Geo-information

Geotechnologies and the Environment

Volume 5

Series Editors:

Jay D. Gatrell, *College of Graduate and Professional Studies and Department of Earth and Environmental Systems, Indiana State University, Terre Haute, IN, USA*

Ryan R. Jensen, *Department of Geography, Brigham Young University, Provo, UT, USA*

The "Geotechnologies and the Environment" series is intended to provide specialists in the geotechnologies and academics who utilize these technologies, with an opportunity to share novel approaches, present interesting (sometimes counter-intuitive) case studies, and most importantly to situate GIS, remote sensing, GPS, the internet, new technologies, and methodological advances in a real world context. In doing so, the books in the series will be inherently applied and reflect the rich variety of research performed by geographers and allied professionals.

Beyond the applied nature of many of the papers and individual contributions, the series interrogates the dynamic relationship between nature and society. For this reason, many contributors focus on human-environment interactions. The series are not limited to an interpretation of the environment as nature per se. Rather, the series "places" people and social forces in context and thus explore the many socio-spatial environments humans construct for themselves as they settle the landscape. Consequently, contributions will use geotechnologies to examine both urban and rural landscapes.

For further volumes:
http://www.springer.com/series/8088

Mathias Lemmens

Geo-information

Technologies, Applications
and the Environment

 Springer

Dr. Mathias Lemmens
Delft University of Technology
Jaffalaan 9
2628 BX Delft
The Netherlands
m.j.p.m.lemmens@tudelft.nl

ISBN 978-94-007-1666-7   e-ISBN 978-94-007-1667-4
DOI 10.1007/978-94-007-1667-4
Springer Dordrecht Heidelberg London New York

Library of Congress Control Number: 2011931685

Printed on acid-free paper

Springer is part of Springer Science+Business Media (www.springer.com)

Preface

This book has been written for students, practitioners and scientists who have no background in geo-information technology but would like to explore the opportunities offered by mobile GIS handhelds and the rich contents of street-view, aerial and satellite imagery, terrestrial and airborne Lidar, and other geo-information using geographical information systems (GIS) or other software tools. The inducement for exploring the powerful capabilities of geo-information technology may be an urban planning project, dike monitoring, road construction or use in environmental, demographic and other studies.

Scope

In course of time a massive amount of books has been published on geographical information systems, land surveying, photogrammetry, geo-databases, remote sensing, web-based GIS and other geo-information technologies. Yet another book . . . . What does it contribute? Compared with other textbooks, the present book distinguishes itself in a number of ways. First of all, this book gives prominence to the geo-scientist rather than technology or mathematical principles. My starting point has been the problems a geo-scientist may be confronted with and from there technology and concepts comes in. I am satisfied when awareness grows among the steadily increasing number of geo-scientists that exploration of geo-data and associated tools is full of pitfalls and snags if not handled with care and with basic understanding of the peculiarities allied to any technology.

Second, it is becoming increasingly seldom that a textbook is written by one or a few authors. Usually the recently published ones are revisions of earlier editions first published way back in the 1950s, 1960s, 1970s or 1980s. Today, it is not unusual when 25 authors, or more, are involved in writing one book. This is a clue as to complexity. The topics covered are complicated and specialized so that the knowledge of one author often falls short to cover the entire spectrum. Writing a book on your own is certainly an endeavour. However, the reader gains a lot in terms of efficacy of communication: repetitions are avoided, the terminology is harmonized, there is clear unity in concept and approach, and the structured outline gives easy access to the basics of technologies and methodologies, their applications and ongoing developments.

Third, the book treats the basic fundamentals of all major geo-data acquisition technologies and demonstrates how they can be used in a plethora of applications. The examples and applications treated are not limited to a specific country or continent but cover the entire globe: Sweden, South Africa, Nigeria, Turkey, China, Siberia, Iran, Nepal, India and Germany, to mention a few countries. In this way the present book demonstrates that all countries struggle with similar environmental issues related to urban, rural and protected areas. The fields of application include environmental management, urban planning, reconstruction of heritage sites, vulnerability assessment of urban areas, assessment of biodiversity and forest biomass, renewable energy sources, census taking, disaster management and land administration. The coverage of a wide spectrum of approaches explored to tackle real world problems all over the globe has been enabled by the fact that I, as previous editor-in-chief of GIM International, to which professional journal I now contribute as senior editor, have published, since 1998, a unique collection of application-oriented short papers written by renowned scientists.

Fourth, although the book treats basic fundamentals, the reader is not bothered with an abundance of formulas and mathematical equations. My basic aim has been to clarify concepts and approaches, not to deepen the nitty-gritty of the mathematics and algorithms resulting from the concepts. They can be found in many textbooks and, furthermore, accessed via internet. When searching for 'collinearity equation' – the central mathematical formula in photogrammtry (Chapter 7) – the first hit reveals a site of the Oregon State University containing lecture notes 'Application of Photogrammetric Techniques to the Measurement of Historic Photographs', written by David A. Strausz, starting with exposing the collinearity equations. When typing RMSE, the abbreviation of Root Mean Square Error, the first hit leads to Wikipedia showing all the formulas.

Fifth, the focal point of this book is an integrative approach. The book aims to bring together land surveying, remote sensing and other geo-data acquisition methods on the one hand and geo-data use in a GIS environment on the other hand in one concise and focussed framework. The majority of other textbooks in the field of geo-information focus either on specific technologies for geo-data acquisition or on GIS technology. The present book does not provide in-depth treatment of specific topics beneficial for the expert specialized in one of the diverse data-acquisition technologies or GIS but gains from centring the diverse topics in one building where the doors are open.

Audience

Professionals, who use a variety of geo-data for solving problems in daily practice, will benefit from this book. After studying this book, students in human and physical geography, mining, agriculture, forestry, demography, urban planning, civil engineering and other geo-sciences will know how they can use geo-information technology for solving scientific or engineering problems and they will better understand the characteristics and limitations of geo-information. This book aims also to address those users of Google Earth and Bing Maps who, fascinated by the opportunities of Earth-viewers, want to explore ways of enriching their professional work. After some months of enthusiastic trials, passionate laymen may discover that the use and processing of geo-data is a discipline and science in its own right. For those non-experts, this book is a perfect entrance.

Contents

The book comprises 15 chapters arranged in three parts. The first part consists of three chapters and gives an overview how geo-information technology has changed over time and how it has gradually moved into the daily lives of the millions through technologies as car-navigation, location aware mobile phones and Earth viewers. Apart from providing an introduction to geo-information technology and its historical development, Chapter 1 treats the basic concepts of geodetic reference systems. Chapter 3 offers an introduction to the concepts of geographical information systems (GIS).

The second part comprises eight chapters (Chapters 4–11) and focuses on geo-data collection and web-based dissemination of geo-information. Global Navigation Satellite Systems (GNSS); mobile GIS systems and Location Based Services (LBS); Terrestrial Laser Scanning (TLS); Photogrammetry; Airborne Lidar; and Earth observation from space are detailed in separate chapters. Together they constitute the primary geo-data collection technologies in use today and the foreseeable future. The content of these chapters is partly based on my research and education experiences, covering the fields of geodesy, remote sensing, photogrammetry, laser scanning, environmental issues and GIS technology, performed at the former Department of Geodesy at Delft University of Technology (TUDelft) as well as my undergraduate course on Geo-information, which I developed for and taught at the Faculty of Civil Engineering and Geosciences (TUDelft). The content further evolved from a module on methods and techniques which I developed for the MSc programme Geographical Information Management and Applications (GIMA) together with Dr. Jan-Jaap Harts, Faculty of Geosciences, Utrecht University and Ms. Ellen-Wien Augustijn, Institute for Geo-Information Science and Earth Observation (ITC), under the auspices of Dr. Stan Geertman. This 'blended learning' MSc programme, introduced in 2003, has been set up to serve both professionals and students. Chapter 10 provides an overview of technologies involved in communicating geo-information via internet and partly evolved from research activities carried out together with my colleague Wilko Quak. The second part finishes with a chapter on quality of geo-information, a subject of which the gravity is often overlooked (Chapter 11). The content is partly based on the introductory part of my graduate course on Quality of Geo-information, TUDelft MSc programme Geomatics.

The third part focuses on applications of geo-information. The first chapter of this part (Chapter 12) focuses, in the form of sightseeing tours and snapshots, on urban planning, reconstruction of heritage sites, vulnerability assessment of urban areas, biodiversity and other uses of geo-information technology. Next, the use of geo-information in Census Taking, Disaster Management and Land Administration is thoroughly considered in separate chapters (Chapters 13–15). The selection of application areas is always affected by an element of arbitrariness. The content of Chapter 12 is primarily based on scientific research and professional experiences published in GIM International throughout the years. The selection of the three subjects, which are deepened in subsequent chapters, has different backgrounds. In the 2005–2006 period, I was extensively involved in the preparation and execution of the 2006 Nigeria population and housing census as a GIS/Mapping advisor and international consultant. The experiences gained on the use of high-resolution satellite imagery and other geo-data collection techniques to establish geo-referenced enumeration areas are reflected in Chapter 13. The focus of Chapter 14 is on disaster management. I have chosen to deepen this subject not only because the role of geo-information technology to support risk and disaster management has been convincingly demonstrated and there is no doubt about its importance but also because my group at TUDelft is comprehensively researching the issues of using (three-dimensional) geo-information for calamity prevention and emergency response. The team headed by Dr. Sisi Zlatanova is internationally renowned. The last chapter covers land administration, which is an essential prerequisite for land management, a subject closely related to environmental issues such as dune erosion, deforestation, soil degradation, flooding of low lands but also climate change. The contents evolved from an MSc course on Land Administration at the Institute for Geo-Information Science and Earth Observation (ITC), since 2011 a faculty of University of Twente, the Netherlands. Under the auspices of Professor Paul van der Molen, I helped to develop a major part of this new course in the period 2007–2009.

Acknowledgements

In writing this book, I am indebted to a number of individuals. First of all I would like to thank all the scientists who were willing to share their knowledge and insights with the readership of GIM International and accepted my, sometimes, tough hand of revision to prepare their in-depth articles for dissemination to a world-wide audience of professionals. In addition to my professional drive and impetus to communicate scientific knowledge to students and practitioners alike, these research activities have been an incentive to write this book. All those given permission to reproduce illustrations are thanked as well. Discussions, throughout the years, with my fellow GIM International editors, Henk Key and Christiaan Lemmen, are gratefully acknowledged. Thanks are due to Prof. Dr. Peter van Oosterom, head of section GIS Technology at TUD, for encouraging finalizing this book. Many thanks too to the series editors for valuable comments on earlier drafts. Finally, I would like to thank my wife Ine and my children Marcia, Karlijn and Fredie for their patience, understanding and accepting, without complaining, huge delays in planned activities.

Delft, The Netherlands   Mathias Lemmens

Contents

List of Abbreviations

AJAX	Asynchronous Javascript And XML
CEN	Comité Européen de Normalisation (European Committee for Standardisation)
DBMS	Database Management System
DEM	Digital Elevation Model
DGNSS	Differential GNSS
DOP	Dilution of Precision
DPW	Digital Photogrammetric Workstation
DSM	Digital Surface Model
DTM	Digital Terrain Model
EDM	Electronic Distance Measurement
EGNOS	European Geostationary Navigation Overlay Service
EM	ElectroMagnetic (spectrum)
EO	Earth Observation
ERP	Enterprise Resource Planning
ESA	European Space Agency
FIG	International Federation of Surveyors
GAGAN	India's GPS Aided GEO Augmented Navigation-Technology Demonstration System
GCP	Ground Control Point
GDACS	Global Disaster Alert and Coordination System
GDF	Geographic Data Files
GDP	Gross Domestic Product
GEOSS	Global Earth Observation System of Systems
GIS	Geographic Information System
GML	Geography Markup Language
GPRS	General Packet Radio Service
GPS	Global Positioning System
GNSS	Global Navigation Satellite System
GPR	Ground Penetrating Radar
GRASS	Geographic Resources Analysis Support System
GSD	Ground Sample Distance

IfSAR	see InSAR
IP	International Protection Rating Code
IMU	Inertial Measurement Unit
INS	Inertial Navigation System
InSAR	Interferometric Synthetic Aperture Radar
INU	Inertial Navigation Unit
ISO	International Organization for Standardization
ISPRS	International Society of Photogrammetry and Remote Sensing
KML	Keyhole Markup Language
LADM	Land Administration Domain Model
LAMBDA	Least-squares AMBiguity Decorrelation Adjustment
Landsat TM	Landsat Thematic Mapper
LAS	Land Administration System
LBS	Location Based Services
Lidar	Light Detection and Ranging
LIS	Land Information System
LoD	Level of Detail
LSA	Least Squares Adjustment
MDG	Millennium Development Goal
MMS	Mobile Mapping System
MPiA	Multiple Pulses in Air
MSC	Multispectral Classification
Navstar	Navigation System with Timing And Ranging
NDVI	Normalized Vegetation Index
NGS	US National Geodetic Survey
NIR	Near InfraRed Band
NMA	National Mapping Agency
OBIA	Object-Based Image Analysis
OGC	Open Geospatial Consortium
OPUS	Online Positioning User Service
PCT	Principal Component Transformation
PDOP	Position Dilution Of Precision
PGIS	Participatory GIS
POB	Point Of Beginning
QA	Quality Assurance
QC	Quality Control
RINEX	Receiver-INdependent EXchange format
RMSE	Root Mean Square Error
RNSS	Radio-Navigation Satellite Services
SA	Selective Availability
SAR	Synthetic Aperture Radar
SBAS	Satellite-Based Augmentation System
SDCM	Wide-area System of Differential Corrections and Monitoring
SDI	Spatial Data Infrastructure
SRTM	Shuttle RADAR Topography Mission

TIGER	Topologically Integrated Geographic Encoding and Referencing system
TIN	Triangulated Irregular Network
TLS	Terrestrial Laser Scanning
UAV	Unmanned Aerial Vehicle
UML	Unified Modeling Language
UN	United Nations
VR	Virtual Reality
VRS	Virtual Reference System
WAAS	Wide-Area Augmentation System
WGS84	World Geodetic System 1984
WLAN	Wireless Local Area Network
XML	eXtensible Markup Language

Chapter 1
Geo-information Technology – What It Is, How It Was and Where It Is Heading to

In a world of climate change impacts, deteriorating physical infrastructure, domestic security threats and a transition to new energy paradigms, geospatial intelligence will be in great demand. All of these will, in my estimation, be far more important drivers than the current boom in consumer interest.

David Schell, former Chairman and CEO, Open Geospatial Consortium

The imperative of measuring and monitoring our environment is increasingly surfacing as the world's population is rapidly approaching seven billion people, natural resources are rapidly being exploited and human activities continue to challenge the quality of land, water and air as well as the Earth's climate. Today's technology, consisting of ground-based, airborne and orbiting sensors combined with information and communication technology enables to collect, process, analyse and disseminate data about a great variety of processes occurring on our planet. The assembly of methods, approaches and devices developed and under development for dealing with the above challenges is called geo-information technology. Geo-information technology is a rapidly evolving engineering discipline, also called geomatics. Which activities does the field of geomatics comprise and how can geoscientists and professionals dealing with the environment benefit from this sophisticated technology? Where is the technology coming from and how did it evolve over time to what it is today? Where is it heading to? This chapter aims at addressing these topics.

1.1 Land Surveying

1.1.1 Famous Land Surveyors

In the past geomatics was simply called land surveying. Famous people have started their career as a land surveyor. For example, every US president who, at some time in his life, worked as a land surveyor has been commemorated by having his face carved into the rock at Mount Rushmore National Monument, South Dakota. Every self-respecting surveyor in the United States of America knows that the first, third and sixteenth presidents had, for a brief period in their lives, been surveyors. The late presidents are George Washington (born 1732), Thomas Jefferson (born 1743) and Abraham Lincoln, and the names of the first two are closely connected to the birth of the United States of America (Rushing, 2006). Indeed, land surveying is often associated with pioneering. But land surveying is also closely connected to science and more specifically mathematics and physics. One of the greatest mathematicians of all time was the German Carl Friedrich Gauss (1777–1855). He was primarily an astronomer and a mathematician and as an astronomer he had to accurately position the observatory in Gottingen, which he headed since 1807. This need for accurate positioning aroused his interest in geodesy. In the period of 20 years between 1828 and 1847, he established a geodetic coordinate system in northern Germany, which in those days belonged to England and was known as the Kingdom of Hannover ruled by King George IV. The most important result of his work as a geodesist was a seminal book on geodesy. To honour his contribution to measurement science, the German central bank issued a 10-Mark note in 1991. The front shows his portrait and his well-known bell-shaped curve and associated equation while the backside depicts a sextant and a triangulation outline consisting of a chain of triangles (Fig. 1.1).

Map making requires a pre-phase in which a geodetic coordinate system is created consisting of a network of fixed points of which the mutual positions in the form of coordinates have been accurately calculated. This framework forms the skeleton over which all other geo-information is draped. Triangulation was the principal method for establishing a geodetic reference frame up to the 1950s and the triangles on the 10-Mark note is a sketch of a triangulation carried out by Gauss. The famous Gauss portrayed as a geodesist demonstrates the key role surveyors play in society. Not only central banks but also local governments have been sufficiently impressed by the achievements of land surveyors to honour the profession by portraying individuals, navigation instruments and surveying methods. When walking through the historical heart of a town one may get suddenly confronted with the statue of a land surveyor as is the case in the Dutch town Alkmaar (Fig. 1.2).

1.1.2 What, Where, When

Literally, land surveying means recording or registering land. The phenomenon that has to be recorded or registered, whether it is a property boundary, topographic feature or the outlines of a building, includes a location component ('where' is the phenomenon located?), an attribute component ('what' is present at that location?) and a time component ('when' was the phenomenon present at the concerning location?). The 'where' component has been considered – throughout the ages – the harder part. No wonder that in many countries, especially on the European continent, the professional is not identified by the term surveyor but as Géomètre (France), Landvermesser (Germany) and Landmeter (the Netherlands), all meaning 'person who measures land'. And hard means here that a substantial part of professional skills of land surveyors, geodesists or geomatics engineers consists of applying

Fig. 1.1 Carl Friedrich Gauss's, astronomer, mathematician and geodesist, portrait on a 10-Mark note (*top*). At the back cover the picture of a sextant and a chain of triangles

mathematics, statistics and physics; just cast a glance at the many textbooks and note the abundance of equations distributed over the pages. Land surveying is thus associated with the accurate measurement of points located on or near the Earth's surface. The connection of points constitutes linear elements, such as roads, or represents boundaries, which, for example, mark property transitions.

Land, in the term land surveying, is interpreted as that the surveyor is standing on with both feet on the ground. However, today the most important sources of geo-information are collected using airborne and orbiting sensors. One type of sensors on board of aircrafts or satellite captures imagery from which the where component is measured using photogrammetric techniques and the what component using remote-sensing techniques. The use of photogrammetry as a means for collecting geo-data started in the nineteenth century (Thomson and Gruner, 1980). The basics and ongoing developments in the field of photogrammetry are treated in Chapter 7. Remote sensing has its roots in space technology, so its history is much younger

Fig. 1.2 Statue of the land surveyor Adriaan Antonisz (1541–1620), erected in 1997, performing a direction measurement to a point outside the picture in the harbour of Alkmaar, the Netherlands

and commences in the late 1950s (Lillesand and Kiefer, 1999). Earth observation from space and how to extract land cover information and identify changes over time from satellite imagery are subjects scrutinised in Chapter 9.

The main activity of the land surveyor or geomatics engineer is to acquire and distribute geo-data, with the emphasis on the geometric component. The data acquisition may involve any Earth-related phenomenon, depending upon the demand of the user or customer. In the past, these phenomena were fairly clear-cut and the tasks of the land surveyor were conveniently arranged and included the following:

- Establishing and maintaining geodetic coordinate systems using the method of triangulation
- Mapping of topographic features such as roads, buildings, railways and canals
- Mapping of natural resources
- Surveys of property boundaries
- Stake-out and measurement work at construction sites

Up to a few decades ago a land surveyor relied on such instruments as measuring chains and tapes, compasses, levels, transits, theodolites and later on electronic distance-measurement (EDM) instruments and total stations to accomplish the above tasks. Levels have been and are still being used for purposes such as calculating volumes of earthwork, providing civil engineers with data for designing highways, railroads, canals, sewers and other infrastructure, and providing geoscientists with data for studying earth subsidence and crustal motion (Wolf and Ghilani, 2006).

1.1.3 Triangulation

Theodolites and their precursors, such as transits, were used to measure the direction from one point, on which the instrument was positioned, to another point in line-of-sight, i.e. visible from the instrument's position. When subtracting the one direction value from the other direction value angles are obtained. Combined with one or a few distance measurements, angle measurements allow the computation of coordinates of points by triangulation, a surveying method developed by the scientist Willebrord Snell, also known as Snell van Royen or Snellius, a Dutch scientist living in the seventeenth century (Haasbroek, 1968). The term triangulation contains the word triangle and the method is based on measuring the angles of a series of connected triangles. When the distance between two points is known, the distances from these points to the third point in the triangle can be calculated. From the end of the eighteenth century up to the 1990s, triangulation was used for accurate large-scale land surveying and for the establishment of nationwide geodetic coordinate systems. For example, during the first decade of the nineteenth century, the Dutch surveyor Krayenhoff triangulated the Netherlands, Belgium and parts of Germany to establish a so-called first-order geodetic reference frame (Haasbroek, 1972) (Fig. 1.3). The majority of the points, which he connected by means of triangulation, were church towers in which a mark was fixed to anchor the exact location of the point. In remote areas special towers were erected. Towers were used as beacons because they are best visible from distances between 10 and 50 km, the lengths of the triangle sides. To transform the angles to coordinates of the points constituting the geodetic network at least one distance between two points in the network has to be known. For this Krayenhoff selected the triangle side Amsterdam–Haarlem, which he determined indirectly by means of, indeed, triangulation using one distance. The measurement of this distance was carried out during the cold winter of 1800 when the inland sea east of Amsterdam, today called IJsselmeer, was frozen so that the surveyors could easily measure the distance between two villages near Amsterdam using a measuring chain. From the distance of 5,650 m and a number of angles the distance between Amsterdam and Haarlem was calculated. Next, after long and lengthy calculations, the coordinates of all points in the geodetic network could be determined, yielding the first first-order geodetic reference frame of the Netherlands. The results were published in 1815. An important prerequisite of this

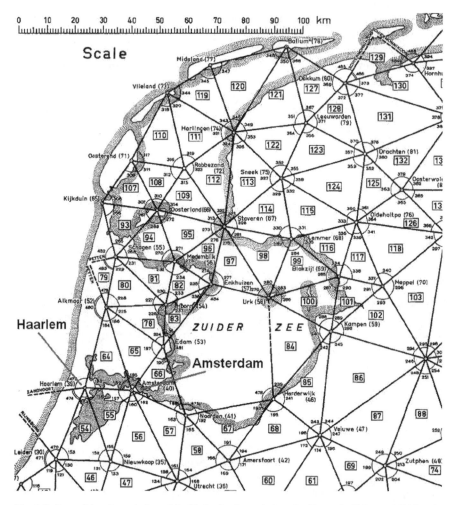

Fig. 1.3 Part of the triangulation of the Netherlands carried out by Krayenhoff for establishing the first-order geodetic coordinate system of the Netherlands

measuring method is that the three points of any triangle are mutually visible while carrying out the measurements.

The period marking the transition of the eighteenth century to the nineteenth century, partly coinciding with the Napoleonic era, was an exiting time for land surveyors as was the sixteenth century for cartographers. Geo-information has throughout the centuries been invaluable for world exploration and exploitation. European expansionist aspirations in the sixteenth century have been an engine for collecting and archiving geo-information of newly discovered territories: 'discovered' here meaning, of course, from the European perspective. To fruitfully exploit their natural resources the specialities of these territories had to be inventoried and archived. The maps published by Gerard Mercator Rupelmundanus and Willem Blaeu represent a summit in terms of artistic and scientific efforts in map making during this hectic explorative period in history (Koeman, 1970; Watelet, 1994). The second part of the eighteenth century and the first part of the nineteenth century was also an era full of exploration, discoveries and pioneering land surveying. It was the time that the exact length of the metre was determined as one ten millionth part of the quarter of the meridian going through Dunkerkque, France, and Barcelona, Spain. The endeavour of measuring its length was carried out by means of triangulation using the recently invented repetition circle and started in June 1792. It took 7 years – much longer than originally planned – and vast financial resources to complete the mission (Alder, 2003).

It was also the time that the East India Company required accurate maps of the Indian subcontinent for the purpose of road construction and visual telegraphy and later on railways and electric telegraph. For the production of the maps the company established the institution of the Survey of India in 1767 in the form of the Bengal Surveys. As stated above map making requires a skeleton, called geodetic reference system, comprising a network of known points, i.e. points of which the coordinates have been accurately determined.

For establishing such a geodetic frame the Survey of India commenced the Great Trigonometrical Survey of India. The activities started on 10 April 1802 and were led by Colonel William Lambton and resulted in the longest arc-of-the-meridian ever accomplished at the time (Keay, 2000). The famous Great Arc runs from Cape Comorin – located in the Southern most tip of the Indian subcontinent where the Bay of Bengal, the Indian Ocean and the Arabian Sea meet – to the Himalayas, almost 2,400 km north. The completion of the arc would take half a century, an endeavour of military proportions both in pecuniary terms and lives lost. From 1818 the project was joined by chief assistant surveyor George Everest, born in Wales in 1790. He became Surveyor-General of India in 1830. Long after his retirement in 1843, the trigonometric survey enabled to identify Peak XV in the Himalayas – known in Tibetan as *Chomo Lungma* (Mother of the Universe) – as the tallest mountain peak in the world. As a tribute to his surveying efforts the peak was renamed Everest in 1856. Indeed, the world's highest mountain has been named after a surveyor. In 1955 the Survey of India determined the elevation of Mount Everest to be 8,848 m (29,029 ft). On 9 October 2005, the State Bureau of Surveying and Mapping of the People's Republic of China officially announced the height of Everest as 8,844.43 m $\pm$ 0.21 m (29,017.16 $\pm$ 0.69 ft). Figure 1.4 depicts a physical elevation model of the Mount Everest area obtained from a digital elevation model. Colonel Sir George Everest died at Greenwich in 1866 without having ever seen the mountain peak named after him. The Great Arc is still India's surveying and mapping backbone and led to the most complex mathematical equations known to the pre-computer age.

The above text emphasises historical developments in land surveying and cartography from a European perspective. But how was China doing, a country where the involvement in producing and using geospatial data is rapidly accelerating in the wake of China becoming a major force in the scientific world and global economy. The history of surveying and mapping in China began more than two millennia

Fig. 1.4 Mount Everest reconstructed as a solid, physical model that bystanders can view and touch as produced by the company Solid Terrain Modelling

ago (Short, 2003). Practice and experience led to formulation there in the fourth century of 'the six principles of cartography': scale, measurement, distance height, angle, curve and line. Using these, the whole of Asia was mapped in the eighth century, taking 16 years. Medieval China also excelled in craftsmanship in survey and mapping, creating masterpieces such as the maps of Africa resulting from nautical voyages. In his book *1421: The Year China Discovered the World* (Bantam Press, London), Gavin Menzies even suggests that the Chinese not only sailed to Africa, but also circumnavigated the globe a century before Magellan, reaching America 70 years before Columbus.

1.1.4 Electronic Distance Measurement

Accurate determination of lengths using measuring chains or tapes is just seemingly an easy task. Obtaining accurate values requires taking a lot of precautions because one operates in a physical environment that may considerably mutilate performance of distance measurements. As simple as they seemingly are, measuring tapes, usually manufactured of steel or other metal alloys, require maintenance and calibration. Every tape has to be regularly calibrated to determine deviations from its actual length. Calibration will take place at a certain temperature, but the temperature in the physical environment will deviate from the laboratory circumstances, sometimes considerably. Temperature fluctuations will cause the length of a metallic chain or tape to shrink or expand even during the same day of measurement. Therefore, accurate measurements, in addition to the actual length measurement, also require temperature registrations. The surveyor's assistants may pull the tape with too much tension, also influencing the accuracy of the length measurement. Stretching a measuring tape along the ground through dense vegetation or over massive rocks will seldom yield a precise measurement. Gravity will cause that the tape bends when it is undesirable to lay the tape on the ground.

Since the 1950s the accurate measurement of distances has become less complicated because of the development of EDM instruments, which measure distances between points using the full and partial waves of electromagnetic energy emitted by the instrument (Burnside, 1991; Reuger, 1996). In the 1970s EDMs were integrated with theodolites into one device called total station (Figs. 1.5 and 1.6). These instruments can measure both distances and horizontal and vertical directions to points of which the coordinates have to be determined (Kavanagh, 2006). Usually a prism pole is positioned over the point, although total stations can also measure reflectorless to light objects up to a few hundred metres. Using a prism reflector the observable distances are in the order of a few kilometres. The calculations are based on trigonometry. Today the measurements are stored digitally and the transformation from directions and distances to coordinates, using trigonometric equations, can be carried out highly automatically by computers on which appropriate software has been installed.

A recent development is the total station equipped with servomotors for automatically rotating the instrument horizontally and vertically, a system for tracking the prism and a communication link that allows just one operator to carry out a survey while he controls the instrument from the prism pole. These robotic total stations also allow unmanned deformation monitoring of dams but the major application is the automated guidance of dozers, graders and other machines. To avoid loss of contact caused, for example, by interruption of the signal by a vehicle driving between base-station and prism, software that predicts the prism's path is used.

Fig. 1.5 A land surveyor at work in Istanbul using a total station

Fig. 1.6 Cadastral surveyors at work in Serbia. The surveyor at the left side holds a prism reflector, which he puts over unknown points on which the surveyor in the middle aims the total station for acquiring distance and direction. The sitting surveying holds maps and aerial photographs on which property boundaries have been outlined, which have to be accurately measured

The most groundbreaking innovation in land surveying since the 1950s is the exploit of the electromagnetic spectrum for measuring distances. In the same decade EDM technology gradually matured, yielding sophisticated measuring instruments, mankind started to rocket satellites into outer space and by the onset of the next decade, on 12 April 1961, the first man – the Russian Yuri Garagin – orbited around the globe enabled by the technology developed by space pioneer and rocket scientist Sergej Koroljow. The birth of space technology combined with EDM technology resulted in the rapid development of a revolutionary way of collecting geo-data: positioning and navigation by means of orbiting satellites. Today, virtually anybody on this planet is using global navigation satellite systems (GNSS) for determining where one is and for finding the way to places one desires to be. GNSS and derived technologies have become essential tools for collecting geo-data and location-based services (LBS). The aim of LBS is to help people to orientate themselves in their vicinity and to find their way to a desired destination. In Chapter 4 we treat the basics of GNSS and in Chapter 5 we elaborate upon devices for collecting geo-data – mobile GIS systems – and LBS.

Many textbooks on the subject of land surveying have been published, some with a long and established tradition covering a series of revised editions. Here we mention a few that have been published in the English language: Kahmen and Faig (1988), Anderson and Mikhail (1998) Bannister et al. (1998), Wolf and Ghilani (2006) and Allan (2007).

1.2 Technological and Societal Developments

From the technological point of view, the main driving forces behind the many changes of the geomatics profession have been the development of new sensors, space technology and the rapid evolvements in information and communication technology (ICT). Only bring to mind digital cameras, global navigation satellite systems (GNSS), high-resolution imagery acquired from the ground, air and space, geographic information systems (GIS), wireless communication and the possibilities of laser technology resulting in airborne Lidar (light detection and ranging) and terrestrial laser scanning (TLS) and the possibility to acquire highly detailed and accurate 3D geo-data. However, technology is not a massive, autonomous process and the evolution of a profession is not only determined by technological developments but the needs of society also create driving forces for change (Lemmens, 2003). Although of a special kind, geo-information differs not at all from other kinds of information. About 80% of all public sector information has a geo-component, either referenced by address or by coordinates specified in a particular geodetic reference frame. Politicians, governors and, in their wake, the military have never underestimated the value of geo-information for achieving their goals. As a result, in many countries the production of topographic maps and other types of geo-data still resides under the responsibility of military agencies. There still exist countries in which topographic data are (heavily) secured from civilian use.

Monitoring and managing of the hallmarks mankind has stamped, and is continuing to stamp, upon the landscape are becoming progressively more important. A comprehensive but non-limitative list of phenomena demonstrating massive expansion over the last decades include the following:

- World Population, resulting in constant migration to cities, so that size and number of mega-cities are increasing and since 2008, worldwide, more people are living in urban conglomerates than in rural areas
- Impact of disasters inducing the need for disaster management and risk prevention
- Mobility, causing the need for establishing and maintaining traffic infrastructure
- Food production, which demands a sustainable approach to agriculture
- Energy consumption
- Exploration of natural resources on and below land surfaces, the marine environment and on and below seabeds

All these developments, including the measurement and monitoring of the environment, have in common the fact that they require detailed descriptions of human activities and earth-related phenomena often in their full spatial (3D) dimensions. They often also require continuous or regular registration (monitoring) by time series measurements (fourth dimension). The plethora of different needs also requires geo-information at different scales; a scale which is suitable for land use planning of rural areas may be insufficient for monitoring of vulnerable biotopes. Representation of reality at multiple scales (called the fifth dimension) is not yet a solved problem and an active research area. Also, the user of geo-information is changing. The majority of traditional clients are cadastres, municipalities, engineers, (landscape) architects, title-insurance companies, real-estate brokers, land developers and construction companies. However, a broad palette of new customers is joining them; roughly portrayed, these are the users of geographical information systems (GIS). Coming from a variety of backgrounds, such as geography, demography, forestry, transport, building maintenance and farming, they have one thing in common: to run their business processes well, to solve their planning issues or to answer their research questions properly, they require accurate, detailed and up-to-date geo-data. As a result, the demand for accurate, detailed and up-to-date geo-data in their full three spatial dimensions and along the time and scale dimensions is increasing throughout all sectors of the economy and geosciences. This development, which is taking place all over the world, steadily increases the importance of the geomatics discipline and those involved in geosciences have to take note of the opportunities and limitations of the technology.

1.3 Geomatics

Today the surveying profession is often referred to as geomatics or geoinformatics. Both terms have been relatively recently introduced compared to their ancestor *land surveying*, which they are meant to replace. Both terms are also used more or less interchangeably to name the science and study of Earth-related information particularly concerned with the collection, manipulation and presentation of the natural, social and economic geography of the natural and built environments. This is the definition as adopted by the Royal Institution of Chartered Surveyors (RICS), which – with 136,000 members across 120 countries worldwide – is one of the most important organisations for professionals involved in land, property, construction and environmental issues (www.rics.org).

Geomatics is a term born in the early 1980s in Canada, where it was coined by political request because it is similar in French and English, as an act of defence against the marching advance of the discipline of informatics. Today, the Department of Geomatics Engineering of the University of Calgary, Canada, defines *geomatics engineering* as (geomatics.ucalgary.ca)

> . . . a modern discipline, which integrates acquisition, modelling, analysis, and management of spatially referenced data, i.e. data identified according to their locations. Based on the scientific framework of geodesy, it uses terrestrial, marine, airborne, and satellite-based sensors to acquire spatial and other data. It includes the process of transforming spatially referenced data from different sources into common information systems with well-defined accuracy characteristics.

Geomatics.nl defines geomatics as follows:

> The discipline of gathering, storing, processing, and delivering of Geographic Information, or spatially refereed information. This broad term applies both to science and technology.

To narrow down the 'broad term' the definition follows with listing the constituting disciplines and technologies: geodesy, surveying, mapping, positioning, navigation, cartography, remote sensing, photogrammetry, GIS and global positioning system (GPS). The possibilities offered by advancements in technologies involved in information processing, (tele)communication, space exploration and sensor development have also dramatically changed working methods, so that the term land surveyor was thought to be not adequate anymore while the terms 'geomatics' and 'geoinformatics' were supposed to encompass more than just traditional land surveying, which focuses on the recording of information about land. However, technology just provides the tools of a profession and it is certainly not appropriate to associate a profession with its tools. A profession is determined by its role in society and in line with this the General Assembly of the International Federation of Surveyors (FIG) adopted a definition on 23 May 2004, which describes a surveyor as (fig.net)

> a professional person with the academic qualifications and technical expertise to conduct one, or more, of the following activities:
>
> – to determine, measure and represent land, three-dimensional objects, point-fields and trajectories,
> – to assemble and interpret land and geographically related information,
> – to use that information for the planning and efficient administration of the land, the sea and any structures thereon, and
> – to conduct research into the above practices and to develop them.

In this definition of the FIG, the 'what' and not the 'how' is rightly placed in the centre.

1.4 Geodesy

1.4.1 Definition and Focus

Another older term associated with geomatics, geo-information technology and land surveying is geodesy, a word of Greek origin containing the words earth (geo) and division (desy). Today many definitions of the meaning of Geodesy are in use over the world. The shortest and most concise one is more than 100 years old and is from the hand of Friedrich Robert Helmert (1843–1917) – coined as the Master of Geodesy and today still known for the development of the three-dimensional similarity transformation – Helmert transformation – which enables to transform a set of points given in one geodetic reference system to another by rotation, scaling and translation. With the two parts of his book on *Mathematischen und Physikalischen Theorien der Höheren Geodäsie* (1880, 1884) he laid the foundations of modern geodesy, while earlier he had published a book on the method of least squares (1872; 2nd edition 1907), one of the most fundamental methods used in land surveying and aimed at minimising the impact of measurement error by making optimal use of redundant observations, that is measuring intentionally more observations than necessary to compute the unknown parameters. In the days of Helmert the observations particularly concerned angles and distances measured with land surveying equipment and the unknowns coordinates of points of a geodetic reference system. The method of Least Squares Adjustment has been introduced into Geodesy by Gauss in 1809. Helmert defined geodesy as follows:

> The science of measuring and mapping the Earth's surface.

This definition is still adequate although technology has dramatically changed. The US National Oceanic & Atmospheric Administration (NOAA) modernised the definition in so far that it changed terms by defining geodesy as (oceanservice.noaa.gov)

> The science of measuring and monitoring the size and shape of the Earth.

Others define geodesy as the scientific discipline that deals with the measurement and representation of the earth, its gravitational field and other geodynamic phenomena, such as crustal motion, oceanic tides and polar motion (en.wikipedia.org/wiki/Geodesy). The Ohio State University has developed a comprehensive definition (geodesy.eng.ohio-state.edu/):

> Geodesy is an interdisciplinary science which uses spaceborne and airborne remotely sensed, and ground-based measurements
>
> – to study the shape and size of the Earth, the planets and their satellites, and their changes
> – to precisely determine position and velocity of points or objects at the surface or orbiting the planet, within a realised terrestrial reference system, and
> – to apply this knowledge to a variety of scientific and engineering applications, using mathematics, physics, astronomy, and computer science.

The aim of geodesy is (Torge, 2001) as follows:

> To determine the figure and external gravity field of the earth and of other celestial bodies as a function of time, from observations on and exterior to the surfaces of these bodies.

The discipline of geodesy may be divided into three hierarchical structured subdisciplines: global geodesy, geodetic surveying and plane surveying (Torge, 2001). Plane surveying is located at the floor level of the hierarchy. Surveyors, who measure outlines of buildings, perform stake-out work at construction sites or collect field data of property boundaries, are involved in plane surveying, that is data about objects are collected at a local level, allowing omitting curvature of the Earth and gravity effects. In plane surveying, the horizontal plane on which a two-dimensional Cartesian coordinate system has been defined is generally a sufficient reference surface. However, plane surveys have to make use of the accomplishments of geodetic surveys, particularly control (reference) points established by geodetic surveys. Geodetic surveyors are carrying out measurements in larger areas and they are not allowed omitting curvature of the Earth and gravity effects. Formulated more precisely, geodetic surveys aim at determining the Earth's surface and gravity field over areas that cover a country, a conglomerate of adjacent countries or an entire continent. The measurements of geodetic surveys have to be connected to the reference networks established by global geodesy. The creation of reference networks requires the determination of the shape and size of the Earth, the orientation of the Earth in space, and the Earth's gravity field and arriving at proper values of these parameters is the basic task of global geodesy (Garland, 1997; Moritz, 2000).

1.4.2 Geodetic Coordinates

The simplest model of the shape of the Earth is a sphere and this model was proposed by the Greek mathematician Pythagoras (582–500 BC) and a century later this model of the shape of the Earth was generally accepted in the Mediterranean region. The dimension of the sphere could be estimated by the founder of Geodesy Eratosthenes of Alexandria (276–195 BC) by carrying out arc measurements. A better approximation of the shape of the Earth is the ellipsoid, which is a figure obtained by rotating an ellipse about its minor axis. The main parameters are the semi-major axis, the semi-minor axis (which approximately coincides with the rotation axis of the Earth), the origin and the orientation in space (Seeber, 1993). Many different ellipsoids have been developed since the early nineteenth century. Each ellipsoid fits best the shape and size of a particular country or other parts of the Earth. For example, the ellipsoid used for mapping Britain, the Airy 1830 ellipsoid (semi-major axis 6,377,563.396 m; semi-minor axis 6,356,256.909 m) is designed to best-fit Britain and is not useful in Africa, Australia or other parts of the world. The position of a point on or near the Earth's surface is given by three coordinates: latitude (φ), longitude (λ) and height above the ellipsoid (h). The triplet (φ, λ, h) is called geodetic or geographical coordinates (Fig. 1.7). Since the introduction of global navigation satellite systems in the early 1980s, the world geodetic reference system (WGS84) is widely used by surveyors. The origin is in the Earth's centre of mass, the semi-major axis is 6,378.137 km and its semi-minor axis 6,356.752 km. Often the WGS84 coordinates are converted into coordinates of a national or local reference system in order to make them compatible with older measurements. The WGS84 ellipsoid is a so-called best fitting ellipsoid, that is the ellipsoid best fits the shape and the size of the Earth as a whole. Textbooks in geodesy develop the equations for transforming one coordinate system into another (e.g. Bomford, 1980). Most GIS systems contain functionality to carry out the transformations by computer.

1.4.3 Height Reference System

Using the Earth's centre of mass as origin a three-dimensional Cartesian coordinate system can be defined. The three orthogonal axes are defined as follows: the Z-axis is defined by the origin located at the Earth's centre of mass and the pole or, in other words, the Z-axis coincides with the rotation axis of the Earth. The X,Y plane lies in the equatorial plane of the Earth and by rendering a line from the origin to the point where the Greenwich meridian intersects the equator, the

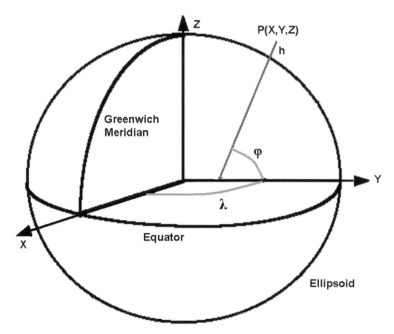

Fig. 1.7 Geocentric coordinates (X,Y, Z) versus geodetic coordinates (φ, λ, h); geodetic latitude (φ) is defined by the angle between the normal of the point to the ellipsoid and the plane of the equator, and it does not refer to the origin

X-axis is defined. The Y-axis is drawn perpendicular to the X-axis to span up a 3D right-handed rectangular Cartesian system (Fig. 1.7). GNSS receivers collect three-dimensional coordinates in this system before they are transformed to WGS84. The advantage of this Cartesian coordinate system is that it provides a universal foundation for incongruent coordinate systems, simple equations, worldwide standardisation and convenient accuracy calculation; the technology is in place and all equations and procedures are available as open source software (Burkholder, 2008). However, such a geocentric coordinate system is not convenient in practice because the X,Y,Z coordinates do give no clue about the difference in elevation between two points. Furthermore, the X,Y,Z coordinates refer to the Earth's centre of mass and are consequently large numbers and thus impractical to handle. Therefore, geodetic coordinates (φ, λ, h) are preferred. However, also the use of geodetic coordinates has practical drawbacks in a number of applications, particularly in the field of engineering. Many people live in flat lowlands situated close to sea. Proper water drainage and flood prevention require accurate elevation data. Dike constructors, dune maintainers and other water managers and engineers should be certain that water flows in the direction as computed from the elevations in their datasets, that is from locations with higher elevation values to locations with lower elevation levels. The flow of water cannot be determined from the heights above an artificial mathematical body called ellipsoid because it does only approximate the actual shape of the Earth and has no physical (gravitational) meaning. The flow of water is determined by gravity. Consequently, to be practically useful elevation values should refer to the gravity field of the Earth. Since water is free to flow in the presence of gravitational forces, surfaces of which the shape is almost entirely determined by the Earth's gravity are sea and ocean surfaces. Other forces acting on seas and oceans are wind, earth rotation and celestial gravity forces. Hence, for engineering works, such as dike construction and prediction of areas vulnerable to flood, elevation values should refer to sea level in one form or another. When no other forces would act on water bodies than the Earth's gravity, their surface would form an equipotential surface. Water above this equipotential surface will flow to this surface. As a result, mean sea level is ideally suited as height reference system. Mean sea level or an approximation of this gravitational equipotential surface is called geoid. This equipotential surface continues underneath land bodies. When carrying out GNSS measurements the resulting heights refer to the ellipsoid. Therefore, for being practically usable, heights above the ellipsoid should be transformed to heights above the geoid. This can only be done when the location of the geoid with respect to the reference ellipsoid is known. The determination of its position with respect to a reference ellipsoid is one of the major tasks of global geodesy. In the pre-satellite era this task was carried out by accurately measuring gravity at a large number of points all around the world both on land as at sea (Heiskanen and Vening-Meinesz, 1958), a time-consuming and expensive job. With the launching of satellites the task has become easier. The orbit of satellites is determined by the gravity field and can be inferred from tracking the satellites from ground stations (Sansó and Rummel, 1989). Another method measures sea level directly using radar-altimeters mounted on satellites (Rummel and Sansó, 1993). From the resulting detailed and accurate gravity measurements of the water surfaces the topography of the ocean floors could be inferred, which is possible because the fluctuations in gravity due to topography affect height of water.

As is the case for ellipsoids, different countries use different geoids and as gravity changes over time due to earthquakes, volcanic eruptions and other tectonic activities geoid anomalies have to be adjusted regularly. Once the location of the geoid is known with respect to the reference ellipsoid, any height above the geoid (H), called orthometric height, can be computed from the ellipsoidal height (h) by a simple subtraction operation (Fig. 1.8).

Expressing the height of the geoid with respect to the reference ellipsoid as N, the orthometric height (H) becomes $H = h-N$, where N is positive when the geoid is above the ellipsoid and negative when below (Heiskanen and Moritz, 1967; Moritz, 1990).

The gravity field and the geoid are uniquely determined by the mass distributions in the Earth's interior. However, the reverse is not true. From the known gravity field the mass distributions cannot be uniquely determined; there are an infinite number of mass distributions that could possibly generate the gravity field as experienced on and above the Earth's surface (Pick et al., 1973). Nevertheless, the shape of the geoid together with seismological data place important constrains on the possible mass distributions and geophysicists have used this insight to intensively investigate

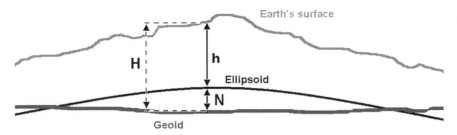

Fig. 1.8 Graphical representation of the relationship between height above ellipsoid, orthometric height (H) and geoid undulation N

mass movements in the Earth's mantle for arriving at an understanding of the driving forces of plate tectonics (Seidler et al., 1983; Jacoby and Smilde, 2009).

1.4.4 Coordinate Awareness

The primary aim of the above discussion has been to demonstrate that the identification of location by means of coordinates is not as simple as it may initially seem when using georeferenced geo-data. For users who merge geo-data, stemming from different sources, in a GIS environment, it is essential to be aware that coordinates can be expressed in different coordinate systems and that geographical coordinates (φ, λ, h) are always defined with respect to a particular ellipsoid. For example, the three geographical coordinates provided by Google Earth while navigating this virtual globe refer to WGS84. Heights or elevations may refer to an ellipsoid and in these cases they are not suited for engineering works. But they may also be defined with respect to a geoid, that is to say a datum close to mean sea level. However, different countries employ different height datums. One should also be aware that large differences exist between various ellipsoidal reference systems. For example, ED50, the reference system used in continental Western Europe excluding Sweden and Switzerland, can differ by as much as 100 m west and south from WGS84. Similarly up to 100-metre differences from WGS84 can be found in the reference systems used by Great Britain and Ireland. What the consequence can be in practice when being unaware of the fact that coordinates may refer to different coordinate systems becomes clear from what Renzo Carlucci, Professor of Geomatics at the University of Roma, communicated to us (Lemmens, 2010):

> Some years ago the Italian highway company Autostrade s.p.a. had to acquire many parcels along highways in order to construct additional lanes. The geodetic reference system of the cadastral maps showing property parcel boundaries did not correspond with construction maps used by the designers. The resulting mistakes caused thousands of legal conflict still pending in the courts of Italy. Maps used for radar control of main airports were based on the ED50 reference system, resulting in a shift of the radar tracks with respect to the WGS84 bearings transmitted by aeroplanes while landing. It does not require much imagination to envisage the precarious situations that could result from wrong manoeuvres. The coexistence of various reference systems is an expensive and dangerous source of error. GIS

operators, not trained in geodetic principles, collate geo-data sources using standard GIS warping operations developed for course fitting only, resulting in mismatches and wrong decisions, the rectification of which may be costly or not possible at all.

In a world that is becoming increasingly globalised, the existence of different coordinate systems side by side also affects cross-border use of geo-data, for example, when investigating the draining of the Danube catchment or flood risk of the river Rhine. Differing coordinate systems confuse emergency workers called upon to help colleagues abroad. Such differences may seem insignificant, but practice shows that they hinder identification of access routes or risk sites such as fuel stations, causing life-threatening delay. So there is an urgent need to eliminate the present heterogeneity of geo-data sets covering border areas and harmonise contents towards seamless integration. To arrive at this goal, the European Union has issued an INSPIRE directive for creating and using shared standards between different countries, which addresses 34 geo-data themes needed for environmental applications. One of these themes concerns shared coordinate reference systems. In the framework of the INSPIRE directive Kadaster, the official Netherlands cadastre, tasked among other things with maintaining and distributing national topographic maps, has together with the National Mapping Agency of North Rhine-Westphalia and Lower Saxony (Germany) released a web service in 2010. The web service seamlessly integrates topographic data on the boundary zone by bringing it all under the same coordinate reference system. Disaster managers benefit from such harmonisation of topographic data, as do water managers and planners involved in creating newly built-up areas, such as business sites.

1.5 Changing Needs

In the second half of the twentieth century, satellite techniques enabled the realisation of the three-dimensional concept of geodesy. Until recently most people would be happy with 2D representations of the world, which worked well, even during eras of ocean voyaging when large areas of the globe were explored, colonised and their resources even exhausted by non-native civilisations. When there was a need to represent the third dimension, such as in hilly and mountainous terrain, lines connecting points at the same elevation above sea level (contour lines) sufficed. But the intensified exploration of the world is drastically changing needs. There is a need for other types of geo-information, a need amplified by the advance of GIS and Internet, which allows use of geo-information, which differs significantly from the use of paper maps. Coping with current governmental issues requires integration of various types of geo-information. The officers and decision-makers using this information differ considerably from those of the past, both in number and background. And governing urbanised areas requires highly detailed representations extending into the full three dimensions.

Resolving all this is thus not just a matter of developing technology to manufacture the same type of geo-information faster and in larger quantities. Technology must also be in place to provide information quickly in the format, scale and at levels of detail (LoD) requested by the user and it will often be necessary to merge various geo-information sets, for example, when a cadastral digital map has to be combined with a road database in designing a road extension (Lemmens, 2006). Satisfying such user demands requires multi-scale representations. Scale is a measure of reduction in representations of reality. A map at scale 1:10,000 has a reduction factor of 10,000: an object of 10 cm length on the map is in reality 1 km. Various issues are involved in multi-scale representations. Simply reducing the scale by zooming out will result in low-quality map representation: too much detail will be visible leading to two main disadvantages:

– The map becomes difficult to read and ugly to look at.
– Transfer of the data by internet is slowed down, even with the broadest bandwidth.

Generalisation and selection of objects are required. This means aggregating several contiguous small objects into one; for example, individual houses, schools, hospitals and other structures within a city quarter are grouped together in one object typed 'built-up area' (Haunert and Wolff, 2010).

Aggregation introduces the issue of semantic consistency; there should be a clear and standardised relationship between objects in the mother database and in derived databases. Objects also have to be represented in less detail. For example, not every curve in a road needs representation at lower scales. This introduces the notion of maintaining topological consistency. An area with several types of buildings located in the mother database on the eastern side of a road should still be on the eastern side after generalisation.

The simplest way to provide maps of various scales is to prepare them beforehand. Scale has to be selected such that the majority of users are served. A main disadvantage here is that all the different representations need storage space. Another solution would be to store just one mother dataset, the most detailed one, from which less detailed scales could be derived at user request by making use of appropriate data structures. This would enable the presentation of real-time generalisation on a computer screen without generating new datasets. However, development of the necessary algorithms and data structures poses fundamental problems, some of which are very hard to solve. This sort of map generalisation, whilst greatly in practical demand, is still the subject of research. The tools currently developed by the main GIS vendors approach generalisation via prior storage of databases derived from the mother database.

1.6 Concluding Remarks

Geo-data is one of the most important components of the digital information scene and is a crucial prerequisite for carrying out research in any of the geosciences and for getting insight into a great variety of Earth-related processes that influence human life. Space technology will make possible the recording of every hook and cranny of the Earth at any required level of accuracy and detail in a highly automatic manner, whilst these data sets will have to be merged with a variety of other data sources, such as topographic databases, to enable the extraction of meaningful information. End-users, including professionals in any of the geosciences, will be increasingly able to produce, to access and to use geo-data themselves.

References

Alder K (2003) The measure of all things. Time Warner Books, London
Allan AL (2007) Principles of geospatial surveying. Whittles Publishing, Dunbeath. ISBN 978-1904445-21-0
Anderson JM, Mikhail EM (1998) Surveying: theory and practice, 7th edn. McGraw-Hill, New York, NY
Bannister A, Raymond S, Baker R (1998) Surveying, 7th edn. Longman, Harlow, Essex
Bomford G (1980) Geodesy, 4th edn. Clarendon Press, Oxford
Burkholder EF (2008) The 3D global spatial data model: foundation of the spatial data infrastructure. CRC Press, Taylor & Francis Group, London, 392 p. ISBN 978-14200-6301-1
Burnside CD (1991) Electromagnetic distance measurement. BSP Professional Books, Oxford
Garland GD (1997) The Earth's shape and gravity. Pergamon, Oxford. ISBN 0-08-010822-9
Haasbroek ND (1968) Gemma Frisius, Tycho Brahe and Snellius, and their triangulations, vol 14, Publication of The Netherlands Geodetic Commission
Haasbroek ND (1972) Investigation of the accuracy of Krayenhoff's triangulation (1802–1811) in Belgium, The Netherlands and a part of North Western Germany, Publication of The Netherlands Geodetic Commission
Haunert J-H, Wolff A (2010) Area aggregation in map generalisation by mixed-integer programming. Int J Geogr Inf Sci 24(12):1871–1897
Heiskanen WA, Moritz H (1967) Physical geodesy. Freeman and Co., San Francisco, London
Heiskanen WA, Vening-Meinesz FA (1958) The earth and its gravity field. McGraw-Hill, New York, NY
Helmert FR (1880/1884) Die mathematischen und physickalischen Theorien der höheren Geodädie. Teubner, Leipzig (Reprinted in 1961)
Jacoby W, Smilde PL (2009) Gravity interpretation: fundamentals and application of gravity inversion and geological interpretation. Springer, Berlin, Heidelberg
Kahmen H, Faig W (1988) Surveying. De Gruyter, Berlin, New York, NY
Kavanagh BF (2006) Surveying: principles and applications, 7th edn. Prentice Hall, Upper Saddle River, NJ, USA
Keay J (2000) The great arc; the dramatic tale of how India was mapped and Everest was named. Harper Collins, London
Koeman C (1970) Joan Blaeu and his grand atlas. Theatrum Orbis Terrarum, Amsterdam, 114 p
Lemmens M (2003) Geo-information technology: changing technology in a changing society. GITC bv, Lemmer, The Netherlands, 192 p. ISBN 90-806205-6-4
Lemmens M (2006) Scale and level of detail. GIM Int 20(9):11
Lemmens M (2010) Lost in Italy. GIM Int 24(11):6–9
Lillesand ThM, Kiefer RW (1999) Remote sensing and image interpretation. Wiley, New York, NY
Moritz H (1990) Advanced physical geodesy. Wichmann, Karlsruhe
Moritz H (2000) The figure of the earth. Wichmann, Karlsruhe
Pick M, Picha J, Vyscoch V (1973) Theory of the earth's gravity field. Elsevier Scientific Publ. Co, Amsterdam, London, New York
Reuger JM (1996) Electromagnetic distance measurement. Springer, Berlin

Rummel, R, Sansó, F (eds) (1993) Satellite altimetry in geodesy and oceanography. Lecture notes in earth sciences, vol 50. Springer, Berlin, NY

Rushing RL (2006) Lasting impressions: a glimpse into the legacy of surveying. Berntsen International, Madison, Wisconsin, USA

Sansó F, Rummel R (eds) (1989) Theory of satellite geodesy and gravity field determination. Lecture notes in earth sciences, vol 25. Springer, Berlin, NY

Seeber G (1993) Satellite geodesy. Walter de Gruyter, Berlin

Seidler E, Lemmens M, Jacoby WR (1983) On the global gravity field and plate kinematics. Tectonophysics 96(3–4):181–202

Short JR (2003) The world through maps: a history of cartography. Firefly books, Toronto, ON

Thomson MM, Gruner H (1980) Foundations of photogrammetry, Chapter 1 in Manual of Photogrammetry. Am Soc Photogramm 1:1–36

Torge W (2001) Geodesy, 3rd edn. Walter de Gruyter, Berlin, NY

Watelet W (1994) Gerardus Mercator Rupelmundanus. Mercatorfonds Paribas, Antwerpen, België, 446 pp. ISBN 90-6153-313-9

Wolf PR, Ghilani ChD (2006) Elementary surveying: an introduction to geomatics, 11th edn. Pearson Prentice Hall, Upper Saddle River, NJ. ISBN 0-13-148189-4

Chapter 2
Earth Viewers

Geo Information has never been so much in use, or benefited so many people. There is increasing exposure of the power of geography to visualise and interpret information over the web, in satellite-navigation systems and in many other aspects of our daily lives.

Vanessa Lawrence, Director General and Chief Executive, Ordnance Survey (GB)

One of the most influential developments boosting the application of geo-information technologies in a wide variety of scientific and professional disciplines has its origin outside the geomatics field although the establishment of the technology heavily relies on recent accomplishments in geo-information technology. The developments referred to concern the emergence of Earth viewers such as Google Earth or Bing Maps accessible by the general public. Earth viewers, also called virtual globes or geo-browsers, allow users to interactively display and explore the information content of aerial and satellite imagery, digital elevation models (DEM), topographic data, earthquake locations, water bodies and many more. The ease with which these viewers may be downloaded on PCs connected to the internet will make people more and more geographically aware. Navigating through Earth viewers can be done rather intuitively not only for those grown up with clicking, dragging and mouse wheels but also for those using computers in their daily work or at home. 'It is like the effect of the personal computer in the 1970s, where previously there was quite an élite population of computer users', Professors Michael Goodchild from the University of California, Santa Barbara says in the scientific journal *Nature* (Butler, 2006). 'Just as the PC democratised computing, so systems like Google Earth will democratise GIS'.

As early as 1993 the first tool for interactive geo-data exploration over the web was developed (Dragićević, 2004). This experimental implementation, developed by Xerox Corporation and named Map Viewer, and other early systems evolved from dissemination static maps only to providing interactive maps with pan and zoom functionality and to viewing the maps using advanced geo-visualisation tools (Kraak and Brown, 2001). With the launch of Google Earth, June 2005, a tool has become available which amplifies geographical awareness as never before. The development

M. Lemmens, *Geo-information*, Geotechnologies and the Environment 5,
DOI 10.1007/978-94-007-1667-4_2, © Springer Science+Business Media B.V. 2011

of (3D) Earth viewers is evidence of how our inherent spatial awareness can be harnessed to provide improved modes of information retrieval and communication (Hennessy and Feely, 2008). Earth viewers have also become a tool for geoscientists from around the world to retrieve geo-data, to visualise their research outcomes and to share data.

The imperative impact of the tool on the scientific, professional and recreational exploration of a broad variety of geo-data has triggered us to continue this book, after introductory Chapter 1, with a survey of opportunities and limitations of Earth viewers.

2.1 Map-Based Searching

To search by location using maps is a very intuitive way of discovering, exploring and collecting information on subjects of interest at or near the surface of the Earth. Suppose you have been lucky enough to rent a small but comfortable room in the very heart of the city you want to continue your academic education or found your first job or new employer. You still have to furnish the room and the next move is to get orientated by visiting a number of showrooms not too far from the heart of the city. A map-based internet search will allow a selection to be made of the types of business to visit, the point of departure and the number of kilometres you want to travel – all with just a few clicks of the mouse. The places selling furniture will then pop up on a map, possibly against the backdrop of an (oblique) aerial or satellite image. A few more mouse clicks will connect you to a route planner and the websites of the shops of interest, or will select alternatives. In densely populated areas, the metropolises, fully three-dimensional (3D) representations of the venue may appear on the computer screen, preventing one from getting lost in and between high-rise buildings after arriving at the destination. Information on the price of occupying a parking place may complete the information dissemination dedicated to your wishes.

Map-based searching enables the collection of all required information from one entry point, avoiding tedious combining of information from several sources stored on different media. So it is not really so surprising that search-engine operators including Google, Yahoo, AOL and Microsoft are demonstrating such great interest in providing map, satellite and aerial imagery-based search possibilities for the mainstream public. The Earth viewers, which can be downloaded free of charge, have broadened public awareness of the possibilities of geo-information. As a result, demand has grown beyond the traditional applications of the users of geographical information systems (GIS). GIS data are being integrated and applied in more business functions, including engineering design, operations management and Enterprise Resource Planning (ERP) systems. Visualisation is a key component of Google Earth's appeal, and it has given GIS providers both the opportunity and challenge to provide more advanced and innovative applications of geo-data.

2.2 Used by the Millions

Earth viewers have been introduced in 2005 and have become increasingly popular ever since. Google Earth, originally developed by Keyhole, a company acquired by Google in 2004, is the most downloaded one. Introduced in June 2005 the most ubiquitous virtual globe was already downloaded over 100 million times just after being 1 year into existence. The software tool is easy in use and has a smooth interface and fast streaming for dynamic geo-visualisation. It allows gaining a rather complete impression of the spatial characteristics of an area of interest and how they can be reached. Although it seems that gigabytes of image data have to be downloaded, the data-explosion problem is trounced by using a tiling structure that transfers increasingly higher resolution data when zooming in. How the tiling structure works can be studied thoroughly when using a slow internet connection. Further, data-compression techniques and other computer engineering methods largely diminish file size. With broadband connection, data transfer and display are performed virtually in real time. The spatial way of accessing information through maps, aerial images and satellite images is a lot more intuitive and convenient than through key word typing in a predefined menu bar. As a result Earth viewers have become so popular that millions and millions of people are using them anyway, anywhere.

2.3 Characteristics

Earth viewers are easily navigable and can be simply downloaded for free from the internet by anybody with access to an online PC; no difficult set-up choices or server settings are required. And it is an extremely user-friendly tool; anybody with just basic computer skills can use it. A news feature in the scientific journal *Nature* reported (Butler, 2006): 'The appeal of Google Earth is the ease with which you can zoom from space right down to the street level'. Indeed, much of the great success of Google Earth stems from the convenient interface of which the use is easy to learn. The user does not need to understand the peculiarities of map projection or geo-referencing and can control the viewpoint from orbiting satellite to low-flying helicopter, avoiding the difficult concept of scale (Goodchild, 2008). Furthermore, Google Earth informs the user when a newer version is on hand, which is immediately downloaded. The viewer also enables to add – just by ticking – other layers to the image background, view road maps, territory boundaries and layers containing information on lodging, dining, shopping and even on pharmacies and grocery shops. Starting from an orbital view of the Earth, you can zoom down from space right to the place of interest, where image resolution allows, even to street level to observe individual cars and to identify houses and other buildings (Fig. 2.1). However, it is not only a tool for accessing and viewing geo-information but it enables also to create oneself geo-information and post it on top of the images.

Immediately after its introduction, exploratory people started to place private documentation of their travels. And all what they brought in could be shared with

Fig. 2.1 Google Earth view on the Three Arms Zone in Abuja, reduced to black and white representation for reproduction purposes. The high-resolution satellite image is produced by DigitalGlobe's Quickbird. Fashioned after Capitol Hill in Washington, DC; the Three arm zones consists of the presidential Villa (*bottom*), the National Assembly (*right*) and the Supreme Court (*top*), all surrounded by a ring road

others, who could interactively explore the result of the home crafts in a visually intuitive interface. Today the adding of own private information has become very easy by the introduction of straightforward tools to marking oneself a location by placing a thumbtack, giving the site a name and adding information in an info window. One can also mark the site of interest by drawing a polygon and one can trail a path, for example, the walk from Grand Hyatt Hotel Beijing to the Forbidden City. Next one can e-mail the personalised image to friends.

2.4 Used by Professionals

The masses use the free version with limited functionality. However, not only the masses but also professionals are fascinated by the Earth view technology and explore its opportunities. The commercial version, Google Earth Pro, is used by many as a marketing tool. For example, national associations of real estate agencies serve potential customers by providing one portal to all houses for sale at national level through maps based on the Google Earth platform using so-called 'mashups'. A mashup is a web application that integrates data from several sources into one application; in the real estate case this is the combination of map and/or image data from Google Earth with photographs and descriptive information, usually placed in 'info-windows', of the houses for sale. The result is a new and distinct web service that was not originally provided by either source.

The use of Google Earth and similar tools for geo-visualisation purposes has been recognised by many and has been widely documented in the scientific and professional literature. Geologists may use Earth viewers as a geological visualisation tool (Lisle, 2006). Demographers may use Google Earth to create animations of urban population growth, as has been demonstrated by Kazuaki Tsuchiya (Google, 2007). Using polygons to represent each of the areas of Tokyo he visualises the realized and expected population change from 1955 to 2025. As soon as Google Maps, Google Earth and MSN Virtual Earth, Microsoft's response to Google's Earth Viewers, became available, the health GIS community recognised their potential for creating custom online interactive health maps of countries and other administrative units (Boulos, 2005). Virtual globes have also been used as a means for information delivery for urban life quality assessment, post-disaster assessment, emergency response and contingency planning (Tiede and Lang, 2010). Google Earth has been studied as a 3D tool for urban planning and renewal, which is often a long-term and complicated process involving decision making at diverse governmental levels as well as public participation (Isikdag and Zlatanova, 2010). Another example, created by web developers, is the geobeijing.info initiative, a portal superimposed on Google Earth for guiding visitors through Beijing during the Olympic Games (Ying et al., 2008). Set up by University of Applied Sciences Stuttgart, Germany, the portal enabled users to search for streets and hotels, and access related information using side panels with hyperlinks (Fig. 2.2).

The reason to build the portal was that information given by commercial geo-browsers such as Google, Yahoo!, Microsoft or MapQuest on streets, venues and so on is not detailed enough to guide Olympic Games 2008 visitors properly through Beijing. The developers used Google's core mapping engine and its map/satellite imagery to add map overlays and to create info windows containing descriptive information. The map overlays capture point features (hotels, sites of interest, and railway and bus stations), linear features (subways, bus lines and more than 1,800 streets) and polygon features (Olympic venues) and were reaped by digitising scanned paper maps. High-resolution satellite images, which can be easily obtained for free through Earth viewers like Google Earth, are well-suited for digitising roads, buildings, walls and trees – geo-information necessary for modelling the sound pressure levels of traffic noise at any location in an urban area (Aditya et al., 2010). Google Earth has considerable potential to enhance methods for teaching geography (Patterson, 2007).

The background data in the Google Earth model basically consist of digital maps and airborne and space-borne images, not all of recent date: some may be up to 3 years old. The literally hundreds of thousands of individual images are derived from over a hundred sources, including Landsat Thematic Mapper (TM) satellite; GeoEye's Ikonos, GeoEye-1 or Orbview-2 satellites; DigitalGlobe's Quickbird and many providers of aerial photographs. Every view shows one, two, three and sometimes even four copyright references. This may give a clue as to how much

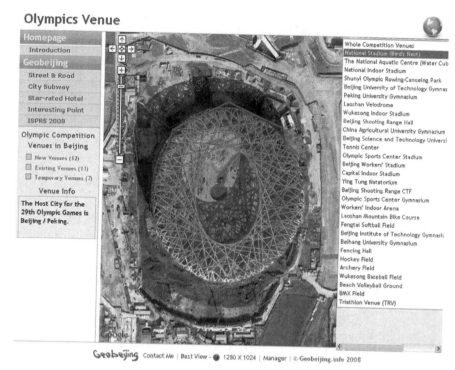

Olympics Venue

Homepage
Introduction
Geobeijing
Street & Road
City Subway
Star-rated Hotel
Interesting Point
ISPRS 2008

Olympic Competition
Venues in Beijing

☐ New Venues (12)
☐ Existing Venues (11)
☐ Temporary Venues (7)

Venue Info

The Host City for the
29th Olympic Games is
Beijing / Peking.

Whole Competition Venues
National Stadium (Bird's Nest)
The National Aquatic Centre (Water Cub
National Indoor Stadium
Shunyi Olympic Rowing-Canoeing Park
Beijing University of Technology Gymnas
Peking University Gymnasium
Laoshan Velodrome
Wukesong Indoor Stadium
Beijing Shooting Range Hall
China Agricultural University Gymnasium
Beijing Science and Technology Universi
Tennis Center
Olympic Sports Center Stadium
Beijing Workers' Stadium
Capital Indoor Stadium
Ying Tung Natatorium
Beijing Shooting Range CTF
Olympic Sports Center Gymnasium
Workers' Indoor Arena
Leoshan Mountain Bike Course
Fengtai Softball Field
Beijing Institute of Technology Gymnasiu
Beihang University Gymnasium
Fencing Hall
Hockey Field
Archery Field
Wukesong Baseball Field
Beach Volleyball Ground
BMX Field
Triathlon Venue (TRV)

Geobeijing   Contact Me | Best View - ● 1280 X 1024 | Manager | © Geobeijing.info 2008

Fig. 2.2 Geobeijing, a portal created on top of Google Earth developed for guiding visitors to Beijing during the 2008 Olympic Games. Shown is the National Stadium, Bird's nest

organisation and negotiation it must have taken to make all this data accessible from one geo-browser. The images are glued together by using one uniform geodetic reference system based on geographic coordinates (latitude and longitude) in WGS84 (World Geodetic System, 1984). The seamless connection of images and maps enables the user to fly over the world and to change altitude with just a movement of the mouse button. Buildings of many major cities, such as Los Angeles, London and Berlin, are represented in three dimensions mostly as rudimentary representation in form of blocks, by experts called Level of Detail (LoD) 1, while some buildings have been enhanced by laminating the facades with photo texture. For Beijing, some tourist attractions, such as the Tian'an Gate – entrance to the Forbidden city – and the granite Monument to the People's Heroes at the centre of Tiananmen Square – the front door of the Forbidden City – are three-dimensionally represented as well as the new National Stadium (Bird's Nest). The stadium hosted the opening and closing ceremonies of the 2008 Summer Olympics, and the main track and field competitions. However, other parts of the world, especially when zooming in on Africa, have to make do with 15-metre resolution imagery as provided by Landsat TM (Fig. 2.3). Resolution here means the size of one pixel in ground units in a digital image and is called ground sample distance (GSD).

Fig. 2.3 Google Earth's view on northern Ghana, reduced to black/white. For Africa the basic image source for rural areas is Landsat TM, GSD 15 m. When available higher resolution satellite images are embedded in the courser ones; these are visible as strips

2.5 Communication Tool with Citizens

Local governments and other governmental organisations are seeking for opportunities to communicate with citizens through Earth viewers.

In an attempt to cut costs on information and communication technology (ICT), Rijkswaterstaat – the government organisation responsible for maintaining and administering main roads and waterways in the Netherlands – tested the suitability of Earth viewers to disseminate geo-information to the general public, an activity which is currently done via an in-house developed geo-information infrastructure called Geoservices (van Asperen and Kibria, 2007). Three Earth viewers were examined: Google Earth, MSN Virtual Earth 3D and NASA World Wind on their performance in disseminating geo-information to the general public (Table 2.1; Fig. 2.4). NASA World Wind is designed for processing and analysing scientific data; its code is open-source, so that scientists and software developers can carry out modifications depending on their needs. The finding of the research was that although Earth viewers share the aim of distributing geo-related information on the web, they differ considerably, making difficult the choice in terms of providing the general public with geo-information. Google Earth proved technically superior but other aspects must also be taken into consideration most importantly compatibility of the viewer with the standards of the Open Geospatial Consortium (OGC).

Table 2.1 Comparison of main properties of three Earth viewers

Property	Google Earth	MSN Virtual Earth 3D	NASA World Wind
Resolution satellite imagery (the Netherlands)	High	Low	Low
Client	Download	Internet browser	Download
Market penetration (presumed)	High	Medium	Low
OGC compliancy	Little	Medium	High
OpenSource	No	Yes (through API)	Yes

Fig. 2.4 Digital Topographic Database superimposed on MSN Virtual Earth in a test conducted to examine performance in disseminating geo-information to the general public through the internet

Suitability for a certain application does not only depend on technologic possibilities but foremost on the setting and general acceptance of standards.

2.6 Getting Geographically Aware

Conventionally mapping the entire world is done through using the Mercator projection, in which the fringes of both hemispheres are extremely exaggerated. When one has only viewed maps showing the entire world in Mercator projection, one would probably belief that Greenland is several times bigger than the Arabic peninsula, but actually their size is similar and that is what Google Earth shows, the world mapped on a sphere, a map showing the true areas of land masses. Earth viewers are increasingly making people geographical aware.

Some people copy the geographical coordinates of their residence or workplace as provided by their Earth viewer and put it on their business card as an additional address line next to their mobile phone number: great evidence that geo-information is becoming part of our collective mind set. As a result demand has grown beyond the traditional GIS applications so that Earth viewers may become a catalyst for the geomatics industry. How will Google Earth and other Earth viewers change the use of geo-information and GIS? Lisa Campbell, Vice-president, Autodesk Marketing Infrastructure Solutions Division, communicated to us (see Lemmens, 2007):

> Today, we're seeing GIS data being integrated and applied in more business functions, including engineering design, operations management and Enterprise Resource Planning (ERP) systems. Visualisation is a key component of Google Earth's appeal, and it has given GIS providers both the opportunity and challenge to provide more advanced and innovative geospatial applications.

Among the millions of users of Google Earth, MS Virtual Earth and other Earth viewers, there will also be a number of people who, inspired and fascinated by the opportunities of Earth-view technology, want to explore ways of enriching and enhancing their own work processes and those of their colleagues using geo-data. But, after some months of enthusiastic trials, they undoubtedly come up against insurmountable problems, discovering that visualisation is the main concept of Earth viewers as functionality for analysing the data is almost completely absent. They will find out that the use and processing of geo-data is a discipline and science in its own right and that expert help is required to accurately extract the relevant information from the mass of geo-data.

2.7 Bing Maps: Microsoft's Virtual Earth Viewer

Above we emphasised the Earth viewer of Google because it is the most popular one worldwide. Bing maps – formerly called Virtual Earth – is Microsoft's platform for bringing maps and images covering the entire digital world to the user's computer screen. The viewer was launched in 2005 as an integrated set of services combining maps, aerial bird's-eye and vertical images, and 3D imagery with search functionality. Bird's-eye images are oblique aerial images taken from an airplane usually flying at an altitude of 1–2 km above earth surface. The oblique images offered by Microsoft are taken with Pictometry's patented camera technology (see Section 6.9). The interest in the use of geo-referenced oblique aerial images for a broad range of applications is widespread. Bing maps is available as an enterprise service and consumer offering, the latter being available through Windows Live Search Maps and an API (application programming interface). This is a free service financed by advertising. The delivery of geo-data services to enterprises is not for free. A wide variety of industries use the data for locating, tracking, geographic analysis and visualisation. In contrast Google's business model is entirely based on selling ads and therefore Google Earth focuses on generating traffic by providing a platform easily accessible to the casual user: the father wanting to plan a family holiday and the businessman to book a hotel. To increase traffic others are encouraged to create their own applications using the platform as a map and image fundament. Google is not actually concerned with how the information is acquired but provides the platform for anybody who has data to share and this business model is uttered by its mission statement to 'organize the world's information and make it universally accessible and useful'. Microsoft acquires imagery with their own digital aerial Ultracam camera, a development of Vexcel, and by contracting image capture out to photogrammetric firms. Next the company processes itself the photographs and offers them as commercial services to enterprises. Another Earth viewer, not mentioned earlier, includes TerraExplorer, which focuses on the professional user.

2.8 Concluding Remarks

The great success of Earth viewers is not just an achievement of the hard work of researchers and developers in the field of information and communication technology (ICT). All the data shown have to be collected and processed. Earth viewers and GIS systems consume huge amounts of earth-related data, including aerial images, satellite images, overview maps, detailed city maps and digital elevation models. Which devices are capable of producing the huge amounts of data necessary? The technologies for collecting geo-data from the ground, from air or from space are treated in the second part of this book. Geomatics professionals are experts in collecting and processing geo-data and determining their quality and fitness for use for certain applications.

The emergence of Earth viewers has also rocketed as a result of political push and campaign. On 31 January 1998, Al Gore, then US vice-president, addressed an assembly at the California Science Center in Los Angeles. In his paper, 'The Digital Earth: Understanding our planet in the 21st Century' (Gore, 1998), he advocated the co-ordinated exploration of all information available about our planet. Since much of this information is geo-referenced, he suggested the creation of a multi-resolution 3D-representation of the planet which he called Digital Earth. He drew attention to how broad and easy access to global geo-information would enable numerous applications to relieve the burden we place on the planet, the extent of which would be limited only by our imagination. And in his wake politicians are recognising more and more how indispensable are land surveyors, or more general geomatics professionals, in helping solve the many problems facing the world, be they environmental, (gender) inequality, sustainable use of land or poverty eradication. However, in many countries the surveying profession is still playing a nominal role and professional organisations, such as the International Federation of Surveyors (FIG), have to 'mainstream the role of surveyors in the national and economic development of these countries', as CheeHai Teo, president of FIG, 2011–2014, stated at his inauguration on 26 November 2010 (Haarsma, 2011). And he continues to state that it is important to show and prove the usefulness of surveying for the benefit of society, the environment and the economy. Before demonstrating by example and by describing geo-data collection technologies in the second part of this book we will first consider the usefulness of geographical information systems (GIS) for bringing geo-data stemming from different sources together at one platform and for analysing the data in an integrated way.

References

Aditya S, Yadav G, Biswas S (2010) Traffic noise mapping: modelling instead of measuring. GIM Int 24(6):29–33

Boulos MNK (2005) Web GIS in practice III: creating a simple interactive map of England's strategic health authorities using Google Maps API, Google Earth KML, and MSN Virtual Earth Map Control. Int J Health Geogr 4(22). http://www.ij-healthgeographics.com/content/4/1/22

Butler D (2006) Virtual globes: the web-wide world. Nature 439:776–778

Dragićević S (2004) The potential of Web-based GIS. J Geogr Syst 6(79):79–81

Goodchild MF (2008) The use cases of digital earth. Int J Digit Earth 1(1):31–42

Google (2007) http://google-earth-kml.blogspot.com/2007_07_01_archive.html

Gore A (1998) The digital earth: understanding our planet in the 21st Century, portal.opengeospatial.org/files/?artifact_id=6210

Haarsma D (2011) Working for the surveyor out there. GIM Int 25(1):12–15

Hennessy R, Feely M (2008) Visualization of magmatic emplacement sequences and radioelement distribution patterns in a Granite Batholith: an innovative approach using Google Earth. In: De Paor D (ed) Google Earth Science, Journal of the Virtual Explorer, Electronic Edition, vol 29, Paper 3. ISSN 1441-8142

Isikdag U, Zlatanova S (2010) Interactive modelling of buildings in Google Earth: a 3D tool for urban planning. In: Neutens T, De Maeyer P (eds) Developments in 3D geo-information sciences. Springer, Berlin, pp 52–70

Kraak MJ, Brown A (eds) (2001) Webcartography: developments and prospects. Taylor and Francis, London. ISBN 0-7484-0868-1

Lemmens M (2007) Committed to open standards. GIM Int 21(2):7–9

Lisle RJ (2006) Google Earth: a new geological resource. Geol Today 22(1):29–32

Patterson TC (2007) Google Earth as a (not just) geography education tool. J Geogr 106(4):145–152

Tiede D, Lang S (2010) Analytical 3D views and virtual globes – scientific results in a familiar spatial context. ISPRS J Photogramm Rem Sens 65(3):300–307

Van Asperen P, Kibria MS (2007) Comparing 3D-Earth viewers: Google Earth, MSN Virtual Earth 3D and NASA's World Wind. GIM Int 21(11):13–15

Ying Y, Behr F-J, Li H (2008) Design and implementation of a portal site for the Olympic games 2008 in Beijing using Google Maps. Int Arch Photogramm, Rem Sens Spatial Inform Sci XXXVII(Part B4):1793–1798

Chapter 3
Understanding Earth-Related Phenomena Through Maps

> *The more projects we completed, the more we learned how using geography as a framework for data integration provided a new dimension in the way people approached problems. GIS allowed them to visualize the problems, which helped provide quicker and better solutions. We became convinced that GIS could truly make a difference in the world.*
>
> *Jack Dangermond, President and Founder, Esri, Redlands, California, USA*

Maps on paper or digital maps alike can make the invisible visible, and thus reveal new insights about the world. Map-making or mapping is one of the techniques to interpret and represent the world and fundamental for representing space and location (Dorling and Fairbairn, 1997). Mapping is only possible if sufficient detailed, accurate and up-to-date geo-information is available. Although availability is an essential prerequisite it is not enough; to be of any value geo-information has to be processed and interpreted by knowledgeable and skilled professionals using the right tools, including Geographical Information Systems (GIS) (Tomlinson, 2007). Today GIS, remote sensing (RS) and Global Navigation Satellite Systems (GNSS) offer a plethora of opportunities for monitoring and managing many facets of our world, which is increasingly considered as vulnerable and affected by human activities. Geo-information technology provides enormous potential for tackling the broad pallet of problems that tend to be branded by pessimists with the tag 'unsolvable'. Induced by increased human population, intensification in agricultural land use and industrialisation, deforestation, soil erosion, degradation in wildlife habitat, loss of biodiversity and pandemics such as HIV/AIDS seem to be a never-ending story. This chapter demonstrates by two historical examples – medical mapping and geological mapping – how insight can be gained about Earth-related phenomena through analysing maps. Next the general steps involved for deriving geo-information from geo-data are considered using a GIS.

3.1 Tabulating Versus Mapping

Suppose the government of some country wants to register data on all the buildings erected in its territory. Such a registration of buildings is, for example, aimed at property-tax collection, granting building permits, distributing petrol-station licences and many more tasks of good governance. A government officer with a background in law, economics, accountancy, sociology or demography, let us call this officer mister Leasd, is probably inclined to define the attributes of the data as 'building identifier', 'address', 'floor area', 'number of storeys' and 'use', and to store all the data in an Excel spreadsheet or an alphanumeric database. The same officer will also want a reference to the document lending official approval to the existence of such an object.

A public health statistician with a background in geomatics engineering, however, would shake his head pityingly. His comment might be 'This dataset will not allow me to assess the health effects of reducing particulate matter in the air or the spread of a contagious disease, even when I combine the building register with the population register and with the registers of the Department of Health'. In reply mister Leasd would sniff, 'Of course you can! You can combine the information through the address and then manually draw everything on our fine, large scale street map'. The geomatics specialist, aware of the importance of georeferencing of data, likely responds, 'Yes, drawing the information on maps would be an option. But how tedious and nerve-wracking that would be, how time-consuming and costly! And it would only allow to compute alternatives at the cost of many man hours, if at all'. He continues, 'Next time you travel to London I will recommend your boss send you on horseback. We are no longer living in the days of Doctor Snow'.

3.2 Medical Mapping

Who was Doctor Snow? Son of a Yorkshire farmer, John Snow (1813–1858), was a London medical physician (Fig. 3.1) and pioneer in the fields of epidemiology and anaesthetics (Vinten-Johansen et al., 2003; Koch and Denike, 2006; Johnson, 2006; Lemmens, 2007). He contributed 82 publications to the scientific literature, including many on chloroform, and proved by assembling maps and geo-statistical (also called spatial statistical) analysis that cholera was transmitted by contaminated water and not by 'miasma' (bad air), as was the prevailing theory at the time. During the cholera outbreak of 1853–1854 he was able to prove his hypothesis by showing that public water supplied to south London residents by private companies was a key factor in the communication of the disease. In contravention of new regulations, Southwark and Vauxhall Water Company was still taking water from local wells and from the River Thames as it ran through central London, while the Lambeth Company had begun taking its water from the Thames about twenty miles upstream of the city. The two companies served different areas, but there was a

Fig. 3.1 John Snow in 1857
when he was 44 years old, a
year before his death

large area of overlap within which the two delivered water through different pipes. Snow outlined the supply areas on a map and started counting deaths from cholera. The statistics thus derived proved a strong correlation between the deaths and water source (Table 3.1).

Although this 'grand experiment', as Snow himself called it, confirmed his hypothesis, he also wanted to acquire evidence on a local scale. For this he carried out research in his own residential district of Soho, served by both water companies.

He plotted on an existing street map the homes of victims of the Soho cholera outbreak of August 1854 and identified a water pump in Broad (now Broadwick) Street as the source of the disease in that neighbourhood (Figs. 3.2 and 3.3). Removal of the handle of the pump and sealing off the water source resulted in an immediate fall in infection rate. It is this map analysis on the local scale of Soho that made Snow famous.

Table 3.1 Geo-statistics enabled Snow to prove strong correlation between cholera and consumption of contaminated water; upstream water, provided by Lambeth Company, was safer than other sources of water distributed by Southwark-Vauxhall Company

	No. of houses	Deaths from Cholera	Deaths/10,000 houses
Southwark-Vauxhall Company	40,046	1,263	315
Lambeth Company	26,107	98	37
Rest of London	256,423	1,422	59

Fig. 3.2 Snow's famous Broad Street map showing distribution of cholera cases. The map has been modified twice, once by Regmarad around 1960 to appear in cartographic books (Johnson, 2006) and again by the current author

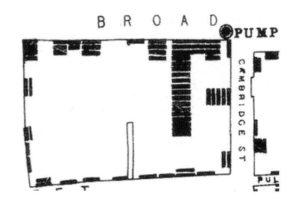

Fig. 3.3 Portion of Snow's original map near Broad Street

In Snow's map, spatial correlation between deaths from cholera and distance to the pump in Broad Street is obvious, and once recognised can be no longer overlooked. Identification is also aided by placing the pump at the centre of the map, and for today's readers by marking this particular pump as a square. However, for a relationship to be readily seen one needs prior notice of correlation. This implies formulation of a hypothesis, or, in more general terms, model. From his work among coal-miners, who 'suffered more from cholera than any other' population, Snow became convinced that the bad sanitation and water facilities in the mines had to do with communication of the disease. 'There are no privies in the coal pits and I believe this is true of other mines: as the workers stay down the pit for about eight hours at a time they take food down with them which they eat, of course, with unwashed hands. As soon as one pitman gets the cholera, there must be great liability of others . . . to get their hands contaminated, and to acquire the malady', he wrote in 1854 in the second edition of his publication *On the Mode of Transmission of Cholera*. The map enabled him to prove that the cause of the disease was water contaminated with excrement from infected people. Since he knew what he was looking for, it was just a matter of plotting deaths from cholera and the source of drinking water, pumps, within their spatial context.

Someone believing that bad air was the cause could have plotted cholera deaths within the spatial context of manure heaps and main wind directions. But nobody did. Although the map convincingly proved Snow's hypothesis, it still took another decade before his germ theory was widely accepted.

Since the twentieth century Snow has been often portrayed as the genius responsible for single-handedly discovering the cause of cholera, and by the GIS community as the mastermind who invented medical mapping. Although some criticise the mythical proportions attributed to his achievements (e.g. Brody et al., 2000), Snow's maps remain the most famous classical example in the field of medical mapping and his study is a landmark still taught today. Koch (2005) has written a comprehensive monograph on medical maps, the history of medical geography and the potential contribution of cartography to understanding and combating disease. The University of California Los Angeles, Department of Epidemiology maintains a John Snow website.

Snow was not the first to use maps to show correlation between spatial phenomena and contagious disease. This was already established more than half a century earlier. In the late eighteenth century Dr. Valentine Seaman, a surgeon at New York Hospital, compiled a map of deaths during an outbreak of yellow fever in what is now the Lower East Side of Manhattan (Stevenson, 1965; Topper, 1997). He was attempting to prove that foul and putrid waste caused spread of the disease. More than a century later, around the fin de siècle, a US commission led by Walter Reed investigated why during the Spanish American War of April to August 1898 so many American soldiers died from yellow fever, more than from fighting. The Reed Commission proved by deliberately infecting mosquitoes and allowing them to feed on volunteers that the mosquito *Aedes aegypti* was responsible for spreading the disease. Most infected people recovered.

3.3 Geological Mapping

The above medical mapping example demonstrates the power of mapping for getting insight in earth-related phenomena. Maps can make the invisible visible, and thus reveal new insights about the world (Lemmens, 2008b). And that is exactly what William Smith did when he published the geological map of England, Wales and parts of Scotland in 1815. The work by Smith, a surveyor by training, has become well known thanks to Simon Winchester (2001). The story described by Winchester goes as follows.

In the late eighteenth century Great Britain was subjected to the incisive mining and construction activities that followed the Industrial Revolution. As a surveyor, Smith had access to pits and dug sites. He observed that throughout the country rock layers were arranged in a predictable pattern, consistently characterised by the same succession of fossils, from older to younger rocks. He collated his findings not only in map form (Fig. 3.4), but also in a book in which he recognised that fossils enabled matching of rock layers across regions. On the map he used conventional symbols to mark features such as canals, copper and tin mines, roads, tunnels and tramways. But how should he represent geological type? He found that the best way was to colour-code them. And these colours depict not only the spots where rock layers surface but also represent time, lots and lots of it, millions and millions of years. The map revealed that the globe must be much older than the 6000 years claimed by the church of that time, so enabling Charles Darwin in the early 1840s to outline his theory of evolution by natural selection as an explanation for adaptation and specialisation. Processes of evolution require many millions of years, and Smith's map proved the viability of such immense time-spans.

This altered view paved not only the way for Darwin's theory of evolution but also enabled Alfred Wegener in 1915 to formulate his 'Continental Drift' theory on the continuous displacement of continents, culminating in the early 1960s in the theory of Plate Tectonics (Hess, 1962; Vine and Matthews, 1963; Kearey et al., 2009). Central in the theory of Plate Tectonics is that the solid part of the Earth's surfacing layer consists of separate plates which are moved by currents in the mantle (Fig. 3.5). While moving with a few centimetres per year they may diverge – resulting in ridges in the oceans and rifting zones on continents, such as Africa's Great Rift Valley – or converge (collapse) resulting in a lifting-up of the border of plates and thus in the creation of mountains. The Andes mountain range and the Rocky Mountains are created by collision of the Nasca plate with the South American and the Pacific plate with the North American plate, respectively. The oceanic plates dive underneath the continental plates and this process is called subduction. The theory led to the unifying insight that earthquakes, volcanoes and derived phenomena such as tsunamis are caused by the same mechanism, i.e. divergence and convergence of plates driven by currents in the Earth mantle. In many countries, such as Japan, India and Iran, the theory of plate tectonics enables scientists to predict – not yet without fault, unfortunately – the occurrence of earthquakes by making use of advance geo-information technology including Global Navigation Satellite Systems (GNSS).

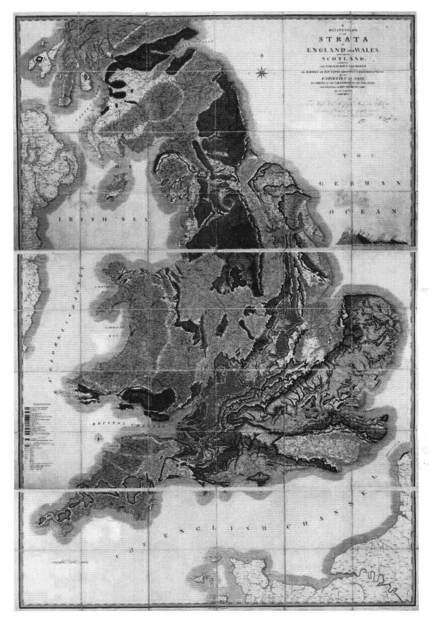

Fig. 3.4 Black and white representation of the map that changed the world: William Smith's geological map of England, Wales and parts of Scotland

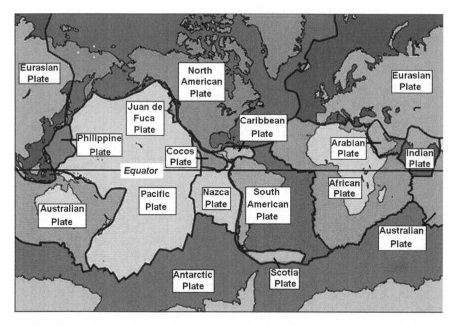

Fig. 3.5 One of the major historical insights gained through maps is the theory of plate tectonics

3.4 From Paper to Bytes

'We are no longer living in the days of Doctor Snow', said the geomatics engineer. What he meant was that mister Leasd just wanted, in the tracks of John Snow, to use address records to geo-reference earth-related phenomena. Combining different geo-datasets with each other and with other datasets by means of address is only feasible when the amount of objects and the number of attributes assigned to each is fairly small. This is because the procedures involved require a lot of manual processing. However, today's datasets are often mammoth-sized. Furthermore, all geo-sciences have made mathematical their understanding of the spatial and temporal dynamics of processes of the parts of the real world they study. Today all real-world processes are modelled as systems of mathematical equations, some more complex than others; and these systems have been transmuted into algorithms. This allows calculations involving such cumbersome and tedious exertion for human beings to be done by computers, fast and free of blunders.

In addition, today's scientists are no longer happy simply proving spatial correlation through visual inspection alone (Lemmens, 2007). Referring to Snow's map, one would today compute all distances from each pump in the district to all the spots where somebody died from cholera, add up the distances per pump and calculate the averages per pump. Then one would determine the minimum of these averages and show that the averaged distance to the pump in Broad Street was significantly smaller than the averaged distance to the other pumps. Or alternatively one would draw circles, all of the same radius, around the pumps, count the number of deaths within each circle and prove by statistical testing that the number in the circle around the pump in Broad Street was significantly higher. In order to carry out these computations by computer the locations of the houses and pumps need to be georeferenced, which means identified in the form of coordinates, all, of course, related to the same geodetic reference system. The computer systems able to carry out such computations are called GIS.

3.4.1 GIS

When computers became available in the early 1950s scientists and engineers started to employ these devices for handling geo-data. These early GIS systems were often developed for specialised applications and required mainframe computers usually only affordable for governments or universities. As GIS technology evolved through multiple parallel but separate applications across numerous disciplines (Pickles, 1999) the resulting approaches and software were dedicated to certain applications in specialised organisations such as Census Bureaus and National Mapping Agencies (NMA) (Coppock and Rhind, 1991; Foresman, 1998). Canada was the first country in the world that produced a map of the entire territory using computers thanks to the pioneering work of Dr. Tomlinson. The computers helped the Canada Land Inventory to establish a map representing a variety of land use types including agriculture and forest. Also in the 1960s the US Bureau of the Census developed software for digital mapping purposes in support of the 1970 Census. The resulting GIS packages enabled to produce the Census TIGER files (Topologically Integrated Geographic Encoding and Referencing system), which are seen today as the forerunner of socio-economic geo-data sets. Another pioneering effort has been delivered by the UK where the Experimental Cartography Unit created software for editing and publishing maps (Goodchild, 2007). By the 1970s all these uncoordinated initiatives and efforts resulted in a focussed vision on how to handle different types of geo-data, culminating in the release of commercial off-the-shelf GIS packages offered by private vendors including Intergraph and Esri (Antenucci et al., 1991).

Many examples of the use of geo-information technology for planning and socioeconomic purposes at a diversity of scales are given in Gatrell and Jensen (2009). The fast majority of applications treated in this book is related to the territory of the US – the volume covers topics as diverse as heat related health disasters; planning and development of peripheral regions; understanding risk in cities; spatio-temporal trends in inequality; tree canopy restoration in urban neighbourhoods; revealing the relationship between local land-use policies and urban growth and many more. Jensen et al. (2007) provide in-depth treatments how geo-information technology can be applied in urban environments. The topics covered by this book focus on urban areas and include the use of remote sensing technology for a variety of purposes such as detecting changes over time, risk assessment and estimating population and leaf area index. Other subjects treated within the book are as follows: mapping, measuring, and modelling urban growth; deer-vehicle collisions along the suburban-urban fringe; and public participation GIS as surveillance tool in urban health.

With respect to medical mapping, today scientists concerned with public health recognise that GIS systems are highly suited for estimating the health risk of people living in a neighbourhood endangered by a hazard or heavy pollution and also that GIS helps to discern trends and correlations difficult to discover in tabular format (Cromley and McLafferty, 2002). GIS allows policymakers to enhance application of resources because the spatial correlation of diseases, health services and terrain characteristics can be easily visualised (Lang, 2000). Since 1993, the World Health Organisation promotes public health mapping using GIS as a tool for a diversity of applications, including (www.who.int/health_mapping/en/):

– Determining geographic distribution of diseases
– Analysing spatial and temporal trends
– Mapping populations at risk
– Stratifying risk factors
– Assessing resource allocation
– Planning and targeting interventions
– Monitoring diseases and interventions over time

Along what lines does a typical monitoring or geo-management task proceed? The general steps involved consist of (Lemmens, 2006):

1. Sampling design of the geo-data collection process
2. Actual geo-data collection
3. Data storage on a computer medium
4. Data processing in a GIS environment to make the data easily accessible and useable in the next step
5. Analysis of the data to arrive at the requested information
6. Visualisation of data and information for communication purposes to others, including policy makers and decision makers
7. Data management
8. Data dissemination

These steps are usually carried out using standard methods and field protocols. But first the geo-management objective has to be specified: without a predefined objective nothing can be done. As stated above, the objectives may cover a great variety of environmental, socioeconomic or planning purposes related to urban or rural areas, developed or developing countries, and cultivated or natural environments.

3.4.2 Essential Data

Key and essential is the availability of all relevant data from which core parameters can be calculated in an appropriate manner. Existing sources are very valuable to start with. Depending on the objective such sources may include topographic maps, forest maps, soil maps, Digital Elevations Models (DEM), contour maps, ortho-rectified aerial photos, archived satellite images, population densities/demographics or average income registered per administrative unit. Frequently a GIS project will result in an abundance of data, and for the smooth retrieval and management of all of this the use of a sophisticated database management system (DBMS), such as Oracle, is an essential necessity. Often existing geo-data are stored in paper-format, so that the essential features first have to be selected, digitised and stored as GIS layers in the DBMS. Here *geo-referencing* comes in, which aims at relating each and every position on the map to the corresponding location on the ground (see Section 1.4). The collection of the coordinates of ground control points, necessary in the geo-referencing process, can be fruitfully done using GNSS technology, in particular the Global Positioning System (GPS).

GNSS technology is also useful for collecting the location of training samples and other ground-truth using mobile GIS systems when multispectral or hyperspectral aerial or satellite images are used. Classification of such imagery requires the availability of a remote-sensing image-processing package. A simple but effective land-use classification method is provided by the Normalized Difference Vegetation Index (NDVI), which is calculated per pixel as a function of the red (R: 0.62–7 μm) and near infrared (NIR: 0.7–1.5 μm) band: NDVI = (NIR – R)/(NIR + R). The value range of NDVI lies between –1 and 1 (minus one and plus one). Chlorophyll in leaves will largely absorb the red band, while NIR is largely reflected. So green areas will show a NDVI close to 1, the NDVI of bare ground will be around 0, and of water it will be close to –1 (Tucker, 1979). When essential features cannot be extracted from existing sources, nothing remains but to abandon one's office chair for the field. Mobile GIS devices, which basically integrate GNSS, GIS, geo-database and mapping in one compact handheld tool (see Chapter 5) are optimised for collecting data in the field as GIS input. GNSS can also be used to test the accuracy of geo-data already stored in the database by field check.

3.4.3 Analysis

When it comes to carrying out analysis, a broad pallet of functionalities is available in today's commercial GIS packages; they can store, retrieve, map, display and analyse geo-data of which the location component have been stored in a common geodetic reference system. GIS software allows a great variety of data, usually stored in different datasets, which can be best envisioned as a stack of layers, the one on top of the other, to be combined and queried (Goodchild, 1987). For example, a cadastral layer, including parcel boundaries, name and address of owner(s), type of ownership, date of achieving ownership, date of creation of the parcel, and land use may be combined with a census layer including district boundaries, average income of households, average age, gender rates and education level. In this way the relationship between size of the parcel and education level or any other quantifiable parameter can be computed in the form of, for example, the normalised cross correlation coefficient, which provides a similarity measure between two functions. If the outcome is +1 there is complete similarity and −1 indicates complete dissimilarity (see any statistical textbook). Buffer and overlay operations appear most useful. For example, when a rural village in a developing country relies on firewood for cooking, one may create buffer zones and overlays of GIS layers to identify forest areas in the vicinity of the village suitable for planting species of wood used for fuel. A functionality, which is fruitful for transportation analysis, is the network operation, from which the shortest routes between a clusters of locations can be computed or the movement of railway passengers simulated. When three-dimensional geo-data in the form of Digital Surface Models are available sight (viewshed) analysis can be performed to identify, for example, the optimal locations for erecting fire watchtowers in a forest area. Availability of a 3D city model allows GIS operators to calculate how buildings, which have been designed by architects or planned by city planners, create shadows throughout the day, enabling to perform sunlight analysis and, if necessary, adjust location, size and shape of residential buildings (see Section 12.4). Also viewshed analysis can be carried out for determining view obstructions when erecting a building or other constructions. Arriving at the most appropriate solution requires communication and collaboration among many actors, including the general public. Communication is best done through visualisations such as images and maps rather than through bare text or spreadsheets.

Often users think GIS is just a mapping tool or a software package as any other. However, it is an advanced analytical tool, the utilisation of which requires in-depth knowledge about concepts and methods of spatial statistics and operations as well as principles of land surveying. A comprehensive introduction to methods and techniques for analysing geo-data independent of software tools can be found in Bolstad (2005).

3.5 Web-Based GIS

The exchange of information between different GIS systems is becoming increasingly important, and this is where web-based GIS comes in. Today communication is done by storing geo-data such as size, shape and location of proposed building designs on a server from where they can be shared with co-workers or viewed by citizens. Using a PC connected to broadband internet professionals and laymen alike can virtually fly-through their city and view buildings, underground constructions and proposed designs from any viewing angle. Communication via internet also strengthens the communication between government and citizens, who can comment on the designs, fill in questionnaires and vote, and in this way (local) government may increase public participation in city planning and other public affairs. Also tourists can access such data and retrieve information and in this way it acts as promotional site where people around the world can take virtual tours, which may inspire them to visit the city.

3.5.1 Architecture and Applications

A Web-based GIS is usually constructed as a three-tier architecture consisting of the client side, the server side and the database component (Huang and Lin, 1999). From the client side users send queries or requests to the server via (wireless) internet. Next the server communicates with the database, where the requested information is compiled by the database engine. Finally the result is stored as, for example, a map in JPG format, and sent to the client. Depending on the GIS functionality present at the client side, the user may carry out zoom, pan, query and other operations on the retrieved map and other data.

When building a web-based GIS system for enabling visualisation and analysis of large datasets stored in a variety of distributed databases one gets faced with a number of technical and research challenges (Flemons et al., 2007). Access should be fast and based on open standards. The map interfaces should be user friendly while sufficient functionality should be available for carrying out analysis on the diverse datasets. Web-based GIS has proven to enable supporting a broad variety of applications including E-Government initiatives (Tsai et al., 2009), disaster management (Mansourian et al., 2006) and visualising, modelling and analysing urban environments (Doyle et al., 1998). Highly dynamic circumstances such as in traffic accidents, fire or other emergencies require quick responses and integration of data stored in a central GIS database and data collected on-site. Adaptive web-based GIS systems are emerging to streamline the information exchange between the diverse actors in rescue teams in order to improve their collaboration according to the needs of the users (Angelaccio et al., 2009).

In web-based applications, digital maps become available to a wide range of users all over the internet without the need that the users purchase any GIS software. The users can access the map dynamically and interactively (Dragicevic, 2004). The number of web-based GIS applications, varying from flood management to providing disease information, which serve worldwide users is increasing rapidly (Holz et al., 2006; Gao et al., 2009).

3.5.2 Web 2.0

The early web was predominantly one-directional, enabling many to access the contents of sites which were solely determined and maintained by the owner (Lemmens, 2010). Web 2.0 allows everybody to contribute to web content, extending the way of communication from one- to two-directional. For example, there exist numerous free online diaries where travellers may post blogs, or web-based journals, as many and as often they want. One can join for free, and setting up a site takes a few minutes, while the only requirement is having an e-mail address. Web 2.0 enables collaboration as never before; one can instantaneously collect, disseminate and share information. People in need of geo-information can choose from a plethora of sources: Google Earth, OpenStreetMap and Bing Maps being the most obvious examples. But individuals can also add information themselves, such as GNSS tracks and georeferenced photographs.

3.5.3 *Volunteered Geography*

Web 2.0 will be the start of a new era of mapmaking. In particular, GNSS in combination with GIS, wireless communication and internet has revolutionised the way geo-information is collected and used. Rapid innovations have triggered the development of mobile GIS systems and Location Based Services (LBS) into work processes requiring data collection in the field or geo-data use while, for example, on the road or on the water (see Chapter 5). New media technologies may change laymen into voluntary collaborators (Goodchild, 2007). In the aftermath of a disaster, various stakeholders can collaborate with each other and interact with those affected to improve rescue and relief activities. In the urban environment, infrastructure planners and maintenance crews can benefit instantaneously and en route from geo-data delivered by civilians via the web. The new technologies also enable augmentation of LBS in the ever-increasing need for navigational assistance in urban conglomerates. People may add their navigational experience to help others find their path to, for example, an Italian restaurant, and provide appraisal of location and food. In such applications it is not algorithms running on computers that guide individuals and groups from point A to point B, but the real-life experience of flesh-and-blood beings.

The latter demonstrates one of the snags of bi-directional communication with low thresholds: the introduction of an uncontrolled level of subjectivity and unreliability. More information is needed and can be produced by exploitation of potentially nearly seven billions sensors: the number of people living on this planet. But expert knowledge and procedures are required to guarantee a minimum level of quality. How to avoid hard disks filled with trash, how purge them of nonsense and blunders and how warrant the same standards as those provided by professionals? The stumbling blocks to this are presented not by technology but, as usual, by the data; and more specifically by its inherently erroneous nature, a feature common to any product of human activity.

Let us take as an example a geomatics activity which is close to the very core of the surveying profession: updating of topographic maps (Lemmens, 2009). While preparing a vacation using maps presented by an Earth viewer or travelling around using a car navigation system the geographically aware traveller may discover discrepancies between the world as mapped and the real world. Some of them would like to inform the map publisher of the flaws they have found. So new technology has created potentially millions of individual information sources from which mapping agencies could benefit for updating topographic databases. However, communicating flaws via e-mail is tiresome and far from optimal in terms of transferring geo-information. And as long as no proper feedback infrastructure has been established, one would be unsure of receiving a response. A feedback infrastructure can be introduced by using today's web-based GIS services. Against a backdrop of existent maps, the co-operative and participating citizen could draw new features and insert GNSS coordinates. The participatory approach would enable an increase in update frequency and give participants the feeling of becoming part-owner of the map they are using. Why should the job be left to professionals alone, when so many laymen have become geographically aware and, given time, know how to control location-based technology? Would not the information provided by laymen be unreliable? To avoid flaws, use can be made here of the statistical law of large numbers. One source of information is no source, two are unsuitable, and with three comes a sense of reliability.

3.5.4 Geosensor Web

Web 2.0 and the continuing miniaturisation of cameras and others sensors, coupled with improvements in distributed computing and the development of open standards that facilitate interoperability, foster the ubiquitous presence of sensors (Lemmens, 2008a). Furthermore, the prices of sensors are rapidly falling. As a result, the attention for web sensors by professionals in a wide range of applications and management fields is greatly increasing. 'Can web sensors partly replace the presence of nurses (who are becoming more and more scarce) in greying societies?' the care manager might think. The observation tasks could be done by sensors and the nurse can jump in after an alarm sounds. (But what about privacy, others might wonder?) 'How can web sensors ease my task to prevent and fight wild fires?' the forester might ask. 'How can I better act in advance to possible evacuation needs of people living in a river delta endangered by floods?' the water board director may ponder.

Configuring and establishing an array of sensors in a network to manage the complex environment or to resolve geo-management problems requires an infrastructure of sensors, (wireless) connections and processing units, consisting of hardware and software (Tao and Liang, 2009). Such an infrastructure has been baptised the sensor web and when location of the sensor is an essential component of the information collected, such a system is called geosensor web or geosensor network (Nittel et al., 2008). Placing sensors in a network may not only support care managers of elderly people but also ease monitoring of phenomena in geographic spaces. Fixed and mobile sensors connected to the web are able to observe many different spatial and temporal phenomena, such as weather, traffic flow on highways, embankments along rivers, earthquakes, volcanism and forests under threat of wild fires. When combining topographic data and satellite/aerial imagery with in situ measurements, a great variety of geo-related applications can be put on ground. A prototype of such an application, developed by Wageningen University, the Netherlands, combines earth observation data and in situ sensor data for daily mapping of vegetation productivity for the Netherlands with a ground sample distance (GSD) of 250 m (Kooistra et al., 2009). To enhance the tempo of disaster response the Meraka Institute, Pretoria, South Africa developed a system combining satellite imagery and in-situ measurements for vegetation fire detection in South Africa (Terhorst et al., 2008).

The road from technological idea to becoming a solid, operational system on the ground is paved with many challenges (Nittel, 2009). The amount of sensors may be vast and together they all produce masses of real-time data that need to be transferred over powerful wireless communication systems. Data are not yet information and thus needs processing to become usable and the highly skilled developer must thoroughly understand all the cogs that pave the pathway leading from the abundance of data collected to usability. He or she should also be able to translate his or her understanding into algorithms and software agents. To stimulate broad use user-friendly applications programming interfaces (APIs) should be developed. As sensors may collect a great variety of data – such as temperature, air particles and satellite data – that have different temporal and spatial resolutions and different data models and formats, a fusion problem arises involving both the processing of the data and their management. Another challenge is that the meaning of data and how data should be interpreted depend on application and context. The data generated in a geosensor network are derived from many and diverse sensors and combining miscellaneous data requires defining ontologies and their mutual tuning (Hornsby and King, 2008). Though the degree of miniaturisation is high and the end is not yet in sight, sensors still consume electrical power. The energy is often hauled from batteries. Progress in the capacity of information technology, such as the number of nozzles on print heads, is often expressed in terms of Moore's law: capacity doubles nearly every 2 years. Unfortunately, the capacity of batteries increases by only about 8%, doubling in only every 10 years.

3.6 Summary and Further Reading

Essentially, GIS supports the collection and up-dating of geo-data, and its storage, retrieval, processing, analysis, visualisation, dissemination and management. GIS enables to retrieve data fast, to compute alternative solution, to include multiple viewpoints and to provide realistic visualisations, which are easily understandable for laymen. Since GIS eases the creation of models of the real world and their interpretation and visualisation it is a perfect aid for all concerned with understanding real world phenomena and processes from geo-data.

In trying to get across these sophisticated topics one may overlook the fact that the crux of successful application of GIS in whatever situation is the availability of accurate, detailed and up-to-date geo-information. All the gigantic number of terabytes of geo-data in the world is of no help in solving Earth-related problems so long as these terabytes fail to reach the hands of knowledgeable and skilled professionals. And with 'knowledgeable and skilled' is meant not 'knowing how to push the button' but thorough understanding of all the facets of the technology and, by far most vital, knowing how to focus all available resources: human, technological, monetary and so on. Or in the words of Roger F. Tomlinson, 'Father of GIS':

> No GIS can be a success without the right people involved. A real-world GIS is actually a complex system of interrelated parts, and at the center of this system is a smart person who understands the whole.

An abundance of textbooks treating principles, technologies and applications of GIS have been published in the course of time. Heywood et al. (2006) and Chang (2007) provide general introductions to GIS. An in-depth introduction to the principles and mathematical foundation of GIS can be found in Burrough and McDonnell (1998). An in-depth treatment of the computational aspects of GIS is given in Worboys and Duckham (2004). For an introduction to the concepts and algorithms of GIS refer to Berry (1993). A recent textbook, treating principles of GIS, edited by Madden (2009) is the Manual of GIS. An integrative approach to GIS, GNSS, digital photogrammetry and cartography orientated towards practitioners is given in Thurston et al. (2003). For those interested in the use of GIS, remote sensing and other geo-information technologies for addressing the major issues in the field of environmental health, refer to Maantay and McLafferty (2011). The on-going developments in Web-based GIS have created a strong interest and academic research emphasis in the area of web mapping. An early exploration of the prospects for web maps using the world wide web, written by multiple authors, has been collected by Kraak and Brown (2001). Introductions to the principles, technologies and applications of web mapping services written for cartographic graduates and professionals are given in Fu and Sun (2010) and Tsou (2011). The basics and possibilities of the many techniques available to collect geo-data are discussed in the second part of the present book.

References

Angelaccio M, Krek A, D'Ambrogio A (2009) A model-driven approach for designing adaptive web GIS interfaces. In: Popovich VV, Schrenk M, Claramunt C, Korolenko KV (eds) Information fusion and geographic information systems. Proceedings of the Fourth International Workshop, 17–20 May 2009. Springer, pp 137–148. ISBN: 978-3-642-00303-5

Antenucci JC, Brown K, Croswell PL, Kevany MJ (1991) Geographic information systems: a guide to the technology. Kluwer, Dordrecht, The Netherlands. ISBN 0442007566

Berry JK (1993) Beyond mapping: concepts, algorithms and issues in GIS. GIS World Books, Fort Collins, CO

Bolstad P (2005) GIS fundamentals: a first text on geographic information systems, 2nd edn. Eider Press, White Bear Lake, MN

Brody H, Rip MR, Vinten-Johansen P (2000) Map-making and myth-making in Broad Street: the London Cholera epidemic 1854. Lancet 356:64–68

Burrough PA, McDonnell RA (1998) Principles of geographical information systems, 2nd edn. Oxford University Press, New York, NY

Chang K (2007) Introduction to geographic information system, 4th edn. McGraw Hill, New York, NY

Coppock JT, Rhind DW (1991) The history of GIS. In: Maguire DJ, Goodchild MF, Rhind DW (eds) Geographical information systems: principles and applications. Addison Wesley Longman, Harlow, pp 21–43

Cromley EK, McLafferty SL (2002) GIS and public health. Guilford Press, New York, NY

Dorling D, Fairbairn D (1997) Mapping: ways of representing the world. Addison Wesley Longman, Harlow. ISBN 978-0582-28972-7

Doyle D, Dodge M, Smith A (1998) The potential of Web-based mapping and virtual reality technologies for modelling urban environments. Comput Environ Urban Syst 22(2):137–155

Dragicevic S (2004) The potential of web-based GIS. J Geogr Syst 6(2):79–81

Flemons P, Guralnick R, Krieger J, Ranipeta A, Neufeld D (2007) A web-based GIS tool for exploring the world's biodiversity: the Global Biodiversity Information Facility Mapping and Analysis Portal Application (GBIF-MAPA). Ecol Inform 2(1):49–60

Foresman TW (ed) (1998) The history of geographic information systems: perspectives from the pioneers. Prentice Hall, Upper Saddle River, NJ

Fu P, Sun J (2010) Web GIS: principles and applications. ESRI Press, Redlands, CA. ISBN 158948245

Gao S, Mioc D, Yi X, Anton F, Oldfield E, Coleman DJ (2009) Towards web-based representation and processing of health information. Int J Health Geogr 8(3). http://www.ij-healthgeographics.com/content/8/1/3

Gatrell JD, Jensen RR (eds) (2009) Planning and socioeconomic applications. Springer, Berlin, Heidelberg, New York. ISBN: 978-1-4020-9641-9

Goodchild MF (1987) A spatial analytical perspective on geographical information systems. Int J Geogr Inf Syst 1:327–344

Goodchild MF (2007) Citizens as sensors: the world of volunteered geography. GeoJournal 69(4):211–221

Hess HH (1962) History of ocean basins. In: Engel AEJ, James HL, Leonard BF (eds) Petrologic studies: a volume in honor of A. F. Buddington. Geological Society of America, Boulder, CO, pp 599–620

Heywood I, Cornelius S, Carver S (2006) An introduction to geographical information systems, 3rd edn. Prentice Hall, Upper Saddle River, NJ

Holz K-P, Hildebrandt G, Weber L (2006) Concept for a web-based information system for flood management. Nat Hazards 38:121–140

Hornsby KS, King K (2008) Linking geosensor network data and ontologies to support transportation modeling. In: Nittel S, Labrinidis A, Stefanidis A (eds) GeoSensor networks. Springer, Berlin, pp 191–209

Huang B, Lin H (1999) GeoVR: a web-based tool for virtual reality presentation from 2D GIS data. Comput Geosci 25:1167–1175

Jensen RR, Gatrell JD, McLean DD (eds) (2007) Geo-spatial technologies in urban environments: policy, practice and pixels. Springer, Berlin, Heidelberg, New York. ISBN 978-3-540-22263-4

Johnson S (2006) The ghost map: the story of London's most terrifying epidemic, and how it changed science, cities, and the modern world. Riverhead, New York, NY

Kearey P, Klepeis KA, Vine FJ (2009) Gobal tectonics, 3rd edn. Wiley-Blackwell, 1st Edition 1990, 2nd Edition 1996, Chichester, West Sussex, UK

Koch Th (2005) Cartographies of disease: maps, mapping, and medicine, 1st edn. ESRI press, Redlands, CA

Koch Th, Denike K (2006) Rethinking John Snow's South London study: a Bayesian evaluation and recalculation. Soc Sci Med 63:271–283

Kooistra L, Bergsma A, Chuma B, De Bruin S (2009) Development of a dynamic web mapping service for vegetation productivity using earth observation and in situ sensors in a sensor web based approach. Sensors 9:2371–2388

Kraak M-J, Brown A (eds) (2001) Web cartography: developments and prospects. Taylor & Francis, New York, NY

Lang L (2000) GIS for health organizations. ESRI Press, Redlands, CA

Lemmens M (2006) GIS, RS and GNSS in whatever. GIM Int 20(3):11

Lemmens M (2007) Medical mapping: importance of coordinates and geo-referencing. GIM Int 21(9):25–27

Lemmens M (2008a) Geosensor network. GIM Int 22(6):11

Lemmens M (2008b) MAPS. GIM Int 22(8):36–39

Lemmens M (2009) Geo-Information from the millions. GIM Int 23(9):65

Lemmens M (2010) Two point nil. GIM Int 24(4):57

Maantay JA, McLafferty S (eds) (2011) Geospatial analysis of environmental health. Springer, Berlin. ISBN 978-94-007-0328-5

Madden M (ed) (2009) The manual of GIS. American Society for Photogrammetry and Remote Sensing (ASPRS). Bethesda, Maryland, USA. ISBN 1-57083-086

Mansourian A, Rajabifard A, Valadan Zoej MJ, Williamson I (2006) Using SDI and web-based system to facilitate disaster management. Comput Geosci 32(3):303–315

Nittel S (2009) A survey of geosensor networks: advances in dynamic environmental monitoring. Sensors 9:5664–5678

Nittel S, Labrinidis A, Stefanidis A (2008) Introduction to advances in geosensor networks. In: Nittel S, Labrinidis A, Stefanidis A (eds) GeoSensor networks. Springer, Berlin, pp 1–6

Pickles J (1999) Arguments, debates, and dialogues: the GIS-social theory debate and the concern for alternatives. In: Longley PA, Goodchild MF, MacGuire DJ, Rhind DW (eds) Geographical information systems: principles, techniques, applications, and management, 2nd edn. Wiley, New York, NY, pp 49–60

Stevenson L (1965) Putting disease on the map: the early use of spot maps in the study of yellow fever. J History Med 20:227–261

Tao V, Liang SHL (2009) Introduction to the spatial sensor web. In: Madden M (ed) The manual of GIS. American Society for Photogrammetry and Remote Sensing (ASPRS). Chichester, West Sussex, UK. ISBN 1-57083-086

Terhorst A, Moodley D, Simonis I, Frost P, McFerren G, Roos S, Bergh F (2008) Using the sensor web to detect and monitor the spread of vegetation fires in southern Africa. In: Nittel S, Labrinidis A, Stefanidis A (eds) GeoSensor networks. Springer, Berlin, pp 239–251

Thurston J, Poiker TK, Moore JP (2003) Integrated geospatial technologies: a guide to GPS, GIS, and data logging. Wiley, Hoboken, NJ

Tomlinson RF (2007) Thinking about GIS: geographic information systems planning for managers, 3rd edn. ESRI Press, Redlands, CA. ISBN: 9781589481589

Topper J (1997) Yellow Fever/Reed Commission Exhibit: this most dreadful pest of Humanity, Yellow Fever and the Reed Commission 1898–1901, University of Virginia

Tsai N, Choi B, Perry M (2009) Improving the process of E-Government initiative: an in-depth case study of web-based GIS implementation. Gov Inf Q 26(2):368–376

Tsou M-H (2011) Designing web map services and network-based cybercartography, 1st edn. Series: Advances in Geographic Information Science, Springer, Berlin. ISBN: 978-3-540-74856-4

Tucker CJ (1979) Red and photographic infrared linear combinations for monitoring vegetation. Remote Sens Environ 8(2):127–150

Vine FJ, Matthews DH (1963) Magnetic anomalies over oceanic ridges. Nature 199:947–949

Vinten-Johansen P, Brody H, Paneth N, Rachman S, Rip M (2003) Cholera, chloroform, and the science of medicine: a life of John Snow, Oxford University Press, New York, Oxford, UK. ISBN: 0-19-513544-X

Winchester S (2001) The map that changed the world: William Smith and the birth of modern geology. Harper Collins, New York, NY. ISBN 0060931809

Worboys M, Duckham M (2004) GIS: a computing perspective. CRC Press, Boca Raton, FL

Chapter 4
Global Navigation Satellite Systems and Inertial Navigation

When one would rank the geo-data collection techniques developed the last three decades or so from most significant to least significant, positioning and navigation by means of Global Navigation Satellite Systems (GNSS) would head the list. GNSS enables obtaining precise positioning and timing information anywhere on land, on sea or in the air, day or night with high precision and reliability and against affordable costs. GNSS does not require cleared lines of sight between survey stations as other conventional surveying procedures, which rely on observing angles and distances between visible ground stations, required for determining two-dimensional or three-dimensional coordinates of points.

Since the turn of the millennium, GNSS has been applied for a broad spectrum of applications, ranging from studying the movements of monkeys through the jungle to precise bombardments. GNSS is not only suited for navigation but is also fruitfully applied by land surveyors and others collecting geo-data in the field for precise positioning at (sub)centimetre level accuracy. Positioning and navigation by

Surveying is often the initial step for exploring natural resources, so surveyors have to operate sometimes under harsh conditions. Here a surveyor carries out GNSS measurements at Antarctica

satellites has now become part of everyday life for many citizens and professionals all over the globe, and a plethora of (commercial) services rely on the continuous well functioning of these systems.

This chapter treats the basic principles of GNSS, its characteristics and ongoing developments. Furthermore we will have a look at the principles of inertial navigation systems as they, combined with GNSS, form the very core of geodata collection from sensors mounted in aircrafts, cars, vessels and other moving platforms. In writing this chapter use has been made of a variety of sources including textbooks, scientific journals, professional journals, Wikipedia, the websites of the two operational providers of GNSS signals (GPS and Glonass) and the websites of the main manufacturers of GNSS receivers (Ashtech, Foif, Hemisphere GPS, Javad, Leica Geosystems, NovAtel, Omnistar, Septentrio, South, Topcon and Trimble). The website of the Department of Geography, The University of Colorado at Boulder, provides an overview of the principles of the Global Positioning System, developed by P. H. Dana (colorado.edu). This website is recommended for those who want to deepen their knowledge on the mathematical principles of GNSS. In the tutorial on the website of manufacturer of GNSS receivers, Trimble, basics are demonstrated using illustrative animations (trimble.com). Where appropriate we will refer to these and other sources.

4.1 Overview

Originally designated the Navstar (Navigation System with Timing And Ranging) the Global Positioning System (GPS) was the first fully operational Global Navigation Satellite System (GNSS). The first GPS satellite (Navstar 1) was launched on 22 February 1978.

Today, when hearing the word GPS, most people think about car navigation that is determining position, velocity and heading of the car. Car navigation systems are becoming an essential facility for many drivers, especially those professionals who commercially transport goods, livestock and passengers and these professional acknowledge that navigation systems ease their daily work and safe arrival at a destination. Based on satellite positioning and micro-electronics, the navigation systems have a high degree of automation, appearances and can be operated by simply button-pushing (Fig. 4.1).

With the broad embracing of car navigation systems by the general public, GNSS technology has become a hit beyond imagining. The success is even such that the man in the street associates the term GPS with navigation, making 'GPS' its synonym, while GPS is actually just one of the navigation and precise-positioning tool presently available or available in the near future. As a mainstream activity GNSS constitutes a formidable growth market: since the early years of this millennium there have been annual growth rates in the satellite positioning industry reaching 25%, resulting in tripling within 5 years. It is forecast that by 2020 annual worldwide turnover will be Euro 300 billion that is 3% of the present Gross National Product of the European Union, and one receiver for every two citizens on the globe

Fig. 4.1 The man in the street associates the term GPS with car navigation (*left*), bike navigation and navigation facilities on a handheld. In some parts of the world they are called Tom-Tom, named after a major manufacturer

(Lemmens, 2007b). However, when a technology goes mainstream complaints soon start being heard from the general public about negative impact on others than the actual users and one of the issues is that an optimal route advised by the navigation system, whilst perfect for the driver, may dramatically interfere with the lives of those dwelling along the road (Lemmens, 2007a). Heavy goods vehicles are relatively more often involved in accidents than cars; they cause approximately 15% of all fatal accidents, whereas they constitute on average around 12% of traffic volume and contribute only 2% to the vehicle pool.

Commercial receivers, build for the consumer market, look very similar to mobile phones and have about the same size. Micro-electronics allows building them in such a compact way, that they even fit in wrist watches. The accuracy is in the 5–30 m range. Instruments built for professionals, such as land surveyors and military, are bigger and accurate to the (sub)centimetre level. Land surveyors and other fieldworkers are increasingly making use of GNSS.

The success of GNSS technology is attributable to a number of mutually reinforcing factors (Leick, 2004). First there is the ability to measure the phases of the carrier signal with a resolution of 1% of the cycle length which equals 1.9 mm for L1 GPS signal and 2.4 mm for L2 GPS signal, while their high frequencies enable to penetrate the ionosphere relatively well. The ionosphere causes delay in travel time, but, since this time delay is inversely proportional to the square of the frequency, carrier phase measurements enable to get rid of most delays introduced by the ionosphere. Furthermore, stability and miniaturisation of atomic clocks have shown significant progress, improving the timing at each satellite. The positions of the satellites, which have to be known, can be accurately determined because they orbit in a stable manner; their high altitude prevents that drag affects their motion which is only determined by the major gravitational forces which are mathematically modelled and thus the position of satellites can be predicted.

Six countries are now involved in the operational provision or development of GNSS services. USA has in place its Global Positioning System (GPS) and Wide-area Augmentation System (WAAS), and the Russian Federation its Global Navigation Satellite System (GLONASS) and Wide-area System of Differential Corrections and Monitoring (SDCM). The European Union (EU) is working hard

Fig. 4.2 First Galileo satellite, launched in the early morning of 28 December 2005; artist perspective

to put its Galileo satellites into orbit (Fig. 4.2) and has the European Geostationary Navigation Overlay Service (EGNOS) on the ground and two satellites in orbit.

The fully deployed Galileo system will consist of 30 satellites and is scheduled for being fully operational in 2014. India and Japan are working on putting navigation satellites into orbit and establishing GNSS augmentation services. The Indian system will consist of seven satellites operating regionally up to 2000 km around the boundaries of the subcontinent and providing an accuracy of better than 20 m as well as a GNSS augmentation system. China's Compass/Beidou Navigation Satellite System (CNSS) is under development. Japan is working on the Quasi-Zenith Satellite System (QZSS), which also operates regionally and will ultimately consist of three satellites. Since the accuracy of its standalone mode is limited, this is actually regarded as an augmentation service for other GNSS systems. All these initiatives facilitate the creation of promising applications in land surveying, land management, sustainable development and many others.

4.2 Basics

Basically, satellite positioning is a trilateration problem (Fig. 4.3). From the known position of three satellites and the measured distances between them and the receiver, the three-dimensional coordinates of the receiver position can be calculated. The distances are determined by multiplying the travelling time of the radio signals, which are in the microwave domain of the electromagnetic spectrum, by the speed of light. The use of microwave signals enables GNSS to work in any weather conditions; the signals can penetrate haze and clouds.

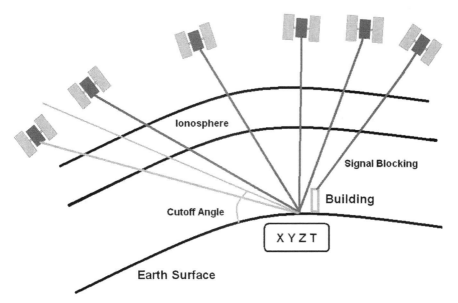

Fig. 4.3 GNSS positioning requires at least four satellites for determining the triplet of position coordinates (*X*, *Y*, *Z*) and the time component (T). The signals from the satellites outmost left and right are not usable due to too close to horizon and signal blocking

GNSS receivers can use two fundamental methods in determining distances (ranges) to satellites: code ranging and carrier phase-shift measurements. The signals continuously emitted by the GNSS satellites consist of a carrier signal on which a code is embedded. The carrier is, so to say, the herald who carriers the message from the sender to the recipient. The code contains the actual message. When using code ranging without making use of GNSS infrastructure (see Section 4.7), i.e. the receiver is operating in the stand-alone mode, the coordinates of a point can be determined with an accuracy of 10–30 m. A great variety of methods has been developed in course of time, both by scientists and manufacturers of GNSS receivers, to improve accuracy and the most important of these is the methods based on carrier phase-shift measurements using Differential GNSS (DGNSS).

The travelling time of a radio signal may principally be determined as the difference between arrival time at the receiver and time of transmission by the satellite. Since radio signals, as all electromagnetic energy, travel at a speed of 300,000 km per second, an inaccuracy in time measurement of 1 ns (one billion part of a second or 10^{-9} s) induces a distance error of 30 cm, whilst with geodetic receivers, making use of DGNSS accuracies at the centimetre or even subcentimetre level can be achieved for the triplet of *X*, *Y* and *Z* position coordinates. To achieve positional accuracy at centimetre level the time difference has to be determined at sub-nanosecond level. To attain such accuracy would require atomic clocks in both receiver and satellites. Atomic clocks are, however, rather too expensive for the everyday consumer. The solution is simple and effective. Because they are few, only the satellites are equipped with costly atomic high-precision clocks having an annual drift of just 2 ns. The receivers are equipped with relatively cheap, quartz clocks accurate up to 10 ns a day. This accuracy is sufficient to allow it to be assumed that time bias between the receiver clock and those onboard the satellites is the same for all satellites. So that just one time parameter is actually unknown: the time bias. This can be determined by measuring the travelling time of radio signals to an additional satellite that means using not three but four satellites. Discrepancies among the calculated coordinates of receiver position (which should, of course, be absent, it concerning just one location) provide sufficient information for accurately determining the time bias. So, at least four satellites should be visible for estimation of the triplet of position coordinates and the time component (Fig. 4.3). However, four satellites represent a minimum and provide quite unreliable results. Every land surveyor knows that for purposes of precision and reliability one should gather more measurements than is actually needed to solve the unknown parameters (Baarda, 1968) – called redundancy – in this case the triplet of position coordinates and the time bias. Therefore high-precision applications will require measurements from more than four satellites.

4.3 Achieving High Accuracy

Capturing the signals of more than four satellites, however, does not warrant high accuracy all the time (Hoffmann-Wellenhof et al., 1994; El-Rabbany, 2002). When the satellites are grouped close together, e.g. in the south-east, the satellite geometry is weak, causing inaccuracy. The satellites should be well distributed over the whole firmament. Interrupted receive of satellite signals can also degrade the reliability of positioning. Interruption occurs when there is no open 'view' between the receiver and the satellite because of presence of obstructions. This will be the case in forest areas where foliage cuts off signals. Also built-up areas with high or densely placed buildings are sensitive to signal interruption while in tunnels and under bridges it will be hard to receive signals. Blocking of GNSS signals frequently occurs in major cities like Beijing, New York and Tokyo. Streets flanked by (highrise) buildings, where receive of GNSS signals is problematic, are called 'urban canyons'.

The combined effect of satellite geometry and obstructions on the positioning accuracy is expressed in a measure called Dilution of Precision (DOP); the lower the DOP value the better the accuracy will be (Langley, 1999). Often DOP is decomposed in a horizontal component (HDOP) and a vertical component (VDOP). The DOP of all three positioning components is called PDOP. A PDOP of one is the ideal situation but a value up to three indicates a geometric configuration which is will suited for surveying purposes. A PDOP value of four or better is suited for car-navigation purposes.

Part of the path travelled by GNSS signals is through the atmosphere. Particularly the delays due to the ionosphere and troposphere cause positioning problems

(Klobuchar, 1991, 1996). The ionosphere is the uppermost part of the atmosphere from 50 to 1000 km above the surface of the earth. The troposphere encompasses the lowest layer and its thickness ranges from 8 km at the poles to 16 km at the equator. The troposphere contains 75% of the mass of the atmosphere and 99% of its water vapour; the temperature decreases with increasing attitude. The accuracy decrease is smallest when the satellite is in the zenith that is above the receiver but the accuracy decreases gradually when the satellite moves from the zenith position to the horizon, since the closer to the horizon the longer the signal will travel through the atmosphere (Fig. 4.3). In course of time several remedies have been developed to diminish the degrading effects of the atmosphere and to arrive at accurate measurements:

– Omitting measurements to satellites which are less than 15° above the horizon.
– Removal of the ionospheric effects by using not one signal transmitted by the satellite (single frequency) but both the L1 and L2 frequencies of the GPS satellites; the delay of microwave signals in the ionosphere is frequency depended. The signal delays caused by the troposphere cannot be eliminated.
– Use of differential GNSS (DGNSS); the simultaneous use of two receivers, one positioned on a point of which the three-dimensional coordinates (planimetric X and Y coordinates and height) are known (base point or reference point).

Particularly, DGNSS allows arriving at computations of coordinates in the National Geodetic Reference System suitable for surveying purposes.

The basic aim of DGNSS is to correct for errors in the measurements of a receiver caused by ionospheric and tropospheric delays, receiver clock errors and orbital errors. DGNSS is carried out by accompanying the use of the user's receiver, called rover, with a second GNSS receiver, called base station, positioned on a location of which the coordinates are known in a geodetic reference system (Figs. 4.4 & 4.5). Since the coordinates of the base stations are known, they can be compared with the coordinates measured with the GNSS receiver located on the reference point. When the atmospheric distortions are more or less the same for the base station position and the rover position, the corrections computed from measurements at the base station are also applicable to the rover measurements. The shorter the distance between base station and rover, also called base-line, the more resemblance between the atmospheric conditions – and hence distortions – the better the corrections will fit. Differential corrections may be transmitted in real-time by radio link or mobile phone to the rover enabling the user instantaneous capture of position. When making use of code ranging solely the achievable accuracy of DGNSS is in the order of 30–50 cm. When carrier phase shift measurements are used in addition to code ranging the accuracy, that can be obtained, can even be at the subcentimetre level, or more formally 5 to 10 mm + 1 ppm, where ppm means parts per million. For example, when the baseline is 12 km, the accuracy becomes 5 to 10 mm + 12 km/1,000,000 is 5 to 10 mm + 12 mm.

Fig. 4.4 GNSS receiver used as rover for stake out purposes. In the background a receiver positioned exactly above a reference point enabling DGNSS and thus high-precision measurements

4.4 Selective Availability

GPS and Glonass are both operational. However, the satellite configuration of Glonass is not yet complete (status 1 January 2011). So, GPS is the only system which is fully operational (fully operational capability was reached in July 1995). All other systems are either incomplete or in a stage of development. Owned by USA and managed by its Department of Defense, GPS got underway in 1973 and was initially intended for military use only (Parkinson, 1996). Nearly 10 years later its continuation became a critical issue as US Congress expressed doubts about the usefulness of the system. However, the trouncing of a civilian Korean airplane (Flight 007) in 1983 over Russian ground changed minds and Congress decided to increase funding and allow its use for civilian aircraft navigation, free of subscription fees or setup charges. Loss of the space shuttle Challenger in 1986 caused further disruption to system plans because these vehicles were designed to carry GPS satellites; delta rockets eventually replaced the shuttle as carrier.

The enormous boost of the use of GPS in many sectors of the economy and society as a whole started May 2000. As a military system, the US government wanted

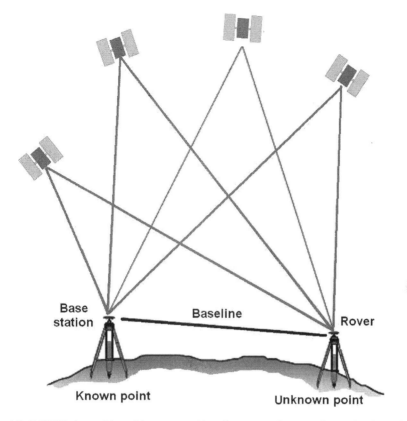

Fig. 4.5 DGNSS: the position of the rover positioned over an unknown point can be determined relative to the position of the base station

that highly accurate GPS signals would only be available to the military and not to civilians; one was afraid of violent abuse. Therefore, GPS signals were intentionally degraded: high accuracy (10 m) was selectively available to the US military only while others had to be content with 100 m accuracy. In the meantime scientists had discovered that astonishing high accuracies could be achieved by combining code tracking with measuring the phase of the signal carrying the code. The ranging codes on the carrier signal, broadcasted by the satellite and received by the user, contain all the necessary information, including satellite position, to compute position. Phase refers to the fractional part of the L1 or L2 carrier wavelength (approximately 19 cm for L1, 24 cm for L2). L1 and L2 are the two signals broadcasted by GPS satellites from the beginning; L1 transmits the Coarse/Acquisition code (C/A) while L2 transmits the restricted Precision code (P-code) reserved for military applications. The codes embedded in L1 signals are freely available to civilian users, but the use of L2 signals is preserved for military applications only.

The US government turned off SA per 1 May 2000 by a Presidential Decision Directive signed by Bill Clinton. The turn-off improved the accuracy of GPS receivers; 'civilian users of GPS will be able to pinpoint locations up to ten times more accurately' as the Directive states. The decision is in line with the 1996 Presidential Decision Directive to 'encourage acceptance and integration of GPS into peaceful civil, commercial and scientific applications worldwide; and to encourage private sector investment in and use of U.S. GPS technologies and services'. The turn off of SA is set down in the Statement by the President regarding USA's decision to stop degrading GPS accuracy. Being pleased to announce that USA will stop the intentional degradation of the Global Positioning System (GPS) signals available to the public, the President states further 'The decision to discontinue SA is the latest measure in an on-going effort to make GPS more responsive to civil and commercial users worldwide'.

4.5 GPS Modernisation

The decision to discontinue SA is often seen as the first step in GPS Modernisation, referred to as GPS III and aimed at improving positioning accuracy, signal availability and system integrity. GPS Modernisation focuses on better serving the civilian users. To get rid of ionospheric delay errors two signals are required but only the C/A code, carried by L1, is accessible to civilian users. The confinement of using single frequency receivers makes today ionospheric delay the largest source of error for GPS receivers used by the general public. The Modernisation program accounts for this by adding a second civilian-use signal to the L1 and L2, called L2C. The satellites equipped with the necessary new hardware are being gradually put in orbit. To benefit from L2C one has to use a dual frequency receiver. To stimulate development of safety-of-life applications for the transportation industry as well as to diminish vehicle fuel consumption and increase capacity in airspace, waterways, highways and railroads a third civilian-use signal has been developed: L5. The transmission power is twice as high compared to the two other civilian-use signals, L1 and L2C, enabling better penetration through trees and other objects which may block satellite signals. On 10 April 2009, an L5 signal was successfully transmitted for the first time from an orbiting GPS III satellite.

4.6 Multipath

Many of the error sources, discussed above, can be eliminated or their effects reduced by taking precautions, however one source of error is difficult to get rid of – multipath. Multipath is caused by multiple reflections of the microwave signal emitted by the GNSS satellite (Fig. 4.6). An electromagnetic or sound signal hitting a surface may interact with it in one of three ways: the signal may be reflected, absorbed or transmitted. Having once been reflected a signal may be reflected a second time or even a third time from another surface. In airborne Lidar multiple reflection results in the recording of longer pulse travel times that ultimately appear

Fig. 4.6 Multipath, part of
the signal reaches directly the
receiver, others along indirect
paths

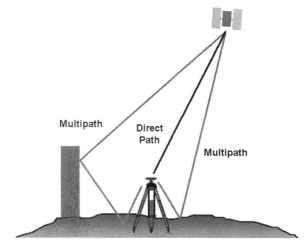

as a dip in the elevation model. Since these are usually individual events, they can be removed by spatial filtering. Similarly, in radar imagery multiple reflections cause objects to produce returns of greater signal strength than might be expected from the size of the object; this results in bright spots.

The precision of GNSS positioning is also affected by reflections from nearby objects such as ground and water surfaces, buildings, vehicles, trees or mountains. Multipath results in the same satellite signal being received at least twice by the GPS receiver via different paths, distorting the measurements. Multipath may even be regarded as the main remaining source of error, since others can be removed by advanced processing methods such as DGNSS (Soubielle et al., 2002). Particularly in urban areas which are characterised by multiple reflecting surfaces, multipath may significantly reduce precision. Hence it is of the utmost importance to detect and/or mitigate multipath error. In contrast, multipath effects are minor in a moving vehicle because a mobile receiver results in solutions from indirect signals failing to converge; only direct signals produce stable solutions.

There are three basic approaches to dealing with multipath. The first, adopting the adagio 'prevention is better than cure', is to avoid measuring in environments where multipath might occur. The second is to physically protect the GNSS antenna from indirect signals using a reverse umbrella, thus safeguarding the antenna from those bad coming from below. The third method is to separate the bad from the good using signal-processing techniques. The first option is obviously impractical and too much impedes operational application of GNSS. But caution during field measurements may help. For example, tracking satellites only when they are more than 15° above the horizon already limits multipath effects.

It is also possible to filter bad from good signals by numerical techniques util-ising redundancy. Reflected signals are always delayed in comparison with direct signals because of their longer paths of travel and these shows as time-dependent variations in measured range to the satellite, which can be detected. In this way

Fig. 4.7   Use of a choke ring to protect a receiver positioned on a reference point for multipath

direct signals can be separated from indirect ones. However, when difference in path length between direct and indirect signal is less than a few metres the signal-processing technique proves ineffective. The remedy is then to entirely ignore the signal from the satellite concerned. This is often quite feasible, as GNSS receivers typically receive signals from eight to twelve satellites while a minimum of four are needed for determination of the three position coordinates and time bias. It will rarely occur that signals from all satellites are being affected by multipath.

A more reliable method is to use a reverse umbrella, or 'choke-ring' (Fig. 4.7). Such a physical device enables rejection of indirect signals hitting the bottom of the antenna, but the device does not stop signals reflected by a building hitting the antenna from above. However, since such indirect signals have a path length of 10 m or even more, they may be mitigated by signal-processing techniques. Numerical and physical protection techniques are thus complementary: by numerical filtering distant indirect signals hitting the top of the antenna can be mitigated, while choke-rings prevent the antenna picking up signals, usually with a path length of a few metres, reflected from nearby ground.

4.7 GNSS Infrastructure

In many parts of the world it is not necessary anymore that surveyors themselves set up a base station when they want to carry out a high-accuracy survey based on DGNSS. Such infrastructures consisting of a network of GNSS receivers positioned

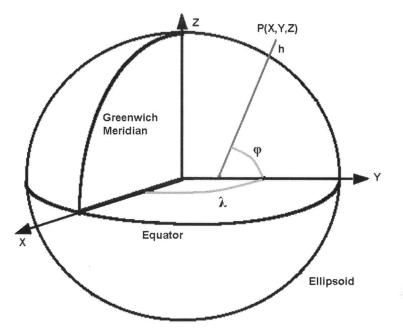

Fig. 4.8 The GNSS triplet of position coordinates P (*XYZ*) of a measured point (P) is converted within the receiver to geodetic (φ, λ, h) coordinates: latitude (φ), longitude (λ) and height (h) above the ellipsoid provided in WGS-84; the coordinates can be internally transformed to other reference systems, such as national geodetic reference system, with the height component referring to the geoid depending on user needs

on reference points (base stations) have been established by public agencies and private companies as added valued services. Such a service releases surveyors from setting up a receiver on a reference point. In addition the coordinates of the points measured can be provided in the National Geodetic Reference System, or another reference system, the infrastructure facilitates the transformation of the WGS84 (World Geodetic System) coordinates delivered by GPS into a local reference system (Fig. 4.8).

The services offered are often free of charge. For example, US Coast Guard maintains a network of base stations particularly around harbours and waterways, and transmits the corrections via radio-beacons covering much of the US coastline in a standard format and in real-time. Many GNSS receivers can accept these corrections and some are equipped with built-in radio receivers.

The Online Positioning User Service (OPUS) is a post-processing service operational since March 2001 and offered by the US National Geodetic Survey (NGS). Through a simple web-based interface users can submit data files recorded with any dual-frequency, geodetic-quality receiver, either in native receiver format or in Receiver-INdependent EXchange format (RINEX). After a sequence of processing steps a report is submitted to the user via email. In partnership with Ordnance Survey, Great Britain, Leica has established a real-time GNSS infrastructure in the

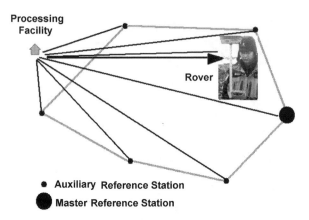

<mark>**Fig. 4.9**</mark> SmartNet GNSS infrastructure. Being connected by mobile phone or other wireless communication the rover transmits its initial position to the network processing facility. Next typically six stations surrounding the rover are selected, the nearest as master station with full corrections and the others as auxiliary correction differences. After computation corrections are returned to the user, enabling positioning at centimetre level accuracy

UK, called SmartNet (Fig. 4.9). Also Trimble has established a GNSS infrastructure in Great Britain to which surveyors can subscript to measure with centimetre accuracy using a high-end receiver and a communication link. The system is based on Trimble's toolkit of hardware and software for easily setting up GNSS infrastructures anywhere in the world, called VRS (Virtual Reference System). Based on Trimble technology, Malaysia completed a GNSS infrastructure covering the entire territory in May 2007. The system includes 27 reference stations and central processing software and the broadcasting to the user is done through the internet. A geometric infrastructure based on DGNSS, called AGRS (Active GNSS Reference System) is operational in the Netherlands since 1997, consisting of five base stations well distributed over the country. AGRS is operated by the Dutch Cadastre and Rijkswaterstaat (the Directorate-General for Public Works and Water Management). Belgium has a similar active system in place (Fig. 4.10).

In addition to ground-based GNSS infrastructures, also satellite-based systems are in place consisting of a constellation of satellites and ground stations that provide positioning corrections to users. These Satellite Based Augmentation Systems (SBAS) support wide-area or regional DGNSS through measurements taken at multiple base stations. Corrections derived from these measurements are sent to one or more satellites. Next those satellites broadcast the correction messages back to enabled receivers. Sub-metre accuracy in real time could be achieved by connecting to an SBAS System. One such system, initially developed for air navigation purposes for all phases of flight, including precision approaches to any airport, is WAAS (Wide Area Augmentation System), the SBAS of USA. It consists of around 25 base stations well distributed over USA. Two master stations, one at the east coast and the other at the west coast, process data from the base stations and broadcast corrections on atmospheric effects, satellite orbits and clock drift to two geostationary satellites.

Fig. 4.10 Distribution of the points of the GNSS reference network covering Belgium

Geostationary satellites circle in an equatorial orbit at such a height that their angular rotation speed exactly corresponds with the angular rotation speed of the earth. Viewed from the earth they preserve a stationary position above the equator; it seems as they do not move. When obstructions, such as trees and mountains are present, the signals may not reach the user, but in open land and on sea WAAS functions very well.

The focal point of use of WAAS is in USA. Europe has developed its own system, Euro Geostationary Navigation Overlay Service (EGNOS) (Fig. 4.11). The ground segment consists of 34 base stations while the broadcasting is done through three geostationary satellites. It provides augmentation services to users all over Europe either in the air, on land or at sea. The Asian continent will be covered by the Japanese Multi-Functional Satellite Augmentation System (MSAS) and India's GPS Aided GEO Augmented Navigation-Technology Demonstration System (GAGAN). The above SBAS systems are public funded initiatives. There are also a few SBAS services offered by private companies. One of them is OmniSTAR offered by the Fugro group of companies and serving both offshore survey applications and land-based applications such as precision farming.

Users active in the survey, geotechnical or geophysical industries are willing to pick-up commercial services instead of using free-of-charge government-funded services, because they rely on accuracy, permanent availability, wide coverage and 24/7 support.

4.8 Improving Accuracy

Two fundamental methods are employed by GNSS receivers in determining distances (ranges) to satellites: code ranging and carrier phase-shift measurements. The distances to satellites can be obtained with better accuracy by acquiring in

Fig. 4.11 EGNOS, the European satellite augmentation system, provides augmentation services to users all over Europe either in the air, on land or at sea

addition to the code information also phase shifts of the GNSS carrier signals. The way such electromagnetic signals propagate through vacuum or the atmosphere is usually modelled as sinusoidal curves. A complete cycle of the L1 signal has a wavelength of 19 cm and is said to have a phase angle of 360°. It will seldom occur that a full cycle will be picked up by the receiver. Usually this will be a part of the wavelength (Fig. 4.12). The determination of the phase of the GNSS signal can be done up to around 1% of the cycle length. Phase measured in degrees can be converted into distance as this corresponds to a part of the wavelength of the signal, yielding a range accuracy for L1 of 2 mm and for L2 of 2.4 mm. This is very precise but ambiguous since the number of phases at the beginning of tracking is unknown but constant over an orbit arc.

The phase observable is thus very precise, at the millimetre level. To benefit from these precise measurements, the unknown integer number of full wavelengths (cycles) needs to be resolved (Blewitt, 1989). The measurement procedure based on phase shifts makes use of the movement of the satellites. When the signal from a satellite is coming in, the GNSS receiver starts tracking of the phase of the signal and counts number of phases while the satellite is orbiting (Fig. 4.13). The number of complete cycles is unknown but constant over an orbit arc. When the number of the full carrier cycles is known with sufficient accuracy a very precise distance measurement to the satellite has been obtained.

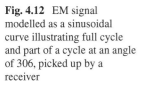

Fig. 4.12 EM signal modelled as a sinusoidal curve illustrating full cycle and part of a cycle at an angle of 306, picked up by a receiver

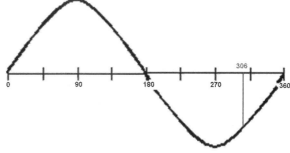

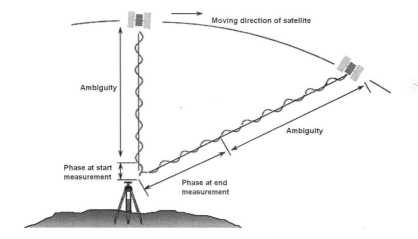

Fig. 4.13 Integer ambiguity resolution of carrier phase-shift measurements

A rough estimate of the number of complete cycles can be obtained by dividing the altitude of the satellite above ground by the wavelength of the GNSS carrier signal. For the L1 GPS signal this will result in 20,000 km, the altitude of the satellite above the ground, divided by 0.19 m, the wavelength of the L1 signal, yielding 105,263,158 cycles. For precise positioning this estimate is much too rough. Other very precise methods have been developed to determine the number of full wavelengths (cycles) that occurred as the signal travelled between the satellite and receiver. This number is called the integer ambiguity or ambiguity. The different techniques used to determine the ambiguity require the collection of additional measurements. Currently, the most efficient method of ambiguity resolution is the LAMBDA method developed at the Delft University of Technology, the Netherlands, by Teunissen and co-workers (see, e.g. Teunissen, 1995; Teunissen and Kleusberg, 1998; Teunissen et al., 1997). LAMBDA stands for 'Least-squares AMBiguity Decorrelation Adjustment'.

4.9 Geodetic Precision

The highest level of accuracy in determining the coordinates of points is achieved when employing carrier phase-shift methods combined with relative positioning methods, which make use of base stations. Since the approaches based on the combination of the two yield the highest accuracy, they are the preferred methods of land surveyors to establish ground control points (GCP) for georeferencing aerial images, to establish or densify geodetic reference networks, or perform other high-precision tasks. A variety of approaches are in use and we will treat some of them to give an impression of types of application and achievable accuracy.

Static relative positioning is used for achieving highest accuracy and is particularly in use for establishing reference networks or their densification. The GNSS receiver, positioned on the unknown point, collects data for a time period of an hour or more depending on the length of the baseline. Usually the measurement duration necessary to obtain subcentimetre precision is expressed as a constant time factor plus a relative time factor per kilometre, e.g. for single frequency receivers the measurement duration is 30 min + 3 min/km, and for dual frequency receivers 20 min + 2 min/km (Wolf and Ghilani, 2006). By slightly changing the measurement procedure the duration of the measurement session can be reduced by 50%. This so-called *rapid static* procedure yields the same accuracy as the static approach, but is only suitable if the baselines are 25 km at maximum and high accuracy can only be obtained when satellite configuration is optimal. The rapid static procedure is often used for establishing a small geodetic reference network, i.e. densification of the existing reference network, when building a bridge, tunnel or other construction work.

Carrier phase-shift methods combined with relative positioning methods are also employed for determining subsequent positions of airplanes, vessels or other moving platforms with high accuracy. When an airplane or helicopters acquires aerial photos, Lidar point clouds, or other sensor data, the position and attitude of the platform has to be continuously determined. The same is true when sonar mounted on a vessel collects sea depth data during a hydrographic survey. The position of the platform can be determined using GNSS *kinematic relative positioning*. This method requires an initialisation procedure for determining the integer ambiguity for each satellite. Several initialisation procedures are in use. The most advanced procedure is on-the-fly ambiguity resolution which requires a measurement duration of a few minutes. While the one receiver remains stationary on the known point, the other receiver moves around. This procedure requires sophisticated software. The kinematic approach can also be used for accurate land surveying purposes such as topographic or cadastral mapping or for construction surveys. This method has received the label *stop-and-go* or semikinematic positioning. The rover's antenna is positioned over points of which the surveyor wants to acquire coordinates. The coordinates of the points surveyed are determined in a post-processing procedure. The accuracy is better than for a true kinematic survey but the PDOP values should be below four. In real-time kinematic GNSS

surveys, the measurements are processed in real time that is the coordinates of surveyed points are immediately available. The accuracy is a little bit less than for
post-processed surveys.

4.10 Atomic Clocks

Without atomic clocks, GNSS would not be possible. Three main types of atomic
clock can be distinguished depending on the element used: caesium, hydrogen
or rubidium (Lemmens, 2005). The caesium-133 atom is most commonly used.
Atomic clocks do not rely on atomic delay and they are not radioactive. The
adjective 'atomic' refers to the characteristic oscillation frequencies of atoms. The
measurement of vibrations is the principle of all atomic clocks. The major difference concerns, in addition to the element chosen, the way of detecting the change in
energy level: caesium clocks separate atoms of different energy levels by magnetic
field; hydrogen clocks maintain hydrogen atoms at the required energy level in a
container of a special material so that the atoms do not lose their higher energy state
too fast, and rubidium atomic clocks, which are the simplest and most compact,
use a glass cell of rubidium gas that changes its absorption of light at the optical
rubidium frequency when the microwave frequency is just right.

Atomic clocks keep time better than any other clock; they are even steadier than
the Earth's rotation. The caesium atomic clock is the most accurate in the long term:
better than 1 s per 1 million years. Hydrogen atomic clocks show a better accuracy
in the short term (1 week): about ten times the accuracy of caesium clocks. The
hydrogen maser oscillator provides fractional frequency stability of about 1 part in
1016 for intervals of a few hours to a day.

The maser's high frequency stability is ideally suited for a variety of space
applications such as Very Long Baseline Interferometry (VLBI) from space, precision measurements of relativistic and gravitational effects and GPS. The hydrogen
maser clock will therefore be Galileo's master clock. As a result of the extremely
high accuracy of atomic clocks, the world's time-keeping system lost its astronom-
ical basis in 1967. Then, the 13th General Conference on Weights and Measures
derived the SI second from vibrations of the caesium atom, which is now internationally agreed as the interval taken to complete 9,192,631,770 oscillations of the
caesium-133 atom, exposed to a suitable excitation.

4.11 Monitoring Space Segment

To perform positioning by satellites the position of satellites must be accurately
known. Arranged on six planes, typically more than 24 GPS satellites are orbiting
at a mean distance from the middle of the earth of 26,560 km that is at a height of
20,200 km above the surface of the earth (Kaplan and Hegarty, 2006). The number

and constellation of satellites guarantees that the signals of at least four satellites can be received at any time all over the world. It takes 12 h sidereal time that is 11 h 58 min earth time to circle the earth. The speed of the satellites is 3.9 km per second.

But how is a continual check to be kept on the positions of these satellites? Their orbits are not completely deterministic but fluctuate due to celestial gravity forces (Lambeck, 1988). The ground segment monitors and controls the position of GPS satellites, called the space segment. The master control station of the GPS constellation is located at Falcon Air Force Base, Colorado Springs, USA. Up to about the middle of 2005 tracking stations in Hawaii, Ascension Island, Diego Garcia and Kwajalein completed the ground segment. They were able to track and monitor the satellites for 92% of the time. For two daily windows lasting 1.5 h, each satellite was out of contact with the ground stations. During August and September 2005 the ground segment was extended with six more remote stations, providing a higher accuracy for the user. Now, every satellite can be permanently 'seen' from at least two stations, allowing calculating more precise orbits and ephemeris data. In the future, the ground segment will be augmented by installing another five stations allowing monitoring of every GPS satellite by at least three stations. The master control station acts as data processing centre for all information, including that collected at the remote stations. Orbit coordinates are continuously determined by triangulation. Comparing the time of the satellites' four atomic clocks with similar devices on the ground provides information on time errors. When a satellite drifts slightly out of orbit, repositioning is undertaken.

The clocks may also be readjusted, but more usually information on time errors is attached to GPS signals as correction factors. The computed corrections, time readjustments and repositioning information are transmitted to the satellites via three uplink stations co-located with the downlink monitoring stations. In this way, all the GPS satellites are able to continuously attach corrections to the parameters they send out, which include ephemeris data, almanac data, satellite health information and clock correction data.

4.12 GLONASS

GLONASS, owned by the Russian government and managed by the Russian Space Forces, was developed by the former USSR at the same time USA was building GPS; launch of the first GLONASS satellites took place on 12 October 1982 (Langley, 1997). The constellation originally consisted of twelve satellites, but by decree of 7 March 1995 the Russian Government opened GLONASS for free-of-charge civil use and the number of satellites was increased from 12 to 24. This constellation was completed in 1997. Since then GLONASS has been designated a 'dual system', available to both civil users and the military. Civil users worldwide are able to make use of the Standard Precision (SP) signal mode, whilst the High Precision (HP) signal mode is reserved for government or military use.

Lack of economic impetus jeopardised continuation of the programme, and by the start of the new millennium Russia had to rely on US GPS signals. By April 2002 only eight satellites were in operation, far too few to act as a global navigation utility. Presidential and governmental decrees issued in 1999 and 2001 were necessary to reverse the downward spiral. An ensuing Federal Target Programme for the 10 years 2002–2011 was to revive the system. At the start of 2006 were in orbit twelve GLONASS satellites with expected lifetime of 3 years, and four GLONASS M satellites with an expected 7-year lifetime. By April 2006 the probability of receiving four or more satellites was 76%, whilst the positioning gap fell from 13.7 h in 2001 to 2.6 h in 2006. In December 2006, three more GLONASS M satellites were put into orbit. According to the goals set by a Presidential Directive of 18 January 2006, eighteen satellites should be in orbit by the end of 2007, sufficient for continuous navigation within the territory of the Russian Federation. The constellation will be further improved by the launch of GLONASS-K satellites. These satellites have half the weight of the M-series (750 kg) and increased operational lifetime of 10–12 years. GLONASS satellites are steadily being launched to bring the system to full capability.

As of February 2009, the GLONASS constellation comprised 20 satellites; 24 satellites are required for world coverage. The ground control segment is located within former USSR; the master control station in the Moscow area, one tracking station in Saint Petersburg, one in western Ukraine and two in Siberia. Initially it was envisaged that GLONASS would be fully operational by 2010. However, the launch, on 5 December 2010, of the rocket carrying the three GLONASS-M satellites that would complete the constellation, failed because the rocket deviated from its course and felt into the Pacific Ocean resulting in the loss of the GLONASS satellites. The aim is to make GLONASS performance comparable with GPS and Galileo. The main goal of the Russian policy is to bring GLONASS to the mass-market. This should be achieved by enabling developers of equipment and applications guaranteed access to the GLONASS civil signal structure by promoting within Russia the combined used of GPS/GLONASS receivers and maintaining GLONASS compatibility and interoperability with GPS and the future Galileo. GLONASS is gaining increasing international attention from partners and users in India, EU, USA and other nations. For example, in December 2004 USA and Russia agreed to shelve any idea of direct user fees for civil GLONASS and GPS services. India and Russia are willing to co-operate on GNSS infrastructure development.

4.13 Galileo

As a joint initiative of the European Commission (EC) and the European Space Agency (ESA) Galileo will be Europe's global navigation satellite system. Several other countries are participating in the project, among which are Israel and China. This system under development will be interoperable with GPS and GLONASS and will operate under civilian control at all times, except in the direst emergency. The

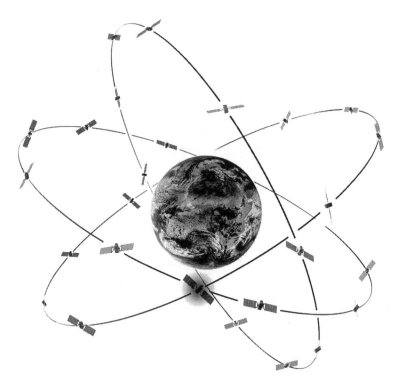

Fig. 4.14 Planned Galileo constellation

satellites will orbit at an altitude of about 23,200 km above the surface of the Earth. The constellation is designed to optimise coverage of the entire planet (Fig. 4.14), a feature not offered by the current GPS and GLONASS constellations.

On the eve of 2006 Galileo became tangible reality. Launched from Balkonur, Kazakhstan, atop a Soyuz-Fregat vehicle, the first Galileo-In-Orbit Validation Element (Giove A) was lofted into orbit in the early morning of 28 December 2005. Two weeks later, 12 January 2006, the satellite transmitted signals from space for the first time. Weighing in at 600 kg, Giove A had during its mission to accomplish three tasks. First, securing use of the signal frequencies allocated by the International Telecommunications Union. The second part was determining radiation characteristics of the three Medium Earth Orbits at altitude 23,222 km to be circulated by the projected constellation of 30 satellites (27 + three active spares). Finally, critical technologies contributing to the future operational constellation had to be examined and checked. These consist primarily of two signal-generation units and two rubidium atomic clocks with a stability of 10 ns per day. The second Galileo satellite (Giove B) was launched on 27 April 2008 (00:16 Central European Summer Time) and began transmitting navigation signals from 7 May 2008. As Europe's biggest project undertaken, the Galileo programme has been subject to ongoing query from the outset, including reservations about the (final) cost. Why add a third Global

Navigation Satellite System to those already operational, GPS and GLONASS? Why should Europe reinvent the wheel? There is nothing wrong with reinventing the wheel and it has been done countless times: why else the many manifestations of the wheel? Galileo is not just a copy of GPS or GLONASS: the differences are significant. In reinventing the global navigation wheel Europe has introduced a number of new features in satellite positioning, bringing with them an incremental step forward in performance. They include greater accuracy, down to a metre, and better and greater integrity through informing users of errors that could compromise performance. Above all the signals emitted by the Galileo satellites will be stronger than those from GPS or GLONASS enabling greater penetration in urban centres, inside buildings and under trees. The constellation is designed to optimise coverage of the entire planet, a feature not offered by the current GPS and Glonass constellations. Furthermore, in tandem use of GPS, GLONASS and Galileo together will enable a further jump in performance.

One big advantage offered by three operational constellations is enhanced integrity and reliability. Triple redundancy as a prerequisite for risk reduction was already recognised over 400 years back by Christopher Columbus, when he sailed away from Europe with three hulls to land on what he thought to be Japan. He sailed back to Europe with just one ship. Due to lack of satellite visibility not all applications one may think of may beneficially use the GNSS technology. An experiment conducted in Copenhagen, capital of Denmark, for example, failed to prove the suitability of GNSS as an alternative car tracking toll system due to the signal blocking by tall buildings and high density of buildings. Signal blocking also occurs in forests and mountains. A major solution to this space dependent deficiency is to put more satellites into orbit. Combined used of GPS, GLONASS and Galileo may thus extend the range of possible applications.

4.14 Beidou

Another system under development is China's Compass Navigation Satellite System (Lemmens, 2007c). China has six navigation-and-positioning satellites; the launch of the sixth satellite took place on 15 April 2009. The system provides regional coverage of China and surrounds. It is named Beidou after the group of seven stars (Septentrio is Latin for 'north') of the constellation Ursa Major, known in many cultures under different designations, in the UK 'The Plough', 'Big Dipper' in USA and 'Big Mother Bear' in Russia. Others see the constellation as resembling a wagon. The first Beidou satellite (1A) was launched on 30 October 2000 followed by 1B on 20 December 2000. Since 2001 China's army and others have thus had access to a domestic satellite positioning system.

Positioning by two satellites? But satellite positioning is a trilateration problem for which at least three satellites are needed and a fourth satellite is necessary to eliminate the time bias. How does China make do with just two satellites? Beidou derives an approximation of one of the three coordinates from a digital elevation model (DEM) and eliminates time bias by dual-way transmission. DEMs can be accurately generated from a multitude of techniques, including InSAR, Lidar and digital photogrammetry, and are abundantly available. Time bias can be eliminated if the signal is shuttled back and forth from satellite to receiver and receiver to satellite. The satellite clock measures travel time, eliminating, most advantageously, the need for extremely accurate atomic clocks.

Just as with GPS and GLONASS each of the Chinese satellites broadcasts signals continuously. Once the user terminal has picked up signals it responds by transmitting them back to the satellite, which in turn forwards the received signals to a central control station. Here the range between terminal and satellite are inferred from travel time; principally by multiplying travel time by speed of light and dividing the result by two. From the ranges to the two satellites and an initial estimate of the elevation coordinate, perhaps taking sea level, an approximation is calculated of user's latitude and longitude. This approximate position is then used to extract an enhanced elevation from an existing DEM, which can then be used to obtain improved estimates of latitude and longitude. In the next round new time measurements are used, or the originals but now with the improved estimate of elevation. This iterative process ends with convergence, and the user receives his position within the Beijing 1954 Coordinate System, claimed accuracy of 100 m and 20 m when using calibration stations. Since the user terminal is not only a receiver but also communicates itself, network capacity limits to 150 the number of simultaneous users. Yet since the signals travel with the speed of light, over half a million users can be served per hour. Dual transmissions also mean a need for more space to accommodate devices, so Beidou terminals are bigger, heavier and more expensive than GPS receivers. China's army uses the two-way communication function to talk to units and monitor their position. However, two-way communication is strategically unfavourable: the enemy may also pick up processed signals and determine from them troop position and movement.

The third Beidou satellite (2A) was blasted into orbit 2.5 years after launch of 1B, on 24 May 2003 to be precise. Nearly 4 years later, on 23 February 2007, the fourth Beidou (3A) was put into orbit and operates as spare. The fifth Beidou, rocketed beyond earth's atmosphere on 12 April 2007, was not like the other four positioned in an approximately geostationary orbit 35,800 km above earth's surface. Instead it is in an orbit of perigee 21,519 km and Apogee 21,544 km. Beidou 1A is positioned north of Irian Jaya at longitude 140E, Beidou 1B south of Sri Lanka at 80E, and Beidou 2A rotates geo-synchronously with the western part of Borneo, Indonesia, at 110.5E. The advantage of geostationary satellites is ease of control. Beidou's ground segment includes the central control station and orbit-tracking stations at Jamushi, Kashi and Zhanjiang. However, the price paid for a system comprising so few satellites and limited ground control is regional coverage.

Regional coverage would seem to constitute something of a disgrace in the eyes of China's leaders, who want a doubling in Gross Domestic Product (GDP) during the first decade of the new millennium. They thus wish gradually to extend Beidou to become a real global satellite navigation system, referred to as the Compass Navigation Satellite System. The ultimate constellation will comprise five geostationary satellites and thirty medium-orbit satellites intended to provide positions with 10 m accuracy, using, like GPS, GLONASS and Galileo, the principle of trilateration.

4.15 Co-operation Among System Providers

To ensure reliable services the adagio among system providers should be co-ordination and collaboration rather than ignoring one another's existence. Co-ordination and collaboration has been put on the cards with the setting up of the International Committee on Global Navigation Satellite Systems (ICG) in 2006. ICG acts under the umbrella of United Nations Office for Outer Space Affairs, meets on a voluntary basis and aims at encouraging universal access to global and regional navigation satellite systems and integrating these services into national infrastructures. ICG also strives to improve compatibility and interoperability among current and future GNSS systems. To provide a means of discussing key technical and operational issues requiring focused input from system providers a Providers Forum has been put in place. The protection of the RNSS (radio-navigation satellite services) spectrum is vital to GNSS services to be assured through domestic and international regulation. The broadcast of open services will be provided free of charge, while many systems also broadcast services to meet the needs of authorised users. To enable manufacturers to design and develop GNSS receivers on a non-discriminatory basis open publication and dissemination of signal and system characteristics by system providers is essential. Also the detection and mitigation of interference with GNSS anywhere in the world is important as is the physical separation of operational satellite constellations and end-of-life disposal orbits.

4.16 Inertial Navigation

To be useful, data collected by moving platforms, either terrestrial, airborne or orbiting have to be georeferenced. For example, transformation of the image coordinates measured in an aerial photo to a ground based reference system requires the three position coordinates of the projection centre of the camera and the three angular parameters (ω, ϕ, κ; omega, phi, kappa) together called exterior orientation (see Section 7.2). The same is true for airborne Lidar (Chapter 8). Here the laser range data have to be transformed to georeferenced point clouds, to which 3D coordinates have to be assigned, which computation also requires the six parameters of the exterior orientation. These six parameters can be determined along direct methods, using GCPs, or indirect methods, using an on board GNSS receiver for determination of the three position coordinates and an inertial navigation system (INS) for determining the three angular parameters. The principles of GNSS have been comprehensively treated above. Here we focus on inertial navigation. An inertial navigation system (INS), also called inertial navigation unit (INU) consists of three

Fig. 4.15 Schematic
representation of INS
principles; measured are
accelerations and rotations
from which other parameters
can be derived such as
position

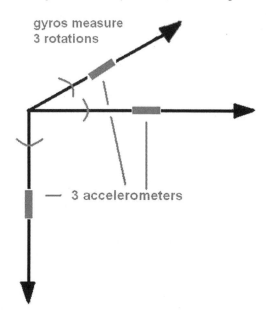

accelerometers and aligned with these three are gyroscopes together with a process-ing unit (Groves, 2008). The accelerometers, placed in mutual orthogonal positions, measure the accelerations of the moving platform (Fig. 4.15). By integration of the accelerations over time the velocity can be determined and by integration of the velocity over time the 3D position coordinates can derived. The basic function of the gyroscopes is to measure the change of attitude of the platform in the form of three angular values along the three orthogonal axes on which the accelerometers are mounted.

INS systems are autonomous systems that can continuously acquire attitude, accelerations, velocity and position of the platform at a rate of at least 50 measure-ments per second (Hertz). In principle, an INS system can provide all six parameters of exterior orientation: position and attitude. However, the accuracy of position determination rapidly degrades with time as inertial navigation suffers from drift, so not only the accelerations are integrated but also their error resulting in discrepancy between the actual position and measured position. Positions determined by inertial navigation are, however, not useless since sampling frequency lies usually in the order of several hundreds of Hertz, while the sampling frequency of GNSS systems is just up to 10 Hz. Hence, inertial navigation positions may fill GNSS gaps due to loss of lock. GNSS provides high position accuracy over long time spans; however, the number of measurements conducted is less then ten per second, and no attitude information is gained. GNSS does not operate autonomously; it depends on satellite signals which may be obstructed by buildings, trees and other objects. In contrast to GNSS, inertial navigation does not require external aid. As an autonomous sys-tem it is immune to any interference from the outside world. So, the advantages and

Table 4.1 Comparison of GNSS positioning and inertial navigation characteristics

Feature	GNSS	Inertial navigation
Positional accuracy	High over the short and long term	High over the short term
Attitude information	Only possible when using multiple antennae arrays; noisy	Accurate
Time dependency	Uniform accuracy, independent of time	Accuracy decreasing with time (drift)
Output rate	Low (~10 Hz), taking into account aircraft speed	High (~100 Hz)
Autonomous	No (requires satellites and GNSS infrastructure)	Yes, but affected by gravity field
Failings	Cycle slip and loss of lock	No signal outages

drawbacks of GNSS and INS are complementary and the integration of both devices yields an ideal couple for determining the six exterior orientation parameters of sensors mounted in moving platform. In an integrated system GNSS measurements can compensate the drift by re-initialising the INS positions and attitude every time a GNSS measurement becomes available. In turn, INS provides intermediate short-term position information and bridges periods that the GNSS receiver losses lock to satellites. Table 4.1 compares the characteristics of GNSS and inertial navigation.

The simplest way of combining INS and GNSS measurements is the uncoupled procedure; the coordinates generated by the GNSS system are used to re-initialise the INS system at regular time intervals. Today, most systems combine INS and GNSS measurements in one single algorithm using Kalman filter technology. The Kalman filter is a mathematical method or algorithm. The basic technique was invented by Rudolf E. Kalman in 1960 and has been developed further by numerous authors since. It provides real-time estimates which are updated while the platform moves. When integrating the various GNSS and INS measurements, the Kalman filter provides an optimal solution giving the system parameters, the measurement and their deterministic and statistical properties. For a more formalised and detailed treatment of Kalman filtering, see one of the many applied mathematics book devoted solely to this subject, e.g. Jazwinski (1970), Gelb (1974), Maybeck (1979), Brown and Hwang (1997), and Grewal and Andrews (2000).

Integrated INS/GNSS systems have been on the market for some time and are installable on any vehicle for direct geo-referencing of geo-data acquired by laser scanners, digital cameras or other sensors collecting geo-data. Their size and price decrease by the year.

4.17 Trends

GNSS receivers, total-stations and other measuring equipment are growing more mobile as it is integrated with vehicles, not only cars, shovels or harvesting machines but also remotely steered mini-helicopters and other unmanned aerial vehicles. The technological trend is towards integration of GNSS with other techniques. Seamless navigation from outdoor to indoor or from gate to gate, for example, which is becoming increasingly important for emergency situations can only be achieved by improving the receiver sensitivity (Tsui, 2005) or by integrating GNSS with other sensors and data. On the consumer market, mobile phones are gradually becoming equipped with a GPS or GNSS chip, and it is true that new applications depend not so much on technological constraints as on the limits of human imagination. Trends in system development can be followed by reading GNSS product surveys, for example, those published annually in GIM International.

References

Baarda W (1968) A testing procedure for use in geodetic networks. New Series, vol 2, no 5. Netherlands Geodetic Commission, Publications on Geodesy New Series. Delft, The Netherlands

Blewitt G (1989) Carrier phase ambiguity resolution for the global positioning system applied to geodetic baselines up to 2000 km. J Geophys Res 94(B8):10187–10203

Brown RG, Hwang PYC (1997) Introduction to random signals and applied Kalman filtering, 3rd edn. Wiley, New York, NY

El-Rabbany A (2002) Introduction to GPS: the global positioning system. Artech House, Boston, MA. ISBN 1-58053-183-0

Gelb A (ed) (1974) Applied optimal estimation. MIT Press, Cambridge, MA

Grewal MS, Andrews AP (2000) Kalman filtering: theory and practice, 2nd edn. Wiley, New York, NY

Groves PD (2008) Principles of GNSS, inertial, and multisensor integrated navigation systems. Artech House, Boston, London. ISBN 978-1-58053-255-6

Hoffmann-Wellenhof B, Lichtenegger H, Collins J (1994) Global Positioning System: theory and practice, 3rd edn. Springer, New York, NY

Jazwinski AH (1970) Stochastic processes and filtering theory. Academic, San Diego, CA

Kalman RE (1960) A new approach to linear filtering and prediction problems. Trans ASME, Ser D, J Basic Eng 82:35–45

Kaplan ED, Hegarty C (eds) (2006) Understanding GPS: principles and applications, 2nd edn. Artech House, Boston, MA. ISBN 1-58053-894-0

Klobuchar JA (1991) Ionospheric effects on GPS. GPS World 2(4):48–51

Klobuchar JA (1996) Ionospheric effects on GPS. In: Parkinson BW, Spilker JJ (eds) Global Positioning System: theory and applications, volume 1, vol 163. American Institute of Astronautics and Aeronautics, Washington, DC, pp 485–515

Lambeck K (1988) Geophysical geodesy. Clarendon Press, Oxford

Langley RB (1997) GLONASS: review and update. GPS World 8(7):46–51

Langley RB (1999) Dilution of precision. GPS World 10(5):52–59

Leick A (2004) GPS satellite surveying, 3rd edn. Wiley, New York, NY

Lemmens M (2005) Know your place from time. GIM Int 19(11):11

Lemmens M (2007a) Car navigation: the nuisance of going mainstream. GIM Int 21(3):45–47

Lemmens M (2007b) A bankrupt PPP. GIM Int 21(6):11

Lemmens M (2007c) Beidou. GIM Int 21(10):11

Maybeck PS (1979) Stochastic models, estimation and control, vols 1–3. Academic San Diego, CA

Parkinson BW (1996) Introduction and heritage of NAVSTAR, the global positioning system. In: Parkinson BW, Spilker JJ (eds) Global positioning system: theory and applications, volume 1, vol 163. American Institute of Astronautics and Aeronautics, Washington, DC, pp 3–28

Soubielle J, Fijalkow I, Duvaut P, Bibaut A (2002) GPS positioning in a multipath environment. IEEE Trans Signal Process 50(1):141–150

Teunissen PJG (1995) The least-squares ambiguity decorrelation adjustment: a method for fast GPS integer ambiguity estimation. J Geod 70(1–2):65–82

Teunissen PJG, Kleusberg A (eds) (1998) GPS for geodesy, 2nd edn. Springer, Berlin, Heidelberg, New York

Teunissen PJG, de Jonge PJ, Tiberius CCJM (1997) Performance of the LAMBDA method for fast GPS ambiguity resolution. Navigation: JION 44(3):373–383

Tsui JB-Y (2005) Fundamentals of Global Positioning System receivers: a software approach, 2nd edn. Wiley, New Jersey

Wolf PR, Ghilani ChD (2006) Elementary surveying: an introduction to geomatics, 11th edn. Pearson Prentice Hall, Upper Saddle River, NJ. ISBN 0-13-148189-4

Chapter 5
Mobile GIS and Location-Based Services

Traditional analogue fieldwork – using map, pen and paper – is increasingly becoming obsolete. Several developments in the ICT realm have firmly stimulated the transition from analogue to digital work flows in many if not all areas of geosciences and geo-information technology. The major driving force steering this transition is that the computer power of handheld devices has been extended to a level that speed of computation and manageable dataset sizes are comparable to what the desktop computers could accomplish just a few years earlier. Together with rapid adoption of the internet, this progress has triggered fitting mobile GIS systems and location-based services (LBS) into work processes requiring data collection in the field or geo-data use while, for example, on the road or on the water.

Mobile GIS and LBS are means of geo-data collection and retrieval using global navigation satellite systems (GNSS) positioning, GIS functionality, wireless connection and mapping, all integrated in one compact device. The capability of GNSS with respect to accurate measuring of the three coordinates of any position, wireless communication and GIS for processing geo-data into has opened a wide variety of new opportunities. The synergy of these technologies enables bringing computer power to field workers for collecting data with increased efficiency and ease (Vivoni and Camilli, 2003; Huang et al., 2005). In addition to cost savings and ease, a fully digital data stream from data collection to the database management system reduces the occurrence of errors (Pundt, 2002). The accurate position capacity of GNSS also triggered the coming up of information and entertainment services, accessible with mobile devices through the mobile network, called LBS. This chapter describes the technology of mobile GIS and LBS, highlights their characteristics and considers fields of application.

5.1 Background

From the technological point of view mobile GIS and LBS systems have much in common and that is the reason why we treat both subjects in one Chapter. Both are based on GIS technology, are equipped with a positioning device, usually a GNSS receiver, which can be constructed small enough to fit within a wrist watch,

M. Lemmens, *Geo-information*, Geotechnologies and the Environment 5,
DOI 10.1007/978-94-007-1667-4_5, © Springer Science+Business Media B.V. 2011

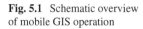

Fig. 5.1 Schematic overview of mobile GIS operation

and are wirelessly connected to a database management system (DBMS) running on a server for downloading (geo-)data from the central GIS to the field and for submitting (geo-) data captured in the field back to the central GIS (Fig. 5.1). The construction of the device is based on chip technology and electronics; unlike traditional surveying equipments, such as total stations, there are no mechanical or optical parts, resulting in very favourable features, including portability, lightweight (typically less than 500 g), small size and handiness (Fig. 5.2). They can be hand-held but they should be ruggedised as they are designed for use in the field, where circumstances can be harsh sometimes. Ruggedised means that they are water, dust and shock resistant while they can be kept fully operational under harsh conditions.

Like mobile phones, mobile GIS systems and LBS systems have many beneficial features that potentially pave the way being exploited in a great variety of application fields, in particular in the still rapidly expanding GIS market, where users vary from public safety agencies, the oil and gas industry to environmental scientists and consultancies. For applications in the area of environmental management and habitat monitoring Tsou (2004) notes the following advantages:

– Field workers can easily carry mobile GIS devices to the field for data collection and validation tasks.
– Wireless communication enables users performing real-time data updates and exchanges between the central GIS and distributed mobile devices.

From a business point of view mobile GIS and LBS aim at serving professionals to better carry out their core business processes. In addition one anticipates –

Fig. 5.2 An example of a mobile GIS handheld: a small, handy, portable and accurate surveying tool (*left*), and an Iphone equipped with Esri mobile GIS software showing Gulf coast oil spill (*right*)

particularly telecommunication companies do so – that LBS will attract a huge consumer base as the services and the necessary technology can be incorporated and integrated in mobile phones.

5.2 Mobile GIS

Many traditional work processes start with dispatching digital geo-data from the central GIS to the plotter next door. In the morning, field workers take paper maps into the field and start pipeline inspection, road maintenance, data collection of railway objects or any other task. Data are recorded by pen on the paper map and in the evening this is shipped back to the office where an operator next day, week or month digitises and keys the data into the computer.

Now that the BlackBerry, iPhone and other smart phones can be purchased at the corner shop, one may question why still sending out crews to the field with paper maps. Why not transmit digital data directly to a laptop, tablet PC or smart phone via the internet (Fig. 5.3)? Field workers can then, on the spot, query and analyse digital maps, view aerial imagery and upload updates to the central GIS (Fig. 5.4 left). When they are missing data they can send a request and receive what they need back from the office wirelessly and instantly. Office and fieldworkers can immediately view and use such updates; a great advance in efficiency and cost cutting.

The technology is in place, and vendors of GIS and survey equipment, well aware that clients do not need tools but solutions, help apply mobile GIS-aided changes in

Fig. 5.3 Carrying GIS into the field using a tablet PC

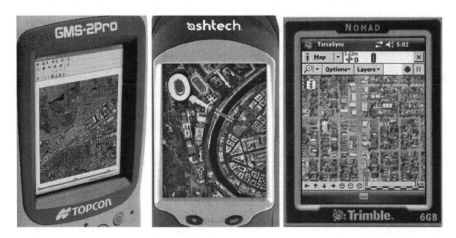

Fig. 5.4 Three handheld mobile GIS devices displaying ortho-imagery as backdrop for geo-data collection

work processes (Lemmens, 2010a). Esri, for example, has modified and extended its ArcGIS for use in the field and provides solutions in which its mobile GIS is embedded in handhelds, usually those from the Trimble stable. These systems can be wirelessly linked with a central server. Since field crews often form a large part of the total workforce, even up to 60%, the units sold per organisation or company may add up to hundreds or even thousands.

A mobile GIS enables thus a user to carry into the field an outdated digital map of a terrain situation in need of updating, or a digital base map on top of which certain features such as street furniture or damaged objects after a storm have to be mapped as a separate layer. To the recorded features of points, lines or polygons, user-definable attributes can be assigned as appropriate to the mapping task in hand. A coding library can be predefined and downloaded prior to going into the field. The user interface is made such that anyone who can operate a mobile phone can operate a mobile GIS. Computer-literate layman can learn to operate the system in less than a day. Researchers with limited programming skills are enabled to tailor the software to the priorities and theoretical needs of the scientific projects they carry out (Tripcevich, 2004). The internal storage capacity, usually expressed in Gigabytes, together with convenient communication ports, enables (down) loading and provides instantaneous access to layers of topographic maps, large-scale base maps, aerial photos, high-resolution satellite imagery and land-ownership boundaries. Furthermore, because of the complete digital storage of data and direct registration of the position in coordinates in any preferred coordinate system, either predefined or user-defined further processing can be done without overmuch human interaction, resulting in a high ratio of automation.

The predecessor of mobile GIS was the pen computer equipped with a GIS. Such a device became available by the mid nineties and appeared to be very useful for on site geo-data collection. The digital map or aerial/satellite images of the site are stored in the computer and projected on screen enabling the operator to manually add features and annotations by writing with a pen on the screen as if he was writing on paper. Compared to using the traditional paper method of capturing data in the field the users soon experienced great advantages.

The data could not only be collected faster but more importantly the conversion from analogue to digital format, which is a tedious, labour-intensive and cumbersome processing step, could be avoided. A next step in the development was to carry the light weight laptop together with a GNSS receiver through the field and to connect the system with a mobile phone to quickly transfer the recorded data to the office GIS. Mobile GIS is a natural step forwards in the progress of this development by integrating all components – GIS, GNSS receiver and wireless communication – in one small device which can be carried along as if it were a mobile phone.

5.2.1 Three Domains

The development of the pen computer to mobile GIS systems stems from three directions: the GIS industry, producers of mapping tools and the survey/navigation industry (Lemmens, 2010b). Coming at it from the GIS side, vendors offer software featuring a suite of GIS functionality including map navigation (zoom, pan, map rotation), geometric computations (distance, area), display, editing, updating and collecting, storing and mapping of GNSS measurements and adding attribute values to collected point, line and area features. Vendors of mapping tools were initially aiming to relieve the work of the surveyor collecting data with a total-station, GNSS receiver or other positioning device. Now these vendors have penetrated the mobile-GIS market this software is usually sold embedded in handheld devices as out-of-the-box solutions for diverse applications such as cadastral surveying, environmental monitoring, defence and utilities. Other software evolved from mapping tools for forestry and farming to become a mobile GIS.

From the surveyor's perspective, the heart of a mobile GIS is GNSS technology and that is why survey/navigation companies produce handheld devices focused on GNSS positioning. GNSS is immune to concerns, such as darkness and rough weather, and allows a person working solo to do the job. Depending on the measuring mode, a high order of precision can be achieved. The fact that surveyors often have to operate in harsh work environments has resulted in shock-resistant and waterproof devices. Usually the software is less sophisticated than GIS software in terms of what it offers but is designed to communicate with GIS packages either by transferring field recordings to the GIS system in the office or by downloading existing maps to the mobile GIS.

5.2.2 Wireless

Wireless connection enables download of geo-data from the central GIS while in the field, and upload of field data either instantly or at set times. Three types are in use: wireless local area network (WLAN), general packet radio service (GPRS) and Bluetooth. WLAN connects two or more devices through an access point, which enables moving around while connected to the wider internet. To ensure effective deployment, the WLAN infrastructure must cover the whole area in which fieldworkers are operating. An alternative is GPRS provided by mobile-phone operators, but these may suffer from delays due to high latency connection, low transfer speed and selective service availability and coverage. Bluetooth is not able to bridge long distances, but is well suited for transferring data from one device to other nearby devices fast and reliably; it can work over 100 m, while no line of sight is required.

5.2.3 Accuracy

All surveyors are trained to accompany their measurements and values derived from measurements, in particular coordinates of captured points, by accuracy statistics. For many applications, such as mapping pipelines, good accuracy is crucial. The GNSS precision features listed by vendors usually come along with recording conditions such as minimum number of satellites, maximum PDOP (Positional Dilution of Precision), minimum (cut-off) angle above horizon, and signal-to-noise ratio. This is a clue as to complexity.

Precision depends on the number of channels, the actual number of signals picked up, software capabilities, GNSS augmentation services used and the facilities to cope with atmospheric delays and multipath. Remedies used to diminish atmospheric degradation include dual-frequency measurements and Differential GNSS (DNSS); the latter an augmentation approach that corrects for errors using a base station with known coordinates. Comparison with continuously measured coordinates enables computation of corrections 24/7, and these are transmitted in real time by radio link or mobile phone. Precision is at sub-centimetre level. In autonomous mode, that is standalone and without making use of augmentation services, precision reads 5–15 m. In addition to ground-based services there exist satellite-based augmentation services (SBAS), which supports wide-area or regional DGNSS through measurements taken at multiple base-stations. Corrections are sent to one or more satellites, which broadcast these to enabled receivers. By making use of various types of differential GNSS, as supported by national or continent-wide GNSS services, accuracies at sub-metre and even decimetre level are achievable (see Section 4.3). So accuracy depends mainly on what is paid for in terms of infrastructure services or, in some parts of the world, on the availability of these.

5.2.4 Operational Features

Mobile GIS systems operate under a variety of weather conditions, terrain characteristics and working constraints. The setting can be harsh, from extreme cold to hot and highly humid. The device may suffer a drop from the bonnet of a vehicle onto concrete, or plunge into water. So it is important enquiring the durability of the system. Several standards are in use, one being the International Protection Rating Code (IP), which indicates durability by IP followed by two digits, for example, IP66. The first digit indicates how the cover protects against intrusion of dust and other harmful particles; 6 is the highest level (complete shield). The second digit indicates the level of water resistance; 5 and 6 mean the device is protected against heavy showers, and 8 is the highest level (survives a plunge into the sea). The IP code can also be extended with a letter and a digit. For example, IPX4 means the device is protected against splashing water. As survey devices often originate in the military, the United States Military Standard (MIL-STD-810) is also in use, addressing much broader operational settings than does the IP standard, including air pressure, lowest and highest temperature, temperature shock, humidity, salt fog, gunfire and random vibration. The qualification is gained in the laboratory where real-world conditions can be only partially duplicated, so that it remains necessary for the professional to exercise their engineering judgment.

So, when purchasing a mobile GIS it is imperative to check such things as whether the device will survive

- tumbling out of the fieldworker's hands towards a crash encounter with paved ground,
- storage in a car standing the whole day in desert sun or polar cold,
- plunging into a ditch full of water, and
- a life-long stay in tropical rainforest.

Further issues concern whether the display is full-colour and readable in daylight, whether the battery can be charged from a vehicle or replaced in the field, and if it can last a whole working day.

5.2.5 Peripherals

Most devices use Windows Mobile (WM), initially developed for the consumer market but gradually Microsoft changed focus of this operating system to the specialised market for rugged mobile devices. A stylus pen or the operator's finger allows tapping commands onto the touch-sensitive screen (Fig. 5.4 right). Equipped with handwriting-recognition software, the device converts written annotations into letter code. When an object is difficult to access, its position can be determined as an offset point from distance and bearing measured with an integrated laser rangefinder and electronic compass. A forthcoming and more accurate alternative to the latter is an inertial navigation system (INS) based on micro-electromechanical system (MEMS) technology. The miniaturisation of accelerometers and gyroscopes has matured to the point where low-cost commercialisation has become viable.

5.2.6 Data Quality and Performance

Mobile GIS tools can support controlling data quality during geo-data collection through functions that check the consistency and integrity of the data directly after entering by field workers (Pundt, 2002). Such functions use the relationships that exist between data components to trace errors.

Before purchase it is a good idea to take any mobile GIS system into the field. As GNSS precision is a multifaceted topic, the best performance test is to check measurements against ground truth. A feasible set-up is measuring a set of known points distributed over an urban area with no obstructing tall objects, so that there are sufficient satellites more than 15° above the horizon. Write annotations on the touch screen, make photos and measure offset points using range-finder and electronic compass. Compute the coordinates in post-processing and compare these with existing coordinates with a precision at least ten times better. To determine sensitivity to multipath, position the instrument under canopy and measure, preferably

Fig. 5.5 A walk along the lines of a car park with two different handhelds (*black* and *grey lines*) clearly shows differences in precision; crosses: ground truth collected with precision GNSS

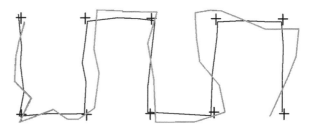

whole day long. Plot the coordinates and use a map or orthoimage as backdrop, check the size and pattern of dispersion and compute the standard deviation. To test real-time performance, walk along some straight lines, such as those demarcating a car park, and judge the jagged nature of the recorded path (Fig. 5.5).

5.2.7 Professional Applications

Fieldwork is needed in many practical applications as well as in many scientific disciplines. In this subsection we look at the use of mobile GIS system in a variety of organisations, which have in common intensive use of geo-data to properly run core business processes based on the inventory made by Lemmens (2005, 2010b). In the next subsection we will have a closer look at scientific applications.

Mobile GIS is used in organisations that map, plan, design, build and operate roads, railways, waterways and other infrastructure (Fig. 5.6). Environmentalists use mobile GIS for mapping invasive species, which may endanger unprotected habitats in even the most pristine natural areas and may have considerable impact on parks and preserves. Fighting invasive species by prevention, early detection and restoration of infested sites requires on-site collection of geo-data at regular time intervals. Mobile GIS provides the technology to do so. Public Health organisations use mobile GIS in the aftermath of a natural disaster, when outbreaks of disease and damage to infrastructure can quickly affect public health and the local economy. During an on-site assessment survey the damaged areas is mapped and documented. Next the data are processed in the office and necessary measures taken. Engineers who want to have quickly available a map of objects, for example, the entire water/wastewater infrastructure also use mobile GIS.

Many land surveyors appreciate using mobile GIS for performing on-site measurements. For example, survey departments are using them for GIS data-gathering for departments of public works, infrastructure and others, and geographic management of all survey and engineering data. Both for mapping and updating, the mobile device enables up to tenfold reductions in surveying time as compared to conventional methods. Using a base-station, so that DGNSS can be applied, allows decimetre accuracy. So that mobile GIS can be real time savers for the land surveyor, whilst also offering good accuracy. Cadastral surveyors use mobile GIS for performing boundary surveying. Especially in developing countries, among the

Fig. 5.6 Use of mobile GIS handheld in utilities

success factors to establish secure land-tenure belongs quick surveying of property boundaries.

In Austria many man-made objects such as roads and buildings are vulnerable to landslides caused by heavy rainstorms and intensified by population growth and changes in land-use, more specifically increase of residential area. With the help of mobile GIS, the Austrian Research Centres, Business Unit Water, conducted a study to determine the characteristics of landslides (Klingseisen and Leopold, 2006). In addition to using a Digital Elevation Model (DEM), colour Orthophotos, national topographic base-map scale 1:50,000 and geological maps data were collected on-site using a Trimble GeoXMTM with ESRI ArcPad 6 software (Fig. 5.7). The accuracy achieved was 10 m.

The above examples refer to terrain of which maps have been produced at feasible scales. However, in many developing regions, such as many countries in sub-Saharan Africa, large and medium-scale maps have either never been made or, if in better times they have been produced, perhaps half a century ago, they have since become desperately outdated and are hence of very little practical value. The suitability of mobile GIS for use in unmapped terrain was demonstrated during bathymetric survey of the Senegal River in June 2004; with a length of 1800 km, the river flows west from Mali to form the border between Senegal and Mauritania. An essential prerequisite for bathymetric surveys is a map of the vicinity of the river. Mobile GIS proved a good replacement for non-existent maps. Two crews drew the map of the river and worked one day ahead of the bathymetric survey vessels.

Another application is related to supporting inventories of specific urban objects such as tree species. Shanghai, for example, created a database of its nine million

Fig. 5.7 Landslide extent geo-data acquisition using mobile GIS with stylus

trees. To inspect their health, fieldworkers use handhelds to navigate to sample trees. There they collect data by filling in a digital form shown on the screen of the handheld which is wirelessly sent to the office, where it is checked right away for completeness. Mistakes are reported back while the fieldworker is still close to the tree. The figures can also provide statistics to serve managers. This application demonstrates that mobile GIS not only makes geo-data acquisition more effective but also that such tools are a good means to control data quality during data collection.

Antalya in Turkey is a rapidly, partially organically expanding urban area populated by over one million souls. To support repair and extension of its drinking-water and sewerage pipelines a GIS database has been created from existing maps. Updates at cm-accuracy are collected with a GNSS receiver which is able to process RTK corrections.

5.2.8 Scientific Applications

The acquisition of data in the field is considered to be a key data collection method in most geo-related scientific disciplines (Gerber and Chuan, 2000). Indeed, a lot of scientific literature focuses on the importance of fieldwork in sciences such as geography, geology, hydrology and biology (Clarke, 2004). Up to around 2005 fieldwork projects were mainly carried out by using paper as the main storage medium while collecting data in the field (Wagtendonk and De Jeu, 2007). The capabilities and functionality of mobile GIS for scientific purposes have been acknowledged by many geo-scientists since the turn of the millennium. Montoya (2003) researched mobile GIS in combination with collecting video for urban disaster management purposes. Recognising that handhelds contribute to scientific studies conducted in demanding field research setting Tripcevich (2004) investigated the opportunities of mobile GIS in archaeological survey and discussed benefits and limitations. Tsou (2004) recognised the great potential of integrating mobile GIS and wireless internet mapping services for regional environmental management programs and natural habitat conservation and developed a prototype GIS application which would allow resource managers and park rangers in the US to access remote sensing imagery and digital maps from a portable web server mounted in a vehicle. Users can conduct real-time geo-data updates and/or submit changes back to the web server via WLAN. Niu et al. (2005) investigated the opportunities of mobile GIS and web-based GIS for the management of coastal land and decision-making, activities which entail great challenges for government agencies.

Other applications fields where mobile GIS is of great help are in

– mapping enumeration areas for national census
– accident mapping: evidence gathering for reconstructing the accident scene
– wild-fire suppression and management for navigating to the scene of a fire, directing fire fighters and ensuring safety at the scene and afterwards mapping the damaged area.

Based on the experiences reported by the above authors and others, Wagtendonk and De Jeu (2007) summarised the most important features and requirements when using mobile GIS systems in scientific data collection. The requirements include that high data *quality* should go hand in hand with minimal data *quantity* that is the samples sizes should be as small as possible. For efficacy purposes, the data acquired in digital form should be transmitted and analysed in digital format using computers and appropriate software. Furthermore, the general criteria of the scientific method are applicable to scientific fieldwork as well. These criteria are: verifiability and repeatability. If an experimental result cannot be verified by repetition, it is not scientific. The keyword of the limitations of scientific fieldwork is *small*, in particular: small research budgets; small application scales but high development costs and complex field methods; small number of qualified field and technical ICT support staff.

5.3 Location-Based Services

Mobile GIS aims at geo-data collection in the field by using support from digital maps stored in a database on a central server. The data are transmitted through wireless communication, both the data collected on-site and the data stored on the server.

The primary aim of mobile GIS is mapping of an area by field visits. Location-based services (LBS) also make use of wireless communication to transfer digital map data from a server to a device in the field. However, the aim of LBS is not to collect data in the field creating a mapping or updating an existing database but to help people to orientate themselves in their vicinity and to find their way to a desired destination. Although the goals of mobile GIS and LBS differ a lot, the architecture of an LBS system is very similar to the architecture of a mobile GIS.

The most often tipped example for demonstrating the magnitude of LBS impact on the commercial market is the father out with his family in an unfamiliar city. The children want pizza for diner. Father takes his mobile device and finds at his fingertips the location of the nearest pizzeria. Within 5 min the restaurant is in view and father is a hero. This example of searching for a restaurant shows that position is the crucial factor to find the desired service and to navigate from the present position to the position where the service is located. Consequently, location-based services can be defined as information services accessible with mobile devices through the mobile network and utilising the ability to make use of the location of the mobile device (Virrantaus et al., 2001).

From a technological point of view an LBS system can thus be defined as a GIS which respond to its location by knowing its own location, which means the information communicated to the users is contextualised to their location. To identify the location of the user, a LBS system should be equipped with a positioning system, and this will usually be a GNSS receiver, although also the cell position of a mobile phone could be used, at the cost, however, of considerably reduced accuracy. Precise indoor positioning is possible by placing pseudolite arrays within the building, which can either operate as an augmentation system of GNSS receivers or as stand alone system. The term pseudolite is composed of the words pseudo and satellite, and is also used to earmark approach landing systems around airports and transponders positioned on sea beds.

A device that contains a GIS, a positioning system and a wireless communication facility may only receive the tag LBS when the system can deliver the following information:

– Showing to the user utilities and their location in his or her neighbourhood; of course the utilities should be an integral part of the LBS service package; in the public safety sector the information may concern the location of fire hydrants.
– How to move from where the user is to where the user wants to be, given a specific means of transport.
– Track the user from where he or she is now to where he or she is going to, particularly to enable others to know where the user is, for example, when clearing a way through a building in fire or a building collapsed due to an earthquake.
– Showing no-go areas, for example, corridors blocked by debris, which is not known to the people on site but which information is already collected by others and stored in the central database at the crisis centre.

Especially when LBS is used in emergency alerts, the devices used by the response teams should be resistant to water, dust, shock and high temperatures while the interface should be easy and the screen resolution high enough to allow complete understanding and decision taking within seconds. To enable users to see both detail and overview of the site in which they are operating the handheld should support flexible zooming in and out in real time which requires advanced generalisation to avoid visual popping effects when changing from a high level of detail to an overview map (Sester and Brenner, 2009). Indeed the increasing availability and use of small mobile computers while in the field or on the road has introduced the need for quickly changing from the one level of detail to the other; as a consequence automatic generalisation has become a very active research area (see, e.g. Foerster et al. (2010); Stoter et al., 2010; Van Oosterom, 2009; Dilo et al., 2009; Meijers et al., 2009).

The LBS market is rapidly growing supported by the success of the mobile web where mapping is among the top ten activities conducted using the internet on a mobile device (Aoidh et al., 2009). One of the main problems induced by the boom in data requests through mobile devices is the abundance of data, which has to be wirelessly transmitted. Reduction of the overload of data transfer through the mobile internet requires techniques which adapt to user needs. Such techniques should consist of efficient algorithms able to handle the complexity and volume of data and combine this with knowledge about the interest of the user. The latter is called map personalisation which is an evolving research issue. The aim of map personalisation is to exclude data that are outside the interest of the user and highlight data relevant to the individual user. This requires development of methods for acquiring, identifying and describing unique user actions (McArdle, 2010).

In complex emergency scenarios saving people, assessing an area and transporting injured to a hospital requires close collaboration among members of the police, fire department and many other emergency team members (Konecny et al., 2010). Collaboration can be greatly improved by utilising mobile devices connected through mobile networks as demonstrated by the EU-funded project Workpad (Catarci et al., 2008). Let us consider a fire fighter who has been instructed to enter a building demolished by an earthquake to rescue a wounded woman. He carries a mobile device that tells him the safest path to the spot using data transmitted from the central GIS to his mobile device. The mobile device also allows retrieving a map of the area with the positions of team members, obstructed corridors and other crucial, run-time information. However, the device allows two-way communication; he may add Point of Interest (POI) information in the form of an annotation that the injured woman is accompanied by a second victim, which information is disseminated to his colleagues through the central GIS; the one who is closest to the victims and available is tasked to give assistance. Such a system is based on the following components:

– Geo-awareness by providing static information retrieved from a central GIS system and run-time information on positions of team members collected on the spot
– Structured annotations describing actual situations as observed by team members

– Task-assignment and co-ordination carried out automatically by a mobile process management system

Such a system greatly supports emergency teams by allowing them to entirely focus on help.

References

Aoidh EM, McArdle G, Petit M, Ray C, Bertolotto M, Claramunt C, Wilson D (2009) Personalization in adaptive and interactive GIS. Ann GIS 15(1):23–33

Catarci T, de Leoni M, Marella A, Mecella M, Salvatore B, Vetere G, Dustdar S, Juszczyk L, Manzoor A, Truong HL (2008) Pervasive software environments for supporting disaster responses. IEEE Internet Comput January/February:26–37

Clarke KC (2004) Mobile mapping and geographic information systems. Cartogr Geogr Inf Sci 31(3):131–136

Dilo A, van Oosterom P, Hofman A (2009) Constrained tGAP for generalization between scales: the case of Dutch topographic data. Comput Environ Urban Syst 33(5):388–402

Foerster T, Stoter J, Kraak M-J (2010) Challenges for automated generalisation at European mapping agencies: a qualitative and quantitative analysis. Cartogr J 47(1):41–54

Gerber R, Chuan GK (2000) Fieldwork in geography: reflections, perspectives and actions. Kluwer, Dordrecht, The Netherlands

Huang B, Xie C, Li H (2005) Mobile GIS with enhanced performance for pavement distress data collection and management. Photogramm Eng Remote Sensing 71(4):443–451

Klingseisen B, Leopold Ph (2006) Landslide hazard mapping in Austria: using GIS to develop a prediction method. GIM Int 20(12):41–43

Konecny M, Zlatanova S, Bandrova TL (eds) (2010) Geographic information and cartography for risk and crisis management: towards better solutions. Springer, Berlin

Lemmens M (2005) Mobile mappers – features and applications: small, portable and accurate devices for GIS input. GIM Int 19(12):16–19

Lemmens M (2010a) Mobile GIS systems: technical and business features. GIM Int 24(12):21–27

Lemmens M (2010b) Mobile GIS systems: hardware and software. GIM Int 24(12):28–31

McArdle G (2010) Personalised map interfaces: understanding mobile users. GIM Int 24(5):26–28

Meijers M, van Oosterom P, Quak W (2009) A storage and transfer efficient data structure for variable scale vector data. In: Sester M, Bernard L, Paelke V (eds) Advances in GIScience. Springer, Berlin, pp 345–367

Montoya L (2003) Geo-data acquisition through mobile GIS and digital video: an urban disaster management perspective. Environ Model Softw 18(10):869–876

Niu X, Ma R, Ali T, Li R (2005) Intergation of mobile GIS and wireless technology for coastal management and decision-making. Photogramm Eng Remote Sensing 71(4):453–459

Pundt H (2002) Field data collection with mobile GIS: dependencies between semantics and data quality. GeoInformatica 6(4):363–380

Sester M, Brenner C (2009) A vocabulary for a multiscale process description for fast transmission and continuous visualization of spatial data. Comput Geosci 35(11):2177–2184

Stoter J, Visser T, van Oosterom P, Quak W, Bakker N (2010) A semantic-rich multi-scale information model for topography. Int J Geogr Inf Sci iFirst:1–25

Tripcevich N (2004) Flexibility by design: how mobile GIS meets the needs of archaeological survey. Cartogr Geogr Inf Sci 31(3):137–151

Tsou M-H (2004) Integrated mobile GIS and wireless internet map servers for environmental monitoring and management. Cartogr Geogr Inf Sci 31(3):153–165

Van Oosterom P (2009) Research and development in geo-information generalization and multiple representation. Comput Environ Urban Syst 33(5):303–310

Virrantaus K, Markkula J, Garmash A, Terziyan YV (2001) Developing GIS-supported location based services. Proceedings of WGIS'2001 1st international workshop on web geographical information systems, Kyoto, Japan, pp 423–432

Vivoni ER, Camilli R (2003) Real-time streaming of environmental field data. Comput Geosci 29(4):457–468

Wagtendonk AJ, De Jeu RAM (2007) Sensible field computing; evaluating the use of mobile GIS methods in scientific fieldwork. Photogramm Eng Remote Sensing 73(6):651–662

Chapter 6
Terrestrial Laser Scanning

Since the early 2000s terrestrial laser scanning has evolved from a research and development (R&D) topic to a geo-data technology, which is commercially offered by a multitude of land surveying companies and other service providers all over the world. The technology is primarily used for the rapid acquisition of three-dimensional (3D) information of a variety of topographic and industrial objects. Cultural heritage, bridges, plants, cars, coastal cliffs, highways and traffic collision damage, all can be accurately modelled and documented with laser technology. Lidar is without doubt the most successful data-acquisition technique introduced in the last decade. As an acronym of Light Detection and Ranging, some prefer to read Lidar as Laser Imaging Detection and Ranging – the term has become a 'proper name' – spelled like your own first and surname with the initial letter the only capital.

Basically, all Lidar scanners measure range and intensity of terrain points hit by the laser beam. The first task is to convert the raw data into positions – three coordinates for each point – in a geodetic reference system. The resulting point-cloud is the basis for further processing, including filtering, visualisation, classification and analysis, or other manipulation.

Lidar instruments can be mounted on a tripod positioned over the ground for capturing, at street view, the surface of objects in the surrounding, such as bridges, dams, building facades, trees or cultural heritage sites (Fig. 6.1). In these ground-based cases the technology is called terrestrial laser scanning (TLS). Lidar instruments can also be placed in moving platforms such as aircrafts, helicopters, cars or vessels. When placed in a flying platform the technology is usually referred to as airborne Lidar or airborne laser scanning (Lemmens, 2007a). When the laser scanner is placed in a car, van or boat, the technology is called mobile laser scanning, terrestrial mobile mapping or, more often, mobile mapping. In recent years mobile mapping has become a rapidly emerging technology, particularly for accurately mapping roads and highways. The biggest advantage of mobile mapping is avoiding closure of the road during the time surveyors would manually measure the trajectory and the extreme rapid acquisition of millions of points. This chapter presents the principles of terrestrial laser scanning and provides an overview of application areas. Mobile mapping systems are also discussed. Airborne Lidar is treated in a

M. Lemmens, *Geo-information*, Geotechnologies and the Environment 5,
DOI 10.1007/978-94-007-1667-4_6, © Springer Science+Business Media B.V. 2011

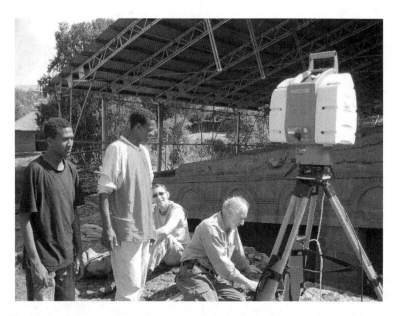

Fig. 6.1 A Leica Geosystem laser scanner at work for heritage documentation in Africa

separate chapter (Chapter 8) after discussion of the basics of photogrammetry, as airborne Lidar and aerial photogrammetry show many similarities especially with respect to geo-referencing and applications. In-depth treatment of theory and principles of Lidar technology can be found in three specialised textbooks, which have been recently published and written by a wide range of scientists and other experts: Shan and Toth (2008), Heritage and Large (2009), and Vosselman and Maas (2010). The technical specifications and development of commercial terrestrial laser sensors are listed on a regular basis in Lemmens (2004, 2007d, 2009, 2010).

Three-dimensional (3D) modelling aims at capturing all geometrical detail of objects, both the exterior parts and the interior ones, and representing these features with high-resolution triangular meshes for accurate documentation and photo-realistic visualisation (El-Hakim et al., 2007, 2008). Terrestrial Laser Scanning makes possible the swift measurement of millions of points by automatically scanning the scene at high speed. In the resulting dense point-cloud objects can be easily identified, allowing the creation of 3D-models of a wide range of objects with a level of detail impossible to achieve within a reasonable space of time using traditional technologies. Practical tests have shown that terrestrial laser scanning compared to conventional land surveying methods, can be up to 15 times faster for creating digital elevation models of open pit mines (Conforti, 2010). The high level of detail with surveying accuracy gives users access to a wealth of information. The products based on dense point-clouds are used by a multitude of professionals such as cultural archaeologists (Fig. 6.1), architects, facility managers, chemical engineers, construction engineers and plant engineers.

6.1 Basics of Laser

A laser is a device that creates a narrow, intense beam of coherent and monochromatic light generated by particles (atoms or molecules) that emit radiation (Siegman, 1986; Wilson and Hawkes, 1987; Koechner, 1992; Silfvast, 1996) The easiest way to understand the working of a laser device is by considering its two-level system: particles have only two energy levels, an upper and a lower state, separated by some energy difference. Electrons in the lower state can be excited to the upper state, usually by heat. Upon returning to a lower state the particle emits light. Usually particles behave independently of each other and the resulting light can be of any wavelength (colour). However, the emission of radiation can be stimulated to produce light of a particular wavelength. This occurs when a particle in the upper state interacts with a photon matching the energy separation of the levels; the particle may decay, emitting another photon with the same phase and wavelength as the incident photon, resulting in two photons emitted. This process is known as stimulated emission. The resulting beam consists of light of a single wavelength; the light is said to be monochromatic. Wavelength is determined by the amount of energy released when the electron falls to a lower state. The beam is also coherent, meaning that the photons move in step with each other; they have the same phase. The divergence of the beam is small, resulting in a very concentrated and narrow ray of electromagnetic energy. A flashlight, on the other hand, releases weak and diffuse light in many directions. Lasers differ greatly in properties, as well as in wavelength, size and efficiency.

Albert Einstein recognised the existence of stimulated emission in 1917, but not until the 1950s were ways found to transfer the theory into practice (Bertolotti, 1983). In 1954 Charles Townes, researcher at Bell Labs US and, independently, Basov and Prokorov in Russia, suggested a practical method of achieving lasing using ammonia gas, which produced amplified microwave radiation instead of visible light (called a Maser). For this the scientists shared the 1964 Nobel Prize for Physics. In 1958 Townes and Arthur Schawlow in a paper entitled Infrared and Optical Masers calculated the conditions needed to produce visible laser light. T. Maiman used a ruby crystal to demonstrate the first true laser in 1960.

Compared to other light sources, the light produced by lasers is in general highly monochromatic, directional, powerful and coherent. A price to be paid for these beneficial properties is that lasers can harm living tissue (Lemmens, 2010). As laser scanners often operate in the presence of people, such as streets in cities and plants, safety is an important issue. Since the early 1970s lasers have been subject to safety classification based on their potential for causing damage to eye and skin. The initial classification system of six classes was slightly revised and extended to include seven classes in 2002. In the old system the US indicated classes by Roman numerals, sometimes appended with letters (I, II, IIA, IIIA, IIIB and IV), whereas the EU used Arabic numerals (1–4), also appended by letter; the latter system is used worldwide in the revised version. The higher the numeral, the more injury the laser can cause: 1, 1M, 2, 2M, 3R, 3B and 4. Most TLS systems are assigned to safety class 3R; in the old system this would be IIIA, meaning 'safe if handled carefully'. Safety depends on maximum output power, laser principle and wavelength. The output power of phase-shift scanners operating in the visible wavelengths should not exceed 5 mW. Other maxima apply for other wavelengths and for pulsed lasers.

6.2 Measurement Principles

Laser scanners are active sensors that emit laser beams for measuring the distances to objects without human/object contact; 'active' means that the sensors themselves emit electromagnetic (EM) energy (Fig. 6.2). Figure 6.3 shows a typical output of a street scene recorded by a laser scanner and derived CAD drawing.

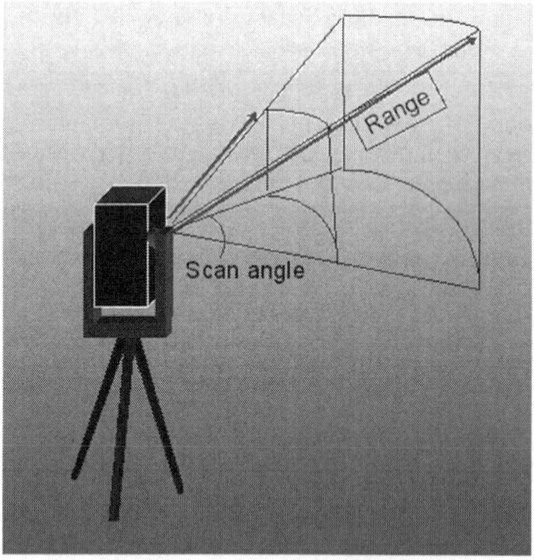

Fig. 6.2 Principle of Terrestrial Laser Scanning; from measured range and scan angle combined with position and orientation of the scanner the three-dimensional coordinates of the points hit by laser beams can be computed, resulting in a point cloud of 3D coordinates

Fig. 6.3 Part of a street in the historical area of Leicester, UK, captured by a laser scanner mounted on a car; point cloud computed from the raw data (*left*) and derived CAD drawing

6.2.1 Time-of-Flight Versus Phase-Shift

Measuring with laser sensors may be based on four principles (Colombo and Marana, 2010):

– Phase shift: waves are modulated in width or frequency; width modulation is sensitive to sharp discontinuities in shape or reflectance of the object, while frequency modulation provides reliable measurements even when return energy is low.
– Pulse measurements, also coined time-of-flight: pulses are emitted and their travel time to and back from the object is measured.
– Optical triangulation, for short-range applications and small objects.
– Interferometry, which offers very high precision and is used in indoor industrial metrology.

The first two laser principles, phase shift and pulse measurements, are commonly used in TLS systems for outdoor applications (Lemmens, 2004). In phase-shift technology the sensor continuously emits beams which are modulated as sine waves (Figs. 6.4 and 6.5). The phase of the reflected part is measured and compared to the phase of the outgoing one, and distance then calculated from the difference in phase (phase-shift). In the time-of-flight technology a pulse is emitted in the direction of the object; the time taken by the part of the pulse reflected back to reach the instrument is measured. Distance is calculated by multiplying this travel time by the speed of light and dividing the result by two.

Fig. 6.4 Two terrestrial laser scanners available on the market based on the range measurement principle of phase-shift, Z&F Imager 2006 (*left*) and FARO LS880

Fig. 6.5 Two terrestrial lasers based on the principle of time-of-flight range measurement: Optech ILRIS (*left*) and Riegl LMS Z420i

6.2.2 Range

One of the most important features of a TLS instrument is measurement range because range determines to a large extent types of application (Lemmens, 2004). A distinction can be made between short-range (up to 25 m), medium-range (up to 250 m) and long-range (larger than 250 m). Phase-based scanners, on the market since the early 1990s, were initially aimed at close-range, high-accuracy industrial applications. They are characterised by a precision ranging from sub-millimetre to sub-centimetre level, and high scan rates of up to half a million points per second resulting in high density. However, these favourable numbers come at the cost of limits to range, which is less than 100 m. In contrast, time-of-flight systems may measure distances up to 1 km and even more, but their precision usually ranges from sub-centimetre to centimetre level, while scan rates are at the 10,000 points per second level.

Another important feature of a TLS instrument is spatial resolution; this parameter determines the level of detail that can be obtained from a laser point cloud and depends on the range (distance to the object), sampling interval and width of the laser beam (Lichti and Jamtsho, 2006). Usually the sampling interval is taken as indicator of spatial resolution. However, a high sampling interval does not necessarily mean high spatial resolution; if the beam width is rather large, a high sampling interval will cause that the footprints of the laser beams as projected on object surfaces do overlap, reducing the effective spatial resolution which one would expect by considering sampling interval alone. Lichti and Jamtsho (2006) propose a new, more appropriate measure: the effective instantaneous field of view (EIFOV).

6.2.3 TLS as a Surveying Instrument

Many TLS systems appear less than perfectly suited to daily survey applications because the equipment is big, difficult to move around and requires an external power supply and computer support for operation (Neubauer, 2007). In many applications, these characteristics may fade into insignificance compared with the benefits; but survey workflow demands versatile set-ups. Therefore manufacturers of surveying equipment such as Topcon and Trimble introduced TLS instruments that can be carried around, set up and operated as a total-station. The battery is internal and the camera is also built in, while the control panel is part of the system. Most TLS systems do have in built imaging functionality, but some manufacturers choose a non-integrated-camera laser scanner, offering instead a camera mounted on top of the instrument. A scanner without integrated camera improves user flexibility. For example, users may want to apply a certain camera type because they are used to it or because their applications require a specialised camera, perhaps a calibrated metric one. It may also be desirable to change lenses depending on type of job (Fig. 6.6).

Fig. 6.6 The Riegl VZ-400 with a Nikon digital camera mounted on top (*left*) emits 122,000 near infrared pulses per seconds, weights 12 kg and measures ranges up to 600 m. Right: Leica ScanStation C10 emits 50,000 green laser pulses per second, weights 14 kg and measures ranges up 300 m green. Both instruments were demonstrated at the European Lidar Mapping Forum (ELMF), December 2010, The Hague, the Netherlands

6.3 Processing Software

The successful use of laser-scanner systems depends not only on the characteristics of the data-collection instruments themselves but also on the capabilities of the processing software necessary to obtain meaningful information after field acquisition of the 3D point cloud. The diversity of characteristics shown by the different, software packages, available on the commercial market, demonstrates that 3D-laser scanning is a promising surveying technology, which is in the process of ongoing and rapid development. Some of the software packages are developed for specialised applications, while others offer a full suite of point-cloud processing tools from scan registration and georeferencing through to feature extraction (Lemmens, 2006). They are specifically developed to support manufacture – specific scanners. There are also packages that offer full support for the entire laser-scanning process, right through to reverse engineering and data comparison, and are not associated with a particular type of scanner. 3D-laser scanning was not initially specifically developed as a technology to support the needs of the land surveyor. However, manufacturers of surveying instruments have recognised the attention being paid by surveyors to the technology. In response, these manufacturers are increasingly adapting the technology, both instruments and processing software, to the needs of the surveyor. As a result, new features are steadily added in support of improving surveying workflows.

6.4 Comparison with Total Stations

Laser scanners are often compared to reflectorless (robotic) total stations and, as far as the measurement principle is concerned, this is fine. Both instruments measure distances using pulsed laser light or phase shifts. This section treats resemblances and differences between TLS systems and total stations, equipment which – together with GNSS receivers – belong to the standard equipment of the land surveyor based on the analysis carried out in Lemmens (2007c).

6.4.1 Measurement Principles and Precision

Total stations can achieve higher precision because many measurements, even up to thousands, to the same point are taken and averaged, and thus improving accuracy, while laser scanners measure each distance only a few times, sometimes just once. To determine 3D coordinates of object points, the coordinates of the position of the instrument have to be known as well as the horizontal and vertical angles of each outgoing beam. The 3D coordinates of each point are calculated in a local or national reference system, from the laser distance, the known X, Y, Z coordinates of the instrument, and horizontal and vertical angles of each outgoing laser beam. The coordinates of the position of total stations are usually determined by centring the instrument above a known point. The position of a laser scanner is usually determined indirectly by placing special targets the three coordinates of which are measured using traditional survey instruments such as total stations. The similarity in measurement technology inspired manufacturers of traditional survey instruments to modify their total stations into a quasi laser scanner able automatically to scan areas of interest at predefined intervals, despite scanning rate being significantly lower than that of laser scanners.

6.4.2 Blind Sampling

At product level, similarity bounces. A land surveyor is used to interpreting a scene as a collection of characteristic points each of which has to be measured individually; connecting the characteristic points correctly allows reconstruction of the boundary of the object. Using a laser scanner, no selection of individual points takes place during scanning; it is a matter of chance which points are hit by the laser beam. As a result, unwanted objects, such as crossing pedestrians, are also captured. The actual measuring is done in the office by fitting geometric primitives such as lines, planes, cylinders and spheres through parts of the point-cloud. Boundaries and characteristic points are computed from intersecting neighbouring geometric primitives. Laser scanners can collect points for hours on end without human intervention, making them particularly suitable for capturing complex objects and operating in hostile environments such as nuclear plants, where placing the instrument should be done pretty smartly.

6.4.3 Time-Efficiency

Laser scanners and total stations are also compared at time-efficiency level. Indeed, surveying with a total station is only feasible when the object can be modelled by a limited number of characteristic points. Laser scanning enables capturing scenes consisting of objects of complex shape, such as chemical plants, cultural-heritage and traffic-accident sites. Some vendors of laser scanners posit the idea that if a laser scanner acquires 8,000 points per second while it takes a survey team 10 s to measure a single point, using the scanner is equivalent to working with not one but 80,000 teams; a faulty and misleading comparison. Surveyors select their points intelligibly, while laser scanners take points blindly, without identification, interpretation and selection. These activities have later to be carried out in the office. As a rule of thumb one may state that the processing of industrial objects, such as plants, will take approximately the same time as the time spend to capturing the data in the field (1:1 field to processing ratio). But as far it concerns heritage documentation a field-to-processing ration of 1:10 is more realistic, that is one day of data capturing in the field requires up to ten man-days data-processing in the office (Rüther, 2007; Lemmens, 2007b). As such, laser scanning bears more comparison with photogrammetry than with surveying. Laser scanning may thus bring land surveyors and photogrammetrists closer together.

6.5 Comparison with Terrestrial Photogrammetry

When comparing the performance characteristics of TLS with terrestrial photogrammetry techniques the pros and cons can be summarised as follows (Lemmens and van den Heuvel, 2001):

1. On-site recording by TLS is independent of the presence of texture on the object. Texture is particularly essential when digital surface models of the objects have to be created from stereo images. The presence of texture in photos is however not required when edges of objects are measured to create for example as-built models.
2. Images can be taken faster and capture both geometry and texture, while TLS systems capture geometry only.
3. By using two or more cameras that can be simultaneously exposed, photogrammetry enables instantaneous recording of a scene. Consequently the recording of dynamic objects is possible. Notwithstanding that the scanning speed of TLS systems is very high, the scene is captured sequentially, point by point, and the recording of a scene may take up to 20 min or even longer. So, in contrast to photogrammetry, dynamic objects, such as human bodies, cannot be recorded by TLS without distortions in the data.
4. Since TLS is an active system, generating its own light, while photogrammetry makes use of light generated by external sources (sun, lamp), light conditions are less critical for TLS.
5. Although manufacturers of surveying equipment are working on instruments that can be carried around, set up and operated as a total-station, today's TLS systems appear less than perfectly suited to daily survey applications because the equipment is big, difficult to move around and uses a lot of power necessitating carrying around heavy batteries. However, their robustness makes them suitable for use in harsh environments, such as deserts and tropical rain forest. Recording scenes by photogrammetric techniques is more flexible because cameras are more portable (less heavy) than TLS instruments, while one does not necessarily have to mount them on tripods.
6. Although the prices of high-precision TLS systems are gradually decreasing as a result of becoming a matured technology, they are more expensive that photogrammetric systems with comparable precision.
7. The creation of digital surface models, necessary for visualisation and rendering purposes, amongst other things, can be done directly and automatically from the recorded TLS data. Photogrammetric images first require a matching step in order to create a set of 3D point, for which specialised matching and editing software is needed (see Section 7.5).

The above shows that the characteristics of TLS and terrestrial photogrammetry do not compete with each other; they are complementary in nature. Since they supplement each other, TLS and digital cameras are often used alongside when capturing complex scenes.

6.6 Integration with Digital Cameras

Technology is moving towards the development of hybrid measurement system consisting of a TLS device solidly integrated with a calibrated digital camera. Such new devices collect 3D point-clouds of objects and, simultaneously, digital images of the same object. They thus integrate surveying with photogrammetry and imaging. These systems are useful both for measuring 3D building points and for directly rendering 3D models with high-definition photo-texture and are useful for architectural surveying, 3D modelling of monumental buildings and heritage documentation. The combination of point cloud and image can be done highly automatically because the position and orientation of scanner and camera are mutually known with high accuracy avoiding manual identifying a set of homologous points between models generated from the laser scans and each image. Colombo and Marana (2007), both from the University of Bergamo, Italy, tested the information content and performance of such hybrid systems by 3D modelling of S. Maria Maggiore, the most important church in Bergamo and found good results (Fig. 6.7).

Fig. 6.7 3D Point models of the interiors of S. Maria Maggiore, Bergamo, Italy (*top*) and orthoimages of decorated intrados of dome and vaults, with line plots in overlay (*bottom*)

Fig. 6.8 Venice palace (Italy) laser-scan data merged with bathymetric point-cloud data

They also conclude that the present manufactory standards for scanners have to be improved as well as the processing software. Although scenes can be captured highly automatically, the in-office information extraction is tedious and labour-intensive because automation level of the processing software is low.

TLS instruments are not only used in conjunction with cameras but also with other types of sensors, such as sonar. The combination of 3D-laser scanning and side-scan sonar can be very beneficial for mapping complicated waterside areas, because the two systems are complementary as Byham et al. (2007) and Bacciocchi et al. (2009) demonstrated (Fig. 6.8).

6.7 Scene Monitoring

Vendors of terrestrial Lidar scanners often proudly present the operation of their instruments in the field as 'a piece of cake'. And they are right; it is convenient. One just needs to place the instrument in front of the object, level it roughly, enter the area to be captured and push the button. It is as easy as that, at least as far as the instrument side of the survey is concerned. But the snag lies not in operating the instrument but in the prelude to this, and in scene monitoring during capture.

6.7.1 Intervening Objects

Before placing the instrument the operator has to ensure that no objects intervene between instrument and scene that might occlude essential parts of it. This means no cars, lampposts, traffic signs, vegetation, people or donkeys (Figs. 6.9 and 6.10). Potential sources of failure may also be hidden in the scene; lapses may result especially from the reflectance characteristics of the surfaces of objects.

Hitting a surface, a laser beam may interact with it in three ways: it may be reflected, absorbed or transmitted (Rees, 2001). Only reflected beams will reach the instrument and thus be of use, but one reflection is better than another. Ideally, a surface behaves as a diffuse reflector, so that the resulting reflections are of like strength in all directions. In this case most of the signal returned from the surface reaches the instrument. But when parts of an object have a specular surface the reflection is deflected and little or no signal is returned to the instrument. Therefore mirrors, shiny metal and brackets of neon lights present in the scene have to be removed or covered. Signal strength, recorded in addition to time of flight of the laser beam, provides the operator with a helpful means of detecting lapses in reflectance during processing of point-clouds. The intensity of the reflections is an important factor determining the accuracy of TLS point measurements. The intensity mainly depends on three aspects: distance from the instrument to the object, reflection properties of the surface of the object and angle of incidence (Lee et al., 2010).

Laser beams may be transmitted through windows; this can happen with buildings, and results in the recording of objects on the other side of the glass. To avoid problems in reconstructing the final scene, windows have to be covered, or special

Fig. 6.9 Trees, fences, and also unmovable donkeys (*bottom left*) may conceal the scene from the laser scanner, impeding capture

Fig. 6.10 Crossing of pedestrians and bikers, while the scene is being captured, causes appearance of spikes and other funny things

caution must be exercised in processing the point cloud. Highway police, when using scanners to record the site of an accident, spray powder on car windows to prevent the scanner 'looking' through them. Pedestrians tend to cross a scene whilst it is being captured, and the same is true of animals, cars and bikes. The surveyor must thus close off the scene area to all traffic before pushing the button. This might sound self-evident, but in practice closing the scene to moving objects often proves much more problematic than it sounds.

6.7.2 Placing Markers

In addition to covering objects and preventing things crossing the scene, the surveyor also has to place and identify objects in it prior to pushing the button. Objects may be markers, such as nails, or marked existing points, such as sharp corners. They are used as control points, necessary for the conversion of range data to 3D coordinates of object points in a national or local reference system. The location of the points has to be selected with care and unambiguously indicated to avoid mistakes. Most commonly, four control points are used at the edges of the scan, although three is enough for geo-referencing; the fourth is redundant and used for reserve and check. Large objects require scanning from several positions, and adjacent scans have adequately to overlap. Once a control point has been identified in one scan, identification in the overlapping scans can be done semi-automatically.

6.8 Applications

Laser scanning has been established as a technique for precise surveying of complex objects and scenes and has become a consolidated practice for many surveying firms and other users. This section discusses a number of applications brought to the footlight during user conferences and discussed in literature.

6.8.1 3D City Modelling

During the fifth user conference of Leica Geosystems HDS held in San Ramon, California, autumn 2007, a broad pallet of applications were brought to stage (see Lemmens, 2008). General Motors Corporation, Detroit, uses 3D as-built models of its plants all over the world created with TLS, giving engineers and suppliers instantly access to the data via the web. The plant sector shows a strong rise in using laser scanning in part driven by the 2006 release of a phase-shift scanner which can measure up to 500,000 points per second. Highly detailed, photo-textured 3D-city models of Glasgow have been created by the Glasgow School of Art, a project funded by the European Union. The data were captured by two scanning surveyors and the 3D models generated by six, sometimes eight, modellers, over about an 18-month period, starting in the summer of 2006. Since Glasgow is an old city with many narrow and curved streets, 3D modelling needs substantially more time than for a new city like Toronto. The University of Rome, Italy, has developed a 3D model of the Coliseum and scanned the interior of today's St. Peter's Cathedral (heritage documentation is not so much an industry as a university research activity). TLS can be even applied to model such complex structures as domes (Fig. 6.7).

6.8.2 Traffic Accidents and Road Safety

The police and the military also benefit from TLS for such applications as traffic-accident survey and forensic research. California Highway Patrol (CHP) uses TLS for accident surveying, since 2006 when they bought no fewer than five scanners. Each accident scene needs to be surveyed quickly, accurately and completely. An extraordinary road accident took place in the early morning of 29 April 2007 on an approach to the San Francisco-Oakland Bay Bridge. A tanker truck carrying over 30,000 litres of gasoline crashed and went up in flames. The heat caused an overpass to collapse onto an interstate (highway) below. A few minutes later CHP's Multidisciplinary Accident Investigation Team was on site and scanned the overpass from ground to top. With conventional means just five hundred to a thousand points can be collected with laser scanning, millions. Unsafe positions can be measured remotely. Also, the road can be reopened sooner and high detail and accuracy allows not only accident reconstruction but is also useful for civil engineers tasked with renovating the road construction.

Three-dimensional road capture is not only a feasible measurement method once an accident has happened but also to study safety of road parts such as crossings in order to prevent future accidents. TLS can become an important measurement technique for this type of applications because (inter)national road safety programmes are gaining growing interest and with these come a need for 3D documentation of roads for sight analysis and traffic-accident reconstruction studies (Pagounis et al., 2007). Also crime scenes can be reconstructed by combining laser-scan data taken after the crime with forensic information. The resulting simulations of the past can be taken to court and seem to be visually convincing, imitating the realism of film, and so often convinces juries that this is how it happened.

6.8.3 Deformation and Heritage

The use of TLS for deformation measuring is gaining increasing interest. Monserrat and Crosetto (2008) developed a new procedure for land deformation monitoring using a sequence of point clouds taken at different moments in time (multi-temporal). The method aims at applicability in a wide range of applications and is based on least squares 3D surface matching as developed at ETH Zurich, by Professor Gruen and his Group. TLS is also applicable for deformation and load test measurements of bridges as Lovas et al. (2008) demonstrate, see also (Lovas and Berényi, 2009). They conducted a test on two Danube bridges, Budapest, one being a suspension bridge and the other a tied-arch bridge. TLS can measure displacements of cables and pylons not possible by traditional methods. Since the accuracy is less than for high-precision land surveying instruments, accompanying use of these instruments remains necessary.

TLS has been successfully applied in the documentation of historic buildings and archaeological features all over the world because this measurement technique enables quickly collecting reliable, high-resolution data and saves up to a hundred man-hours over a typical 1-month excavation (Neubauer, 2007). In addition to documentation the data can be used for creating virtual-reality models, restoration planning or virtual reconstruction and other products attractive for the general public. Combining terrestrial laser-scanning, magnetic field and electrical resistance data may also reveal anomalies at heritage sites that remain concealed during fieldwork and in aerial photos, which was demonstrated during preservation of Por-Bajin fortress, Siberia (Anikushkin and Kotelnikov, 2008). Figure 6.11 shows the Trimble laser scanner in action to capture the Por-Bajin fortress site. Section 12.2 in Chapter 12 provides a detailed overview applying TLS in recreating the past.

6.8.4 Other Applications

Foresters can use TLS for determining stand value and log products (Murphy, 2008). TLS has been successfully applied in a wide variety of civil engineering settings.

Fig. 6.11 Trimble GS200 at work at the Por-Bajin Fortress heritage site, Siberia

6.9 Mobile Mapping

A mobile mapping system (MMS) consists basically of a car, van or boat equipped with a positioning system – comprising a GNSS receiver integrated with an Inertial Measurement Unit (IMU) – and laser scanners (Tao and Li, 2007). Digital cameras, thermal sensors or other geo-data capturing systems may also be mounted on the roof rack. Figure 6.12 shows an example of a car equipped with GNSS, IMU, laser scanners and cameras. For inspection of the condition of construction surfaces, such as road asphalt, ground penetrating radar sensors may be attached to the vehicle too. Mobile mapping systems are sometimes preceded by the prefix 'terrestrial' to distinguish them from airborne platforms equipped with similar geo-data acquisition sensors. The positioning system comprising a GNSS receiver integrated with an Inertial Measurement Unit (IMU) enables to geo-reference the data collected by the sensors. In urban areas GNSS satellite signals may be blocked by high-rise buildings and other tall objects (urban canyons). To bridge the time GNSS satellite signals are blocked attaching odometers to the vehicle wheels is beneficial for improving the geometric quality of the data. These three redundant positioning technologies – GNSS, IMU and odometers – enable to obtain accurate positions of the laser and other sensors while they move around and acquire data. All devices are connected to a control unit from which the operator steers the capture of 3D coordinates of millions of points.

To save costs it is important that the time necessary for mounting, installing and setup of all devices is reduced to minimum. The cost per kilometre of surveying by means of MMS is falling by the year. Although the diverse manufacturers built a wide variety of systems, a typical MMS configuration may consist of the following components (Fig. 6.13):

Fig. 6.12 Mobile mapping system; note the odometer at the driving wheel at the rear. At the other side, not visible, a second odometer is mounted

Fig. 6.13 Mitsubishi MMS-X640 mounted on top of a car. The configuration comprises from *left* to *right*: laser scanner, digital camera, GNSS antenna on top of an IMU, digital camera and laser scanner

– One GNSS receiver, measuring 10 times per second the position of the car while being locked to L1 and L2 GPS signals and Glonass signals
– An IMU with a data rate of 100 measurements per second
– Two odometers
– Two digital cameras
– Two laser scanners

After collection of the data in the field the trajectory of the vehicle has to be computed first and next the point clouds, images and other sensor data have to be georeferenced. To compute the trajectory all data from GNSS receiver, IMU and odometers have to be integrated. After georeferencing measurements can be performed on the 3D point clouds or on overlapping images using a digital photogrammetric workstation to determine the 3D coordinates of, for example, traffic signs by conventional photogrammetric intersection (Fig. 6.14). The images can also be used to attach colour to each laser point. Using special software it is possible to select points in the images and retrieving the coordinates of these point from the DEM created from the laser points. This is made possible because both the images and the 3D laser point clouds have been accurately georeferenced.

The coming out of MMS is accelerated because practitioners, such as highway managers, are in need of fast and cost-effective data capture systems, while, in turn, the availability of MMS reshapes the Lidar mapping market allowing the introduction of new applications and end-customers (El-Sheimy, 2005). The main application with a high potential of commercial success is mapping of highways and other roads, necessary for construction, maintenance, safety and environmental purposes. Surveying of roads by traditional means is a time- and resource-consuming task. Mobile mapping systems provide an enormous time saving in capturing road networks and their surroundings (Cheng et al., 2008). MMS is able to acquire up to one billions points per hour together with tens of gigabytes of imagery. Field trials have shown that accuracies up to two centimetres can be achieved (Sukup and Sukup, 2010). So, high-quality geometric information of objects on, above and alongside roads can be extracted from dense point clouds. Data capture while the

Fig. 6.14 Measuring 3D coordinates of a traffic sign from two overlapping images captured during a mobile mapping survey

vehicle is moving with a speed of 100 km per hour is just the first link in the chain of providing proper information to users. The huge volumes of data require dedicated software to manage and analyse the billions of points. To speed up the cumbersome manual extraction of features much is done for developing software for extracting road signs, roadsides and the like automatically; research institutes are doing a lot of work is this realm. To allow access to the data anytime, anywhere by a wide range of professionals and other users a web-portal consisting of at least a web-based viewer should be employed.

References

Anikushkin M, Kotelnikov S (2008) 3D-model of Siberian fortress. GIM Int 22(7):13–15

Bacciocchi M, Byham P, Conforti D (2009) Data integration for coastal surveying: combining laser scanner and bathymetric sensor. GIM Int 23(3):13–15

Bertolotti M (1983) Masers and lasers: an historical approach. Adam Hilger, Bristol

Byham P, Bacciocchi M, Conforti D (2007) Waterside mapping in Italy: integrating 3D laser scanning and side-scan sonar. GIM Int 21(8):47–49

Cheng W, Hassan T, El-Sheimy N, Lavigne M (2008) Automatic road vector extraction for mobile mapping systems. Int Arch Photogramm Remote Sens 37(Part B3b):515–521

Colombo L, Marana B (2007) Camera laser scanner: geo-knowledge for preservation of St Maria Maggiore, Italy. GIM Int 21(8):15–18

Colombo L, Marana B (2010) Terrestrial laser scanning: how it works and what it does. GIM Int 24(12):17–20

Conforti D (2010) Measuring open pit mines: testing static and mobile laser scanners. GIM Int 24(8):17–19

El-Hakim S, Gonzo L, Voltolini F, Girardi S, Rizzi A, Remondino F, Whiting E (2007) Detailed 3D modelling of castles. Int J Archit Comput 5(2):199–220

El-Hakim S, Remondino F, Voltolini F (2008) 3D Modelling of castles: integrating techniques for detail and photo-realism. GIM Int 22(3):21–25

El-Sheimy N (2005) An overview of mobile mapping systems. FIG Working Week 2005, Cairo, Egypt, pp 1–24

Heritage GL, Large ARG (eds) (2009) Laser scanning for the environmental sciences. Wiley-Blackwell, Chichester, West Sussex, UK

Koechner W (1992) Solid-state laser engineering, 3rd edn. Springer, Berlin. ISBN 0-387-53756-2

Lee S, Lee JO, Park HJ, Bae KH (2010) Investigations into the influence of object characteristics on the quality of terrestrial laser scanner data. KSCE J Civ Eng 14(6):905–913

Lemmens M (2004) 3D Laser mapping. GIM Int 18(12):44–47

Lemmens M (2006) 3D Laser scanner software. GIM int 20(9):49–53

Lemmens M (2007a) Lidar. GIM Int 21(2):11

Lemmens M (2007b) Laser scanning technology challenged: shortcomings in spatial documentation of heritage sites. GIM Int 21(3):25–29

Lemmens M (2007c) Merging professions. GIM Int 21(8):11

Lemmens M (2007d) Terrestrial Laser Scanners. GIM int 21(9):41–45

Lemmens M (2008) Leica geosystems, HDS: fifth user conference and HQ, San Ramon, USA. GIM Int 22(2):37–41

Lemmens M (2009) Terrestrial laser scanners. GIM int 23(8):62–67

Lemmens M (2010) Terrestrial laser scanners. GIM Int 24(8):30–33

Lemmens MJPM, van den Heuvel FA (2001) 3D Close-range mapping systems; detailed and accurate 3D object models from laser. GIM Int 15(1):30–33

Lichti DD, Jamtsho S (2006) Angular resolution of terrestrial laser scanners. Photogramm Rec 21(114):141–160

Lovas T, Barsi Á, Dunai L, Csák Z, Polgár A, Berényi A, Kibédy Z, Szöcs K (2008) Terrestrial laser scanning in deformation measurements of structures. Int Arch Photogramm Remote Sens 37(B5):527–531

Lovas T, Berényi A (2009) Laser scanning in deformation measurements: acquiring 3D spatial data of bridges. GIM Int 23(3):17–21

Monserrat O, Crosetto M (2008) Deformation measurement using terrestrial laser scanning data and least squares 3D surface matching. ISPRS J Photogramm Remote Sens 63(1):142–154

Murphy G (2008) Determining stand value and log product yields using terrestrial Lidar and optimal bucking: a case study. J Forestry 106(6):317–324

Neubauer W (2007) Laser scanning and archaeology: standard tool for 3D documentation of excavations. GIM Int 21(10):14–17

Pagounis V, Tsakiri M, Palaskas S (2007) Road safety analysis: terrestrial laser scanning to improve road safety. GIM Int 21(8):52–53

Rees WG (2001) Physical principles of remote sensing, 2nd edn. Cambridge University Press, Cambridge, UK

Rüther H (2007) Laser-scanning and heritage: academic and manufacturer face challenges. GIM Int 21(5):14–17

Shan J, Toth ChK (eds) (2008) Topographic laser ranging and scanning: principles and processing. CRC Press, Boca Raton, London, New York. ISBN: 978-1-4200514-2-1

Siegman AE (1986) Lasers. University Science Books, Mill Valley, CA, USA. ISBN 0-935702-11-3

Silfvast WT (1996) Laser fundamentals. Cambridge University Press, Cambridge, UK. ISBN 0-521-55617-1

Sukup K, Sukup J (2010) Mobile mapping: accurate enough for mobile mapping. GIM Int 24(6):17–21

Tao CV, Li J (eds) (2007) Advances in mobile mapping technology. ISPRS book series, vol 4. Taylor & Francis, London, UK

Vosselman G, Maas H-G (eds) (2010) Airborne and terrestrial laser scanning. Whittles Publishing, Dunbeath, Caithness, Scotland, UK. ISBN 978-1-904-445876

Wilson J, Hawkes JFB (1987) Lasers: principles and applications, International series in optoelectronics. Prentice Hall, Upper Saddle River, NJ, USA. ISBN 0-13-523697-5

Chapter 7
Photogrammetry: Geometric Data from Imagery

Practitioners who wish to construct and maintain an urban, local or national geographic information system (GIS) or Land Information System (LIS) face the difficult and complex problem of data capture. Images have for a long time been major information sources for topographic mapping of large areas and creating base maps. Taking images is the fastest and most reliable way to capture reality.

This chapter treats in greater detail accurate and detailed information extraction from imagery. The professional fields dealing with information extraction from images are photogrammetry and remote sensing. Once the images have been captured with cameras mounted in airborne, ground-based or spaceborne platforms, the measurements are done in the office using software running on PCs, which are slightly modified to enable stereo viewing. Photogrammetry is mainly concerned with the geometric aspects: *where* are features of interest, such as buildings and roads, located. Remote sensing, on the other hand, focuses on the extraction of thematic information; this implies measuring *what* is present at certain locations. The importance of the *when* component, i.e. at what time the geo-data have been captured, is steadily growing. This is because scientists, managers and governors want to know how urban-fringes, sea level, dune erosion and many other earth-related phenomena evolve over time. The methodology to follow earth-related processes over time is called *change detection*, a topic we will treat in greater detail in Chapter 9. Extracting the 'where' component requires other techniques, methods and algorithms than does the 'what' component. This chapter focuses on the where component: photogrammetry. A conduct on the 'what' component is provided in Chapter 9.

7.1 From Analogue to Digital

Photogrammetry was already in existence more than 150 years ago, although, judging by current standards, only in its infancy. The main issue during all the long history of photogrammetry has been to reduce human involvement by automating parts of the complex process to arrive from the planar coordinates of points in images to (three-dimensional) object coordinates in a national or international geodetic reference system. The degree of complexity appeared so massive that it led to photogrammetrry evolving as an autonomous discipline. During its history Photogrammetry has passed through the phases of plane-table, analogue and analytical photogrammetry to digital photogrammetry (Torlegård, 1988) and so today the extraction of geometric information from images is done along fully digital paths, enabled by digital storage of reflectance values (pixels) and the use of Digital Photogrammetric Workstations (DPW). The technology, gradually refined over time and now ripened into maturity, means that today reliable, accurate and detailed geodata can be acquired of large areas by taking measurements from imagery recorded by film or digital camera onboard an aircraft, satellite or ground-based station.

After the invention of a new technology its first and primary application is often for serving the geo-information needs of the military. In the field of photogrammetry Almé Laussedat, a colonel in the Engineer Corps of the French Army and referred to as the Father of Photogrammetry, from 1849 onwards made many efforts to prove that terrestrial photographs could be beneficially used in the preparation of topographic maps (Thomson and Gruner, 1980). Not only the birth of terrestrial photogrammetry but also of aerial photogrammetry originates in France and the history of 'air surveying' is marked by three important dates: 1783, 1837 and 1855. On 4 June 1783 the first demonstration flight of a hot-air balloon took place in Annonay, France. It was by the Montgolfier brothers, since known as the inventors of the first practical balloon (Gillispie, 1983). In 1837 Jacques Mandé Daguerre, by profession a painter of diorammas, obtained the first practical photograph using the process named after him (see www.daguerre.org). In 1855 Gaspard Felix Tournachon better known as Nadar for the first time in history combined the two technologies by climbing into a balloon with a camera in his hands and taking the first aerial photograph from eighty metres above earth's surface. The first images from balloons were 'birds-eye' view snapshots and Emperor Napoléon III immediately recognised the potential for images taken from artificial 'high ground'; funds soon became available for developing airborne photogrammetric techniques to produce topographic maps for military purposes.

The earliest known application of photogrammetry in the United States was by the Union Army in 1862. Theodor Scheimpflug, a well-known name in the field of photogrammetry, was himself a captain in the Austrian Army. After World War I, photogrammetry was intensively applied for mapping remote areas. For example, Dutch mapping agencies carried out, under the passionate supervision of Prof. Willem Schermerhorn, who became after World War II Netherlands' first prime minister, massive efforts to collect and archive detailed topographic and geographic information of Indonesia. In the 1960s and 1970s, huge parts of the Amazon area were mapped from the air for exploitation purposes, using radar technology (Fig. 7.1).

A few more significant dates in the history of photogrammetry follow here. The International Society for Photogrammetry (ISP), predecessor of ISPRS, was founded by E. Dolezal in Austria in 1910. In 1921 came the introduction of the Autocartograph, the first universal analogue plotter. In 1932, Schermerhorn, famous founder of the International Institute for Geo Information Science and Earth Observation (ITC) in the Netherlands, began systematic tests on aero-triangulation

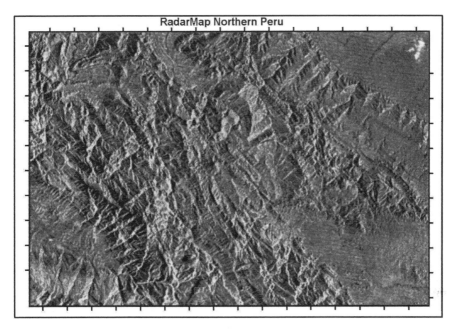

Fig. 7.1 Radarmap of Northern Peru, South America, surveyed from the air

techniques. These he applied in 1936 to mapping Irian Jaya, then under Dutch East Indies colonial rule and today part of the developing world. The purpose was oil exploration. Since its inception, over a century ago, the major application of aerial photogrammetry has been the production of detailed topographic maps covering entire countries. Furthermore, it has been used for a plethora of other tasks including environmental monitoring, urban and rural planning and road, railway and oil pipeline planning, construction and monitoring. Close-range photogrammetry enjoys the interest of archaeologists, architects, car engineers and many more who want detailed and accurate geometric information of objects in three dimensions (3D). Other well-known photogrammetrists who pioneered to introduce developing technologies in the photogrammetric data acquisition process, which we want to mention here, are Uki Helava and Fritz Ackermann. In the 1950s Helava equipped analogue photogrammetric mapping instruments with encoders and servomotors and connected these to computers and so invented the analytical plotter. Professor Ackermann's primary contribution to photogrammetry is to rapidly bring scientific methods into practice through his company Info GmbH, which was acquired by Trimble in 2007. Based on sound scientific fundaments, developed at his Institute at Stuttgart University – renown for excellent photogrammetric research and education – the software is most sophisticated, robust and reliable. Ackermann's name is further associated with bundle block adjustment and automated stereo matching.

Many aerial images are still taken with film cameras. This is done for several reasons. A major one is these photogrammetric cameras are specially developed for aerial surveys. They are metric cameras build for a specialised market, which is actually a niche market; they are precision instruments of which the internal geometry is precisely known. They are insensitive to the intense vibrations of an aircraft in air. The focal length is fixed and remains stable over time. The lenses have been carefully polished resulting in nearly absence of lens distortions enabling the realisation of a near-ideal central projection. They are voluminous and heavy; it takes two men to drag the camera to the airplane and to mount the instrument in the hole specially cut in the floor. The size of the exposure part of the film is 23 × 23 cm, from which image coordinates can be obtained with an accuracy of 0.005 mm (5 μ) or better; at photoscale 1:1,000 this image measurement accuracy means an object-space accuracy of 0.5 cm. This is the same accuracy as can be obtained by total stations, DGNSS and other high-end ground-based instruments used by professional surveyors. Accordingly, aerial photogrammetric cameras are very expensive and robust; they can last for decades. So, once a photogrammetric company has purchased a camera, they want to take the best out of it, money-wise. But economics is not the only reason for using film cameras as discussed in Section 7.10.

To convert the photos into digital images at the front-end of the photogrammetric process, a high-precision, calibrated scanner is needed which possesses the capability to maintain the geometric and radiometric characteristics of the original image. Purpose-built scanners are necessary to transfer accurately the analogue film to a computer readable format (Fig. 7.2).

Fig. 7.2 Photogrammetric scanner for scanning entire film roles installed at COWI, Denmark

Both geometrical and radiometrical scanner resolutions may affect the results substantially. Contrary to digital satellite images, which have a predefined (fixed) resolution, aerial images may be scanned with basically any resolution. Typical pixel sizes range from 7.5 to 50 μ.

The equipment developed for extracting geometric information from images evolved from analogue instruments, via analytical systems to digital systems. Analogue instruments are robust systems in which the transformation from image coordinates to object space coordinates is done along optical-mechanical lines. Up to the 1970s such instruments constituted the main processing utilities. Equipped with encoders to enable conversion of analogue coordinates into digital coordinates, they still can be found as work horses in topographic mapping agencies in some countries. Making use of the advances of computer technology, which rapidly emerged after WWII, analytical systems were developed in the fifties; the first analytical plotter was introduced in 1957 by the Finnish Uki Helava, who supplied many leading edge photogrammetric developments to the commercial marketplace often as responses to US government contracts (Miller and Seymour, 1995). From the 1970s onwards, analytical plotters gradually replaced analogue systems, to be surpassed themselves by Digital Photogrammetric Workstations by the turn of the millennium. Analytical plotters are measuring instruments specially constructed to process film-based stereophotos using the processing power of computers. They consist of a stereo-plotting unit, which are basically two high-precision digitisers, stereo-viewing optics and a computer provided with specialised software (Fig. 7.3).

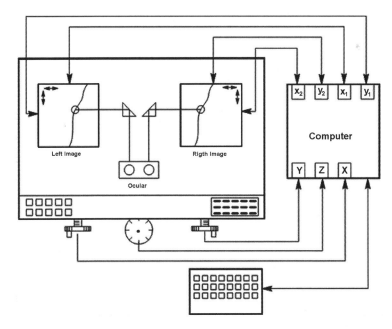

Fig. 7.3 Schematic overview of an analytical plotter

Digital photo-coordinates of corresponding points in both images of the stereo-pair are manually measured in stereo mode. Software installed on the computer transforms the image coordinates in 3D coordinates of points in an object-space-related coordinate system. During measuring or in an editing stage, the operator attaches labels to points, lines connecting points and polygons made-up by a sequence of lines. Examples of labels are building, road and railway.

The major difference between an analytical system and a digital system is the availability of the image in digital form, enabling the automation of, in principle any photogrammetric workflow; in practice automation is limited to a few tasks: determination of the image coordinates of tie points as a pre-processing stage for aerotriangulation, creation of Digital Elevation Models (DEM) and production of orthoimages. The basic enabler of automating these tasks is stereo image matching (see Section 7.5). Instruments for processing analogue and digital images can be very advanced and accordingly expensive, or more rudimentary and accordingly cheap. Digital Photogrammetric systems are treated in a separate section of this chapter (see Section 7.7).

When a discipline exists already for such a long time, would it not be critically endangered that means at risk of becoming extinct? The reply of Remi Jean, President DVP-GS and Vice-president and Partner Groupe ALTA DVP-GS, Canada to this assumption reads (see Lemmens, 2006):

> On one side, Google and Virtual Earth are currently creating a never before seen interest and demand for geospatial data by the 'good enough' large public. On the other side, major manufacturers have developed very efficient captors and digital cameras that produce data of exceptional metric and radiometric quality: 1+1 = 2. It is only a matter of time before we see a significant increase in the demand for quality data. Photogrammetry has definitely come out of the shadows (or the dark room!) and is active again.

7.2 Basics

The basic task of photogrammetry is to transfer the coordinates of points as measured in images to 3D coordinates in an object-space related reference system. The fundamental assumption is that each and every point in the image uniquely correspondents to one point in the terrain, or more generally, object space. The transformation involves application of a wide spectrum of mathematical principles which is thoroughly described in many textbooks, including the *Manual of Photogrammetry* published by the American Association of Photogrammetry, of which the fifth edition was released in 2004 and which serves as the comprehensive reference for photogrammetrist and others working with extracting information from aerial images. The principles of extracting information from images taken from the ground – Close Range Photogrammetry – are described in Luhmann et al. (2006) and Atkinson (2003). A volume specially dedicated to the calibration and orientation of cameras in computer vision, the sister discipline of photogrammetry, has been written by various authors brought together by Gruen and Huang (2001). Many textbooks providing a general introduction to the basics of (aerial) photogrammetry have been published, including Kraus (2007), Mikhail et al. (2001), Wolf and Dewitt (2000), and Konecny (2003). The principles of digital photogrammetry are treated in Schenk (1999) while Linder (2006) provides an introduction to digital photogrammetry from a practical point of view. The basic essentials of aerial photography, emphasising on the survey aspects are presented in Read and Graham (2000). Some of the above books acted as a source for writing the present chapter together with our own lecture notes and publications.

The geometry of a camera is modelled as a central projection. That means the rays of sunlight reflected by each point in the scene pass the camera exactly through one point, the projection centre. Next the ray of light meets the image plane and leaves its mark there. Whether the mark is dark or bright depends on the intensity of the ray. The basic notion in which all the mathematics in photogrammetry is founded is that object point, projection centre and image point constitute one straight line; they are collinear. Mathematically this notion is expressed by the collinearity equation, which is the central mathematical expression in photogrammetry and can be found in any photogrammetric textbook. To compute object-space coordinates from image coordinates, first of all the planar position of the origin of the image coordinate system has to be known. In the camera system also the focal length, by photogrammetrists called camera constant and abbreviated to 'c', has to be known. Principal point and camera constant are the elements of the interior orientation. Ideally the principal point would coincide with the origin of the image plane, spanned up by the perpendicular lines connecting two opposite fiducial marks but due to construction limitations a slight shift will be present in the order of a few microns (Fig. 7.4). The two coordinates of the principal point and the camera constant are usually determined together with the parameters describing lens distortions

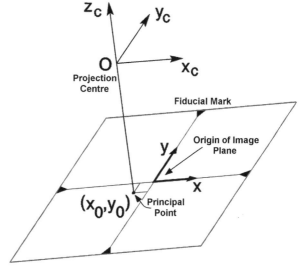

Fig. 7.4 Elements of the interior orientation: origin of the image plane is determined by the fiducial marks. The principal point is defined by the intersection of the perpendicular from the projection centre with the image plane and slightly deviates from the origin of the image plane. The distance from the Projection Centre O to the principle point (x_0, y_0) is the camera constant

by calibration in a laboratory. However, they can also be determined from known geometry present in the scene applying a self-calibration procedure.

In addition to the three elements of the interior orientation, also the orientation of the camera in space has to be known. The three position (X, Y and Z) coordinates of the projection centre and the three angular parameters (ω, φ, κ; omega, phi, kappa) together are called exterior orientation. The XYZ coordinates describe the position of the projection centre of the camera in the object-space related reference system at the instantaneous moment of taking the image. The angular parameters describe the direction of the optical axis and the image plane of the camera with respect to the three axis of the coordinate system of the object-space (Fig. 7.5). The three position parameters and three angular parameters sum up to six parameters and these six parameters of the exterior orientation can be determined along direct or indirect methods (Jacobsen, 2001). The indirect determination is the oldest method and consists of using a set of well-distributed Ground Control Points (GCP). The direct method uses on board GNSS receivers for determination of the three coordinates

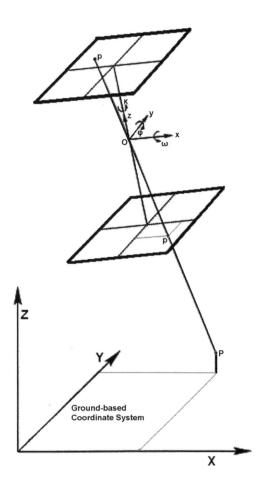

Fig. 7.5 Linking the coordinate system of the camera with the ground-based coordinate system, through the six parameters of exterior orientation: three coordinates of the projection centre (O) and ω, φ and κ

of the projection centre. In addition an inertial navigation system (INS) is used to determine the three angular parameters (see Section 4.16).

Over time the systems have evolved such that GNSS receiver and INS system are integrated, both physically and along software lines. Onboard GNSS and INS enable a substantial reduction in the number of GCPs without loss of accuracy. Since the collection of GCPs is a costly and time-consuming activity, direct methods did emerge rapidly once GNSS became reliable and accurate in the mid nineties. It has become standard procedure to determine the exterior orientation parameters of the hundreds of images recorded during an aerial survey by aerotriangulation and bundle block adjustment (see Section 7.4).

When the three parameters of the interior orientation and the six parameters of the exterior orientation are known, the image coordinates of an object point of which the *XYZ* coordinates are known can be computed by using the collinearity equation. However, the reverse is not true; from measuring the planar coordinates of a point in one image it is not possible to determine the three coordinates of a point in object space. At least three observations are necessary to compute three unknown parameters. Therefore, overlapping images taken from different viewpoints are necessary (Fig. 7.6). From the two coordinates of a point in the one image and the two coordinates of the corresponding point in the other image, the three coordinates of the object point can be computed using the collinearity equation. Overlap between the images is created during air survey by recording the images in a series of strips, together constituting a block of which the shape is usually more or less rectangular. Usually during one flight hundreds of images are recorded. In theory 50% overlap

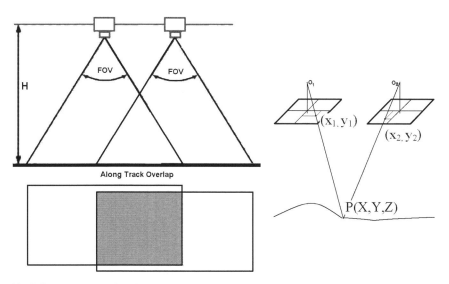

Fig. 7.6 From two overlapping images (stereopair) the three-dimensional coordinates of any point in object space can be accurately computed, provided that the interior orientation parameters and the six parameters of exterior orientation are known

between the images in a strip would be exactly sufficient to measure all points in the scene in stereo.

However, between theory and practice there is air causing turbulence so that aircrafts will drift away from nominal course, while also rolling, pitching and dipping will contribute to non-covered patches (gaps) or areas with too much overlap. Too avoid gaps and to guarantee sufficient overlap, the overlap between adjacent images in the strips has been standardised and set to 60% (forward overlap or along-track overlap), while the overlap between adjacent images between strips is set to 20% (side, or across track, overlap). Sufficient forward and side overlap is also a prerequisite for bundle block adjustment. To optimise the measuring process use is made of the stereo viewing capability of human being.

7.3 Stereoscopic Viewing

When two images of the same scene but taken from different viewpoints are presented to the human eyes – one image to the left eye, the other image to right eye – the human visual system is able to combine the two images into one representation of the scene in which the third dimension is preserved. The complete separate presentation of the two images allows the brain to create a realistic three-dimensional model of the scene. This stereo ability of humans is extensively used in the information extraction process from images. The advantages of measuring in stereo-models are plenty and the two most important ones are as follows:

– Height is an important visual clue. Use of the third dimension enables to better interpret the scene and to see more detail than in a single image.
– The extraction of 3D coordinates (XYZ) requires digitising a point in the left image as well as the corresponding point in the right image. Using stereo mode the measurement is done in just one track. The two digitising marks – one for the left image, the other for the right image – are fused in the human brain to one mark which can be landed on the ground.

It should be noted that 3D coordinates can also be achieved by using single images alone. This is, however, only possible when the images as recorded (raw images) have been converted into orthoimages using a digital elevation model (DEM). The combined use of orthoimages and a DEM enables to extract 3D object-space coordinates from single images; the planar coordinates from the orthoimage and the height coordinate from the DEM. This measuring procedure is called *monoplotting*.

The equipment available on the market to view stereo images can be separated into four categories:

– Optical solution in which the operator looks through oculars; the left image is positioned underneath the left lens system and the right image underneath the right lens system. This is the method used by analogue and analytical instruments. During the infant years of digital photogrammetry oculars were also mounted in front of the computer screen; the left image was projected on the left part of the screen and the right image on the right part resulting in split screen stereo. In the meantime this method of stereo viewing on computer screens has vanished because it is physically stressful, inconvenient and is considered unhealthy because the operator's eyes have to be positioned close to the computer screen.

– Anaglyphs: one of the images is projected in red tones and the other in blue tones. Covering one of the eyes with a red transparency and the other with a blue allows stereo viewing. This is a rather inexpensive approach but only black-and-white images can be processed so that one of the most important visual clues for image interpretation – colour – gets lost.

– A method making use of passive glasses, as does the anaglyph solution, but which allows the view of colour images is based on different polarisation of the left image and the right image (Fig. 7.7, right). The glasses used for observing the images in stereo are polarised accordingly. As with the anaglyph method the glasses are comfortable but a substantial loss of light is incurred due to polarisation.

– Image shuttering using liquid crystal glasses (Fig. 7.7, left). The left and the right image are projected alternatively on screen, with frequency of 50 or 60 Hz, depending on country (use is made of alternate current voltage of the regular power supplier). When the left image is on screen, the right glass of the liquid crystal glasses becomes opaque and vice versa. Although no loss of light occurs, operators may experience a flickering effect which is constraining to the eyes.

Since operators work with DPW equipment the whole day long, the performance of stereo viewing facilities in the personal experience of operators forms an important checking criterion when purchasing such equipment.

Fig. 7.7 Two methods of stereoscopic viewing, active shutter glasses (*left*) and passive polarised glasses, which just look like sun glasses

7.4 Aerotriangulation

Without the availability of accurate exterior orientation parameters of each image (six in total) no coordinates in a ground-based reference system can be calculated from coordinates of image points. Basically, the determination of the exterior orientation parameters of an image along the indirect method requires at least two GCPs for each image. The three coordinates of each GCP ($2 \times 3 = 6$) are exactly the number necessary to compute the six exterior orientation parameters. However, to improve cost effectiveness photogrammetrists have always been keen to discover methods to determining the exterior orientation parameter of each and every image in a photogrammetric block using as few GCPs as possible (Ackermann, 1981). Not only is the measuring of GCPs expensive but also their monitoring between the moment of measurement and the day of air survey when they are measured at forehand. By using the forward and side overlaps of the images in a photogrammetric block the number of GCPs can be artificially densified. In the overlaps tie points are defined and these bridge the space between GCPs. The aim of aerotriangulation is to compute the coordinates of the tie points and from these the parameters of exterior orientation of each image. The procedure is developed such that more coordinates of tie points are measured than strictly necessary for determining the unknowns; a redundant data set is thus created. This improves reliability and accuracy. The apparatus generally applied to process redundant data sets is based on least squares adjustment (LSA). In the aerotriangulation the use of LSA is called bundle block adjustment. The selection and determination of the image coordinates of tie points can be done fully automatically without human intervention using image-matching techniques (see Section 7.5). An optimum use of GCPs is obtained through locating them at the perimeter of the block (Fig. 7.8).

The number of GCPs can be further reduced, even substantially by using on board GNSS and INS (Ackermann, 1992). Practical considerations for topographic mapping how onboard GNSS can reduce the number of GCPs are described in Milanlak and Majdabadi (2005). Commercial photogrammetric software also supports block adjustment of high-resolution satellite images, such as Ikonos, QuickBird and Orbview imagery, hyperspectral images and radar images.

7.5 Image-Matching Techniques

Automation in photogrammetry is largely made possible by stereo-matching techniques that enable (semi-)automatic aero-triangulation and creation of DEMs (Ackermann, 1983; Pertl, 1985; Hannah, 1989). The aim of matching is to identify corresponding phenomena in two or more sets. In photogrammetry the sets are the left and right image of a stereo pair, and the problem in corresponding them is to trace and locate in the right image the conjugate of a point in the left image. Ever since images could be stored in a computer as pixels stereo-matching algorithms have been developed; there are many and new ones emerge on a regular basis. The

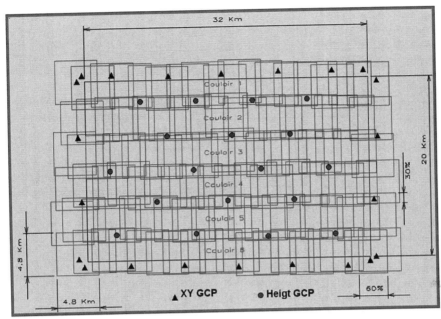

Fig. 7.8 Planar or 3D Ground Control Points (GCP) are located at the border of a photogrammetric within the block Height GCPs act as pillars to prevent 'bending' of the photographs, which are tied together by corresponding points in overlaps

methods may be categorised into two broad classes depending on how the image is approached (Lemmens, 1988; Heipke, 1996). From the signal-processing perspective an image is regarded as a set of grey values or colour values representing the intensity of reflected electromagnetic signals. But an image may also be seen as a representation of features present in object space where each feature, such as the corner of a building, a road crossing or tree, is represented by an irregularly shaped group of pixels. The evaluation of the similarities of the features extracted in the one image with respect to the features in the other image can be done using sophisticated techniques developed in the field of Artificial Intelligence based on relation matching (Vosselman, 1992). An algorithm for extracting features in digital stereo images, which has been widely implemented in DPWs, has been developed by Foerstner (1986).

In the signal-based approaches, also called area or intensity-based, correspondence is sought using intensity values in a regularly shaped patch. A target patch the size, for example, of 9 × 9 pixels, is defined and shifted over a search patch in the other image the size of which depends on how well the approximate location of the conjugate point is known. This process is schematically outlined in the bottom part of Fig. 7.9. For each position a similarity measure is computed, for example, the normalised cross-correlation, resulting in a connected set of similarity measurements, one for each pixel. The highest similarity value determines the corresponding

Fig. 7.9 Image pyramids used for stereo image matching

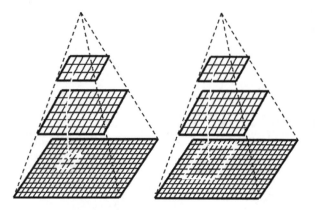

patches, their centre pixels selected as corresponding points. Acceptance of the match depends on whether the similarity value exceeds a predefined threshold; the value can be selected in a heuristic way or, when using normalised cross-correlation, on a statistically sound basis using Student's t-testing. Sub-pixel accuracy can be achieved by fitting a function, for example, a second-order polynomial, through the correlation values and then determining the maximum of that function.

Correlation techniques allow at best a linear difference (gain and shift) between the intensity values of the left and right image, but a shift only in geometry. Since geometric differences do exist as a result of differences in exterior orientation and presence of relief, these are tackled through an iterative least-squares approach (Grün, 1985). These usually model the geometric differences as affine transformations. However, gain comes at a cost: the approximate location of the conjugate point has to be known accurately in advance, even down to the level of a few pixels. This problem can be coped with along two lines. The first by establishing an approximate match with feature-based matching, whereby first points, line or areas are detected in both images using differential operators such Marr-Hildreth, Sobel, Moravec or Förstner. Next, attributes are assigned to the features, such as average and variability of grey values. Knowing the search range, corresponding features in the left and right images are found by comparing attributes.

A consistency check is then performed, based on the assumption of smooth object surfaces, to remove faulty assignments. The location of features serves as an approximation for least-squares matching. Another way to tackle the approximate value problem is by adopting a multi-resolution approach in which an image pyramid is created with at its base the original, full-size images and at subsequent higher levels images generated from uniting 2×2 pixels (Fig. 7.9). This may be repeated until an image of just one pixel remains. By selecting a hierarchical level that best reflects the approximate position of the conjugate point least-squares matching is carried out at that level. The resulting correspondence is now used to track the matching down through the image pyramid until the original image has been reached.

High computation load requires reduction of search space. This is achieved by using information on relative position and orientation of the stereo pair and

characteristics of the terrain topography. The first enables use of epipolar geometry, which reduces matching to a one-dimensional problem; the latter might avoid
starting at too high a level in the image pyramid. The characteristics of automatic
matching require the resulting DEM be edited; if editing is not performed correctly
flat or sloped areas may appear in the orthophoto as breaks and cuts. Editing is
usually done by removing errors after visual inspection of the DEM.

7.6 Scale and Orthoimages

The average scale of an image depends on the focal length of the camera (camera
constant) and the distance from the camera to object space, in aerial images this will
be the terrain. When the camera constant is 15 cm and the flight height is 987 m,
then the overall scale of the image is 1:987/0.15 = 1:6580 (Fig. 7.10). In just one
very rare case the scale is everywhere the same in the image and this is when the
terrain is flat and the image is taken exactly parallel to the terrain. This will hardly
ever occur in practice and to give an image the same scale characteristics of a map
that means the scale is everywhere the same, corrections have to be applied: the
image has to be transformed from a central projection to an orthogonal projection.

  When the terrain is flat, as may be often assumed in river delta areas, correction
for tilt suffices. The correction process, called rectification, requires four GCPs for
each image. When the terrain is hilly, relief distortions have to be eliminated for
which elevation data are required, usually presented in the form of a DEM in which
the terrain heights are stored on a raster. The image is said to be orthorectified; the
image is cleared from scale distortions due to terrain height and camera tilt; the
scale is the same everywhere in the image. Or, in other words, an orthoimage is

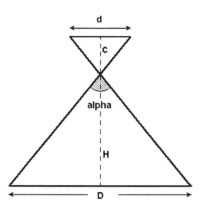

Fig. 7.10 If the terrain is flat and the image is taken parallel to the terrain, the scale is exactly
determined by the ratio between flying height (H) and camera constant (c). The angle alpha determines the field of view. The size of the image plane for aerial photogrammetric frame cameras has
been standardised: 23×23 cm and hence smaller the c bigger the alpha. If $c = 300$ mm the camera
has a normal angle, if $c = 152$ mm a wide angle and if $c = 88$ mm a super-wide angle

an image transformed from perspective projection to orthogonal projection by the performance of corrections for tilt and relief displacement. Orthoimages, often also called orthophotos, can be used in a GIS as a backdrop on which other layers, such as a topographic map, are superimposed. This enables an operator to update the map by digitising new features and to remove the ones which have disappeared. Other applications of orthoimages include urban planning, cadastral mapping and disaster management. The combined use of orthoimages and DEMs enables extracting 3D coordinates of points (monoplotting).

Although an orthoimage will be geometrically correct as a result of transforming the central projection to an orthogonal projection, the image still may contain failures with respect to content. Especially for built-up areas the results will be poor. Due to the central projection roofs of buildings are incorrectly positioned, facades of buildings and the edges of other high objects are visible in the image; the effect increases from the centre of the image to the borders; the buildings seem to lean over. At one side of the building too much is visible (facades) while at the other side objects are occluded by the lean-over of the building (Fig. 7.11). When using aerial photos of 60% along track and 20% across track overlap, as captured by convention in traditional photogrammetry, in combination with coarse DEMs the visual appearance of orthoimages in urban areas will be far from realistic; a feature not much appreciated by GIS operators, who apply orthoimages as a backdrop, because these effects cause that their vector maps do not exactly correspond with the outlines of objects, severely impeding data extraction. Also operators creating 3D city models are in need of more realistic images. For these applications the technique of creating true orthoimages has been developed. In true orthophotos distortions have been largely removed by using greatly overlapping imagery and a dense DEM. The resulting images show real near-vertical views for every position (Fig. 7.12).

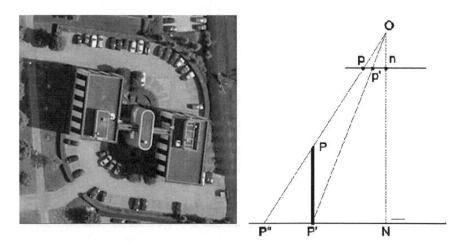

Fig. 7.11 Aerial photograph of an office area in Rotterdam, the Netherlands (*left*). Lean over of buildings and associated occlusion

Fig. 7.12 Traditional orthoimages make buildings appear to lean (*left*) while in true orthoimages the view is near-vertical

7.7 Digital Photogrammetric Workstations

Since the early 1990s, developers in the field of photogrammetry have focused on the design of computer-based systems – Digital Photogrammetric Workstations (DPW) – which can process digital images and also to automate parts of the information extraction process. A digital photogrammetric system is defined by the International Society of Photogrammetry and Remote Sensing (ISPRS) as:

> Hardware and software designed to derive photogrammetric products from digital imagery using manual and automated techniques.

The trend is to take the hardware component from off the consumer shelf, whilst specialised software has become the distinguishing component. The hardware and software components of current DPWs are thus often de-coupled. As a result, DPWs have become software products that run on off-the-shelf consumer PCs to which may be attached special mousses, stereo-viewing hardware, and sometimes hand and foot wheels or other devices to support specialised photogrammetric tasks (see Lemmens, 2005, 2007b, 2009a, 2011a).

In the mid-1970s computer vision systems became available which operated successfully in industrial settings and other highly restricted scenes, where illumination conditions, types of object and camera positions are rigidly constrained. These achievements encouraged photogrammetrists to go in search for computer algorithms which could automatically extract roads, houses and other objects from photogrammetric images. The dream was loading digital aerial images into a computer and within minutes retrieving – without human intervention – a fully detailed map. Soon, however, the multifaceted nature of image content became evident, and visual interpretation, so easy for human beings, proved too complex for transfer into computer algorithms. Scientists had to admit that the visual cognitive abilities of human beings are so highly developed that one had overlooked to envision that such a process might be difficult at all to implement on a computer. In the early 1990s the Artificial Intelligence community, which was very active at that time, recognised that the standard computer vision paradigms fail to provide a means for reliably recognising any of the object classes common to the natural outdoor world. The dream evaporated. In response to user demands for higher automation and more flexibility some manufacturers have implemented automatic line-extraction modules for mapping features like road edges and shorelines. As a result, to date, commercial DPWs support at best semi-automatic extraction of points, lines (mostly roads) and objects (mostly buildings). A human operator approximately locates and identifies a feature, and then comes edge detection and matching algorithms or line-following techniques before the speed of the computer is utilised to extract the feature in its accurate position, in real time.

DPWs are cost-friendly compared to the opto-mechanical mastodons of the past, designed to avoid manual computation (Lemmens, 2009a). The most costly component of a DPW today is its software. Prior to 1980 the main photogrammetric products were orthophotos and line maps. The introduction of GIS systems had by the 1980s created a wolverine, and DPWs enabled alleviation of the hunger for geodata; purchasing a DPW requires only a fraction of the investment once needed to buy its analogue or analytical counterpart. The same budget can be used to spectacularly expand the instrumental capacity of an organisation. Of course, low price would be of little benefit in enhancing production capacity if one still had to rely on highly skilled people with good stereoscopic vision. But DPWs enable the job be done by laymen. Precise pointing has been consigned to history by matching algorithms that keep 'the floating mark on the ground'. The need for stereo vision is rendered similarly redundant. Hence a few days training is enough to teach the uninitiated the nitty-gritty of photogrammetric mapping, resulting in a potentially dramatic increase in human resources.

A second reason for images having become the main source of geo-data for GIS input is that low-price DPWs have made possible mapping outside specialised environments, where once analogue and analytical instruments had to be staffed around the clock for their cost-effective exploitation (Lemmens, 2009b). Cost, together with ease of use, has removed equipment as an impediment to widespread use of photogrammetry. Users outside photogrammetric organisations are increasingly carrying out mapping on an occasional basis. However, one can put a great deal of knowledge into software, but a basic level of understanding of theoretical concepts is required to fix any job well. If the result of a mapping task deviates from expectations, ignorance of the basics could inspire the layman to point the finger of blame at technology.

It has often been advocated that one of the main advantages of Digital Photogrammetry (DP) over Analytical Photogrammetry (AP) is flexibility; AP requires specialised instruments for every product, whilst DP needs just one instrument, the computer. With DP the actual product is generated along software-supported workflows. However, this does not mean that all manufacturers offer general-purpose systems. The capabilities differ a lot, and end-user scope and intended types of application are the decisive elements in their choice. The distinguishing element even is that the various systems are dedicated to different applications and specific end-users. Some focus on a heavily map-producing environment, with high productivity and performance, where the work is done by highly trained end-users; others have close-range applications or the occasional GIS user in mind. Most DPWs are able to process, in addition to the traditional central projective images of aerial cameras, spaceborne imagery, close-range imagery and radar imagery.

For many applications it is important that the system be able to adapt to the frequent release of various new, high-resolution airborne and space-borne sensors with their own data formats. Advanced mapping systems are able to handle many different sensor models and to carry out, aerotriangulation, DEM extraction and orthoimage generation in a (semi)-automatic way. Which DPW is the best depends on end-user scope and intended application. A topographic mapping organisation which uses aerial and satellite images for creating from scratch large-scale maps requires a steam engine whilst a municipal public-works department might make do with a simple system.

7.8 Digital Aerial Cameras

The first digital aerial cameras were presented to the photogrammetric community at the 2000 ISPRS congress in Amsterdam. Z/I Imaging (today Intergraph) and LH (today Leica Geosystems) were the two companies responsible for this innovation. Back in the 1990s, when experimentation began on developing digital aerial cameras, the basic design problems involved getting enough pixels into the focal plane to capture an adequate level of detail qua ground coverage, and how to acquire colour images. The basic solutions were either to place linear CCD arrays, in the focal plane or to use several, area CCD chips (Lemmens, 2008c). The camera system shown in Fig. 7.13 is based on area array technology. The linear-array architecture, also called the pushbroom scanner, principally employs a single lens head, while colour (or multi-spectral-band capture) is obtained by placing three (or more)

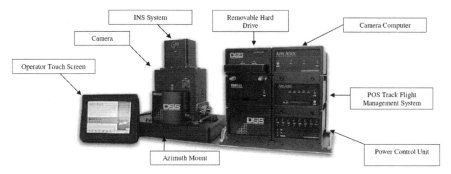

Fig. 7.13 Applanix Digital Sensor System (DSS) consists of integrated medium-sized digital camera (system sensor heads weight about 7 kg), direct georeferencing system (POS AV) and flight-management software. POS AV provides the six exterior orientation parameters in both real-time and post-mission mode. Applications vary from updating and maintaining cadastral databases to rapid response for disaster management

linear arrays in the focal plane, upon each of which are projected different parts
of the visible and possibly near-infrared electronic-magnetic spectrum; enabled by
beam-splitters. The area-CCD-array solution is a camera consisting of several cones
(multi-head).

The general benefits of digital aerial over film cameras include better radiometric
performance and elimination of film processing and scanning costs. Availability of
image content in digital format enables a highly automated workflow, creating the
possibility of generating photogrammatric products such as orthophotos/mosaics
and DEMs with little delay between capture and end-product. This might allow,
for example, for rapid response in the aftermath of a natural disaster. Twelve-bit-
per-pixel radiometric resolution, or even higher, ensures good light sensitivity. The
use of image enhancement techniques means details can be made visible in parts
of the imagery made bright by reflections or overcast and dark due to shadow or
cloud. This relative insensitivity to unfavourable light conditions enables extension
of the daily time-span during which images can be taken, and of flight season. It also
allows data collection on overcast days, thus optimising aerial survey productivity
and area coverage. For detailed specification of the diverse digital aerial camera
systems, available on the market, refer to Lemmens (2008b, c, 2011b).

For some applications, such as topographic mapping, the prompt availability of
end products is not a decisive factor and for such projects still film cameras are in
use, such as was done for mapping the Maldives, a group of islands scattered across
an area of 100,000 km^2 (see Section 7.10).

7.9 Oblique Aerial Imagery

Pictometry is an aerial image acquisition and data processing technology developed and patented by US-based Pictometry International Corp, headquartered in
Rochester, New York. The essential difference with conventional airborne photogrammetry is that in addition to vertical also oblique images are taken, which
is enabled by a sensor system consisting of five cameras, one directed nadir (image
plane approximately parallel to terrain), the others viewing forward, backward, left
and right (Fig. 7.14). In aerial photogrammetry a vertical image is an image which
is approximately parallel to the terrain. This is the prevalent approach when creating topographic or cadastral maps. The viewing angle for all sideward looking
cameras is approximately 40 degrees off-nadir. The (mutual) geometry of the five
cameras is accurately calibrated. The dynamic range of the grey values is 12 bits
enabling to carry out surveys under unfavourable light conditions. The present standard approach stems from US homeland security purposes and includes a flying
height for neighbouring images of 3,000 ft (1,000 m) and for community images
6,000 ft (2,000 m) and pixel size 6 inches (15 cm) and 1 foot (30 cm), respectively.
In oblique images the pixel size varies from 10 cm at the bottom to 18 cm at the
top of the image. The standard products acquired by Blom Group of Europe are
usually neighbourhood images. Up to 16 km^2 can be acquired per hour and every

Fig. 7.14 Simultaneous recording of vertical and oblique images with the Pictometry digital camera system

1.5 seconds photos are taken. Each image consists of 6 MB of data while each km^2 is covered by around 50 views, that is around 310 mb of data. Each point on the ground is visible in up to 18 oblique images, provided absence of occlusion (point is not visible because it is concealed by another object in the line of view). Direct georeferencing is enabled through the onboard, integrated GNSS and Inertial Navigation Systems.

What does Pictometry technology add to traditional aerial photogrammetry? In principle just oblique images and that is nothing new. Oblique images of urban areas are useful for both professionals and non-professionals because facades of buildings become visible, giving a different visualisation experience than that offered by traditional orthophotos. Oblique imagery shows parts of buildings and other constructions, which remain hidden in vertical aerial photos. The number of storeys and height of buildings can easily be determined. Such information combined with DEM information is also useful for land administration purposes, for determining property taxes and providing building permits.

In the past extracting accurate geometric information from aerial images was technologically restricted and could only be done on a production scale with vertical images. Much emphasis was on using vertical images measuring stereoscopically and the whole image acquisition and measuring process was adapted to this. Today the geometry of the sensors can be calibrated accurately, direct georeferencing can be done through GNSS and IMU while complex geometric transformations can be carried out by computer and additional information sources, in particular high-resolution, accurate DEMs, can be computationally incorporated. As a result, extraction of accurate geometric data from oblique images is now possible.

The main advantage of oblique images compared to vertical images is better and more intuitive interpretation. Interpretation of vertical images requires training and craftsmanship but tests have shown that the interpretation and taking measurements in oblique images can be done after a short training (Lemmens et al., 2007). Oblique images make aerial information thus accessible to a large, non-professional user group, such as officers at municipalities to support their actual tasks. Poul Nørgård, head of the Mapping and Geodata Department, COWI, Denmark, who has over 10 years experience with in-house build oblique systems, details the benefits of oblique images as follows (see Lemmens, 2007a):

> Without leaving the office, municipal officials can now register how many floors a building has and determine the shape of roofs and volume of buildings. The photos also give architects and the public a good impression of how a new building fits into the neighbourhood, or the appearance of a town quarter after changing facades. In combination with orthophotos they give fire brigades and sanitation workers information on how to access building blocks through back gardens.

What does a license for using Pictometry products, which is rather expensive, add to using the same images available at Microsoft's local.live.com website for free? Actually, the two scenarios are not comparable. First, images may be available but not yet included on the website. Furthermore, and more importantly, only the oblique images are available, not the orthoimages so that no stereo-images will be available. Users are not enabled to carry out measures on the images and to use them as a navigation tool, for example, to access other data sets. In short: Microsoft's website local.live.com enables just viewing of weakly georeferenced oblique images.

The Netherlands' Cadastre, Land Registry and Mapping Agency (Kadaster) performed a research to determine the potentials of Pictometry technology for cadastral purposes in close co-operation with the GeoTexs company from Delft, The Netherlands and Blom Info, Copenhagen, Denmark, part of the Blom Group (Lemmens et al., 2007). The Pictometry images used were captured in 2006 (Fig. 7.15 shows an example of an oblique image used in the research). The tests reveal that the accuracy of taking location measurements in orthoimages is 19 cm and in orthoimages 86 cm, expressed in terms of root mean square error (RMSE). The accuracy of the elevation component depends on the accuracy of the underlying DEM. Although Pictometry technology has been announced as a visualisation tool, not as a surveying tool, the above measures demonstrate that photogrammetric surveying accuracy can be achieved. Within a cadastral context Pictometry may serve as aid in splitting parcels and carry out parcel formation. Furthermore, it appears to be a suitable tool for (1) preliminary boundary determination via notary, (2) building registration and (3) communication from government to citizen.

Georeferenced oblique images are also useful for highly automatically creating 3D city models. From airborne Lidar, terrestrial laser scanning and planar maps of buildings and streets, the block structures of buildings can be derived. Since the

Fig. 7.15 Headquarters of the Dutch Cadastre, Apeldoorn, the Netherlands, viewed from an oblique perspective captured with Pictometry technology

oblique images are well-georeferenced the outlines of the buildings can be accurately projected into the oblique images, and next the image patches can be glued on the facades enabling automatic augmentation of block 3D building models with textures. This is not only more precise and cost-effective but also faster than the manual process; cost savings are up to 80%.

In summary, oblique aerial imagery is complementary to traditional photogrammetry and Lidar and cannot replace photogrammetry for accurate mapping nor replace Lidar for creating DEMs. Pictometry allows users to easily and efficiently view and observe any point chosen from five angles (Nadir, North, West, South and East). This allows them to enhance the visual appearance of models created from photogrammetry or Lidar. Because oblique imagery shows previously hidden parts of buildings and other structures the technology enables creation of a much richer database. It can be viewed in existing GIS or included software with standard features such as zoom, pan, measure (height, distance, area and altitude) and walkabout. The number of storeys and height of buildings can easily be determined. Such information on building structures and facades, combined with DEM information, is useful for existent cadastral applications, for determining property taxes and providing building permits. Lidar can be a very good complement to oblique images as it gives very accurate and detailed elevation information, allowing highly automatic creation 3D city models.

What market can one expect for Pictometry, both in terms of public/private segments and regions? The answer is given by Arne Saugstad and Nils A. Karbø, Blom Group, who communicated to us (Lemmens and Lemmen, 2007):

> The market is very varied and ranges from municipalities to media and from blue lights to telecom. Customers appreciate that they are buying a standardised, up-to-date product that has the same features across the country and across Europe. The plug-ins to major GIS packages also offer comfort [. . .]. The possibilities of the technology will reach millions of people in Europe and open up a big potential market, which is, we think, spread fairly evenly across Europe . . .

7.10 Aerial Photogrammetry in Practice

How is aerial photogrammetry applied today for mapping extensive areas? In answer we can look at a project aimed at digital mapping of the Republic of Maldives (Lemmens, 2008a; Raghu Venkataraman et al., 2008). The Maldives is geographically tricky territory, scattered and spread as it is across an area of 100,000 km^2 between 1°S and 8°N latitude, 72°E to 74°E longitude. Ninety-nine percent of the country is water and the landmass consists of 26 atolls and 1,192 islands, only 33 of them exceeding 1 km^2 in area. The project was realised through a collaborative effort between the Government of India and the Republic of Maldives, and carried out by the National Remote Sensing Agency of the Department of Space, Govt. of India, Hyderabad. Fronting such a project stands the requirement for a nationwide geodetic reference frame for obtaining consistent coverage with homogeneous accuracy. At the tail end comes product specification, and for the Maldives this was the creation of 1:25,000 digital line maps of the entire country from 1:40,000 aerial photos, and 1:1,000 maps from 1:6,000 aerial photos of sixteen islands. The 1:25,000-scale maps needed to demarcate all land and water boundaries, atolls, built-up areas, roads and vegetation. Figure 7.16 outlines the methodology used.

Thirteen GNSS reference stations and 41 ground control points signalised with 5 × 5 m targets established the reference frame in WGS84 reference system for the air survey of the entire Maldives. Compared to film a digital aerial camera allows, for example, rapid response in the aftermath of a disaster. But prompt availability of end products is not a decisive factor in mapping large areas, and the Zeiss RMK 15/23 camera used provides metric-quality photographs with wide swath, so reducing overall flying time. During February, 2004 – the ideal cloud-free season in the region – a total of 4,456 photos were taken using the metric film camera integrated with onboard GNSS and international navigation systems. The airplane was a Beech Super King Air B-200 and the flight was guided by a Computer Controlled Navigation System. Male International Airport was the base of the operations. An optimal flight plan was devised in terms of flying time and avoiding wastage of aerial

Fig. 7.16 Diagram of the
mapping the Republic of
Maldives using aerial
photogrammetry

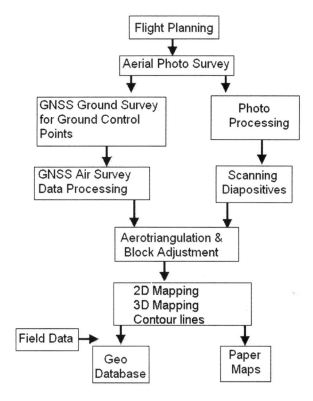

film. Account was also taken of the limited window of cloud-free days, and the flight
plan was prepared using IRS 1C/1D satellite images, existing atlases, old maps and
so on. To prevent heat degradation, rolls of film were kept in an air-conditioned safe.

After aerial survey, 2,377 B&W stereo 1:40,000 photos for preparing 1:25,000
digital maps were scanned using a precision Zeiss SCAI photogrammetric scanner pixel size 21 μm, corresponding to 85 cm Ground Sample Distance (GSD).
Scale 1: 6,000 photos for preparing 1:1,000 digital maps were scanned with 20 μm
pixels (12 cm GSD). Digital photogrammetric techniques were used for automatic
aero-triangulation and block adjustment using Socetset software and Orima block-
adjustment software. Next, 1:25,000 scale line maps were created by first producing
digital orthoimages using a DEM and then capturing vectors from the orthophotos using PC-based, low-cost systems. The 1:1,000 scale maps were created using
Digital Photogrammetric Workstations, resulting in 3D-maps and contours.

The 85-cm GSD of scanned photographs is similar to what very high-resolution
satellites can deliver (Ikonos: 1 m; Quick Bird: 60 cm, Worldview 1 and GeoEye
1: 50 cm). However, their use would require non-selective purchase of 800 scenes
over the entire area, whilst aerial photos enabled selective data acquisition covering
only those areas of significant landmass and provided stereo images for generating
DEMs. More than a thousand small, scattered islands covering just 1% of the area
made aerial photography preferable to satellite imagery.

7.11 UAV Photogrammetry

Which data collection method is best suited for surveillance of pipelines, dams and flooded areas, or for monitoring volcanoes and vulnerable land? How, fast, efficiently and cheaply, to collect in-situ data on a road accident or building on fire? Such questions frequently arise in our ever more urbanised and motorised world. Ongoing miniaturisation of electronics means that the answer increasingly sounds: 'use Unmanned Aerial Vehicles (UAV) equipped with colour, thermal or near-infrared cameras, airborne Lidar systems or other sensors'. Re-usable aircrafts flown without onboard human crew have been in existence since 1916 and the bulk of applications concern reconnaissance and other military missions too shady, dirty or dodgy for human lives to be jeopardised. The recent introduction of lightweight UAVs enables to perform low-cost aerial photogrammetric surveys over small areas and, compared to manned alternatives, UAVs can access calamity areas while flying at low altitude without endangering human life (Eisenbeiss, 2009). UAVs are also suited for cost-effective geo-data acquisition for environmental purposes; in July 2010, researchers from the University of Calgary used a UAV to survey a region in Canadian Arctic (Whitehead, 2010). Within 16 flight lines 148 photos were recorded from 300 m altitude enabling to create a detailed DEM and a 10-cm GSD orthomosaic. The level of detail suffices to create a 1:1,000 map.

Compared to manned photogrammetric surveys the keywords associated with UAVs, suited for photogrammetric purposes, are lightweight and small and so should be their payload. Small cameras have a smaller field of view compared to digital photogrammetric cameras. Consequently, more images are required to cover the same area. Smaller cameras are also less stable, which negatively affects image quality. The GNSS/INS unit, which is necessary for direct georeferencing purposes, should also be small which limits accuracy. The small engine reduces range and altitude. However, the decisive factor limiting range is line-of-sight to the UAV, or pilot ability to blindly adjust position; they do not carry systems for collision avoidance. UAVs are literally a toy in the wind and sensitivity to air movements limits their use to conditions of windless, sunny weather. They do not carry systems for air traffic communication and regulations of air security authorities do not cover UAVs.

UAVs appear in manifestations as diverse as aeroplanes and helicopters of all sizes and shapes, motorised para-gliders, blimps, kites and balloons (Fig. 7.17). What sort of UAV is suitable as a platform for digital cameras and other remote sensors? The answers depend on application. The minimal set of criteria to be considered include cost, size and weight of payload; stability and vibration; number of people needed for launch and control; level of piloting skills; flight time; range; minimum airspeed (the lower, the less blurring of images); minimum size of take-off and landing area; and safety. An additional design criterion concerns whether the UAV be 'dedicated', that is, equipped with specific sensors constraining the range of applications, or 'general-purpose', capable of carrying a diverse set of sensors exploitable for many applications. Kites, blimps and the like are affected by wind and thus ill-controllable. Fixed-wing UAVs are more stable and, depending on wingspan, can carry a payload of up to a few kilograms. They have to maintain a

Fig. 7.17 Top (from *left* to *right*): lightweight, small fixed wings on runway during a demonstration; Scout B1-100 helicopter equipped with Riegl airborne Lidar scanner; Fulmar developed by Aerovision in operation. Bottom: RQ-7 shadow in a military operation (*left*); Inview fixed wing

minimum speed to stay in the air and they cannot hover. An expert has to handle the remote control to avoid crashes, and good flight preparations are vital. They are suited for mapping areas of a certain extent, conveyed in square kilometres. But data processing has to be largely automated, while there should be no need for ground control points, otherwise expensive manual post-processing negates the advantage of cheap data collection. Compared to fixed-wing UAVs, helicopters offer more flexible manoeuvring, vertical take-off and landing and hovering, but they require high-level piloting skills. Helicopters need blades of at least several metres in length in order to carry a payload of up to a few kilograms while remaining stable in the air. Larger, single-blade helicopters powered by petrol engines offer one option, but the rotor may cause injury, so again the pilot has to be well trained. Quad-roto helicopters are stable enough, and less dangerous, but battery-powered flight time is limited, while vibration may blur the imagery acquired.

A solution is guidance without human intervention, enabled through small sized and lightweight boards integrating GNSS and inertial navigation and the design of more stabilised systems. With the maturation of miniature autonomous systems the usability of UAVs as a new lost-cost alternative to traditional aerial photogrammetry will increase, particularly when employed for small projects covering areas limited in extent.

References

Ackermann F (1981) Block adjustment with additional parameters. Photogrammetria 36(6): 217–227

Ackermann F (1983) High precision digital image correlation. Proceedings 39th photogrammetric week, Institut für Photogrammetrie, Stuttgart, Heft 9, pp 231–243

Ackermann F (1992) Kinematic GPS control for photogrammetry. Photogramm Rec 14(80): 261–276

American Society of Photogrammetry (2004) Manual of photogrammetry, 5th edn. American Society of Photogrammetry. Bethesda, Maryland, US

Atkinson KB (ed) (2003) Close range photogrammetry and machine vision. Whittless Pub, Dunbeath, Caithness, Scotland, UK

Eisenbeiss H (2009) UAV photogrammetry. PhD thesis, ETH Zurich

Foerstner W (1986) A feature based correspondence algorithm for image matching. Int Arch Photogramm Remote Sens 26(part B3) Rovaniemi.

Gillispie C (1983) The Montgolfier brothers, and the invention of aviation. Princeton University Press, Princeton, NJ, USA

Gruen A, Huang TS (eds) (2001) Calibration and orientation of cameras in computer vision. Springer, Berlin

Grün A (1985) Adaptive least squares correlation: a powerful image matching technique. South Afr J Photogramm Remote Sens Cartogr 14(3):175–187

Hannah MJ (1989) A system for digital stereo image matching. Photogramm Eng Remote Sens 55(12):1765–1770

Heipke C (1996) Overview of image matching techniques. In: Kölbl O (ed) OEEPE workshop on the application of digital photogrammetric workstations. OEEPE Official Publication No. 33, pp 173–189

Jacobsen K (2001) Direct georeferencing. Photogramm Eng Remote Sens 7(12):1321–1332

Konecny G (2003) Geoinformation. Taylor & Francis, London

Kraus K (2007) Photogrammetry, geometry from images and laser scans, 2nd edn. Walter de Gruyter, Berlin

Lemmens MJPM (1988) A survey on stereo matching techniques. Int Arch Photogramm Remote Sens 27(part B8):V11–V23

Lemmens M (2005) Digital photogrammetric workstations. GIM Int 19(5):41–45

Lemmens M (2006) Photogrammetry out of the dark room. GIM Int 20(6):7–9

Lemmens M (2007a) COWI Consultation: subsidiary Kampsax flourishes. GIM Int 21(11):36–39

Lemmens M (2007b) Digital photogrammetric workstations. GIM Int 21(12):22–25

Lemmens M (2008a) Mapping the Maldives. GIM Int 22(4):11

Lemmens M (2008b) Digital aerial cameras. GIM Int 22(4):18–25

Lemmens M (2008c) Digital aerial cameras: system configurations and sensor architectures. Prof Surveyor 28(5):66–72

Lemmens M (2009a) Digital photogrammetric workstations. GIM Int 23(12):32–35

Lemmens M (2009b) Layman's photogrammetry. GIM Int 23(12):57

Lemmens M (2011a) Digital photogrammetric workstations. GIM Int 25(12): in preparation

Lemmens M (2011b) Digital aerial cameras. GIM Int 25(4):35–42

Lemmens M, Lemmen C (2007) Pictometry: long-term impact on GI market. GIM Int 21(4): 7–11

Lemmens M, Lemmen C, Wubbe M (2007) Pictometry: potentials for land administration. 6th FIG regional conference, San José, Costa Rica, 12–15 Nov 2007

Linder W (2006) Digital photogrammetry: a practical course. Springer, Berlin, Heidelberg, New York. ISBN 3-540-29152-0

Luhmann Th, Robson S, Kyle S, Harley I (2006) Close range photogrammetry, principles, methods and applications. Whittles Publishing, Dunbeath, Caithness, 510 p. ISBN: 1-870325-50-8

Mikhail EM, Bethel JS, McGlone JC (2001) Introduction to modern photogrammetry. Wiley, New York, Chichester

Milanlak A, Majdabadi MGh (2005) Optimal GCPs with onboard GPS: BLOCK adjustment for photogrammetric production of map of Iran. GIM Int 19(9):46–47

Miller SB, Seymour RH (1995) Uuno Vilho Helava contributes to photogrammetry in the United States. ISPRS J Photogramm Remote Sens 50(6):19–24

Pertl A (1985) Digital image correlation with an analytical plotter. Photogrammetria 40(1):9–19

Raghu Venkataraman V, Srinivas P, Rao J (2008) Aerial survey of the Maldives: total 1:25,000 Topographic Map Cover. GIM Int 22(8):17–19

Read RE, Graham RW (2000) Manual of aerial survey. Whittles Publishing, Dunbeath, Caithness, Scotland, UK, 408 p. ISBN 1-870325-62-1, 0-8493-1600-6

Schenk T (1999) Digital photogrammetry, vol 1. Terra Science, Laurelville, OH

Thomson MM, Gruner H (1980) Foundations of photogrammetry, chapter 1. In: Manual of photogrammetry. American Society of Photogrammetry, Falls Church, VA, USA

Torlegård K (1988) Transference of methods from analytical to digital photogrammetry. Photogrammetria 42(5–6):197–208

Vosselman G (1992) Relational matching. Lecture notes in computer science, vol 628. Springer, Berlin

Whitehead K (2010) Unmanned aerial vehicles for glaciological studies: airborne survey of Fountain Glacier's Terminus region, GIM Int 24(10):26–29

Wolf PR, Dewitt BA (2000) Elements of photogrammetry with applications in GIS, 3rd edn. McGraw-Hill, New York, NY. ISBN 0-07-292454-3

Chapter 8
Airborne Lidar

The year 1997 marked the start of creating a highly detailed nationwide Digital Elevation Model (DEM) of the Netherlands, up to one height point per 16 m^2, making use of airborne Lidar technology. The accuracy specifications of the so-called AHN were: 15 cm root mean square error (RMSE) and 5 cm systematic error. At that time, commercial Lidar was still in its infancy and many operational hurdles had to be overcome. The AHN project was completed in 2003. Figure 8.1 shows a part of a raw airborne Lidar data set used to create AHN. 'The Netherlands is flat as a coin. Why you need such a detailed DEM?' some foreigners laughed, especially those living in mountainous areas. The answer is quite simple: when 40% of a country's territory is situated below sea level, every decimetre counts in the struggle to keep feet dry. In the meantime, error sources have been understood and remedies developed to avoid them or to get rid of them.

Today, many countries have created accurate, detailed, nationwide DEMs generated from airborne Lidar surveys or are in the process of creating DEMS covering the entire territory of the country. For example, airborne Lidar is being used to establish a new national elevation model of the entire territory of Sweden (450,000 km^2), one point per two m^2, in a project spanning 2009–2013. The DEM is particularly created in the framework of climate change, which entails rising sea levels, heavier rainfall and higher temperatures; occurrences which make estimating the risks and effects of flooding, landslides and other eventualities vitally important (Petersen and Burman Rost, 2011). The project is carried out by two internationally operating Scandinavian geospatial product service providers: Blom, headquartered in Oslo, Norway, and Denmark-based COWI. To stay in pace with planning up to five Lidar Surveys are carried out simultaneously. Figure 8.2 gives in on board impression of such a survey.

The first commercial airborne Lidar systems appeared on the market in the mid-1990s. Figure 8.3 shows three airborne Lidar systems available on the market in 2011. This active remote-sensing technology rapidly evolved to become the laser scanners which can today emit 400,000 pulses per second and record the full waveform of the backscattered signal. Soon they will be even able to record half a billion full waveforms per second (Lemmens, 2011a). Airborne Lidar has matured, as witnessed by the development shown by product overviews given in Lemmens (2007a, 2009a, 2011b) and earlier.

M. Lemmens, *Geo-information*, Geotechnologies and the Environment 5,
DOI 10.1007/978-94-007-1667-4_8, © Springer Science+Business Media B.V. 2011

Fig. 8.1 Raw airborne Lidar data used to create AHN, note the high level of detail. *Black square* marks cross roads of two highways (Ouderijn); *right* to *left* Enschede-Amsterdam, *bottom* to *top* Maastricht-Utrecht

Fig. 8.2 On-board view on an airborne Lidar Survey carried out by geospatial service providers Blom and COWI

This chapter details the basics of Lidar technology and sketches applications. An overview of basic relations and formulas concerning lasers, laser ranging and airborne Lidar can be found in Baltsavias (1999a). Lemmens (1997, 1999a, 1999b) provides an overview of the factors influencing accuracy of DEMs created by airborne Lidar. A playa how a country or region may benefit from using digital surface models collected by means of airborne laser scanning is given in Lemmens and Lohani (2001).

Fig. 8.3 Three commercial Airborne Lidar systems. From *left* to *right*: Optech ALTM Gemini (23.4 kg, 26 × 19 × 57 cm); Riegl VQ580 (12 kg, 36 × Ø 22.2 cm); Leica Geosystems ALS 70

8.1 Overview

Lidar is a most successful data-acquisition technique. As an acronym of Light Detection and Ranging – some prefer to read Lidar as Laser Imaging Detection and Ranging – the term has become a 'proper name', spelled like your own first and surname with the initial letter the only capital. Although the rise of Lidar as an operational system began just at the end of the 1990s of the previous century, its history dates back to the 1960s when it was first tried out (Ackermann, 1999). In the 1970s experimental systems were developed. Accurate positioning remained a bottleneck until, in the early nineties, the Global Positioning System (GPS) became a reliable, stable and precise positioning technology. In the meantime Lidar has been recognised as an advanced technology with many beneficial characteristics: high rates of data capture (up to 100 km^2/h) and levels of automation, right up to detailed 3D-reconstruction of the real world. Lidar features high accuracy and precision and a high level of detail (up to millions of points per square kilometre). As a result airborne Lidar has gained widespread recognition as a major source for the reconstruction of real-world surfaces. Its many applications include the creation of accurate Digital Elevation Models (DEM) and 3D-city models for planning purposes. But also outside the realm of DEM creation, airborne Lidar data have proven to be a valuable data source. It offers benefits for planning, design, inspection and maintenance of infrastructure. Dunes, beaches, dykes and floodplains can be captured accurately and with a high level of detail, enabling coastal-erosion modelling, monitoring and flood-risk management. The goal of surface reconstruction is to approximate geometry, topology and features of a surface using a finite set of sampled points. The result of a Lidar survey is a dense cloud of irregular, distributed 3D points characterised by coordinate triplets associated with attributes, in particular intensity of return signal.

Compared to photogrammetric creation of DEMs using digital stereo imagery and automated image matching techniques, airborne Lidar has the following advantages (Baltsavias, 1999b; Wehr and Lohr, 1999; Lemmens, 2001; Romano, 2004; Meng et al., 2010):

- Lidar facilitates the recording of the first and last parts of backscattered pulses enabling mapping of the bare ground surface even in forests and other regions with dense vegetation.
- In contrast to stereo images no texture is needed to create a DEM; so generation of high accuracy DEMs without missing data (gaps) of dunes and other areas without vegetation is rather easy.
- Weather and visibility conditions only slightly affect flight surveys, making the technique fairly independent of season and daytime.

8.2 Basics

Airborne Lidar devices are multi-sensor systems, consisting of a reflectorless laser sensor, a positioning system based on GNSS and international navigation and usually further equipped with a digital camera or other additional recording device. As the aircraft flies over the area the laser sensor emits pulses towards the ground (up to 250,000 pulses per second, depending on device). The pulses travel with the speed of light (300,000 km per second). After having hit the ground, the pulses are reflected and part of the reflected energy will reach the receiver part of the laser sensor, depending on the reflectivity characteristics of the surface. Basically airborne Lidar works similar to Terrestrial Laser Scanning, discussed in Chapter 6. However, airborne Lidar is more demanding as the six parameters of the exterior orientation have to be determined in a dynamic environment that is while the system is moving through the air. A rotating or nutating mirror, possibly supported by fibre optics, enables scanning perpendicular to flying direction. To compensate for mechanical instabilities and guarantee constant alignment, fibreglass optics may be mounted in front of the mirror. At the same time, scan angle is measured with a scan angle encoder. A positioning system for determining the six exterior orientation parameters, as in aerial photogrammetry, is required to transform range measurements and scan angles into 3D terrain coordinates either expressed in XYZ coordinates or the triplet (φ, λ, h), that is Latitude, Longitude and Elevation above reference ellipsoid or, preferably, (φ, λ, H), where H is the orthometric height, i.e. height above the geoid (see Section 1.4). The positioning component of the exterior orientation is determined by a GNSS receiver installed on-board and one or more receivers installed on reference stations, which is necessary for obtaining high-precision measurements using Differential DGNSS (see Section 4.3).

The determination of the attitude of the sensor is done with an inertial navigation unit, which uses a combination of accelerometers and gyroscopes to detect rate of change in acceleration and attitude; the latter usually defined as pitch, roll and yaw (see Section 4.16). Figure 8.4 provides a schematic overview of the working of airborne Lidar. Figure 8.5 schematically outlines an airborne Lidar workflow.

Fig. 8.4 Schematic overview of the working of an airborne Lidar system

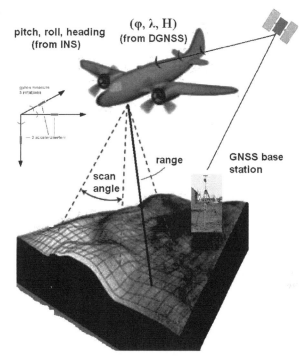

The generation of the 3D coordinates of point clouds from integrating GNSS-, INS- and Lidar data is done in post-processing. However, methods have been developed for the integration of these data sets directly in flight with sub-decimetre accuracy (Skaloud et al., 2010). Real-time determination of the 3D coordinates of points with better than decimetre accuracy is appropriate for a wide range of applications requiring short response time such as post-disaster damage inventory.

8.3 Pulse Characteristics

Depending on system, flying height, speed and number of flyovers, point densities up to some dozens per m² can be acquired. Helicopters are better suited for high-resolution coverage because they can easily limit their speed. Since Lidar is an active system, data acquisition is independent of sun illumination, while no shadows are generated. Weather and visibility only slightly affect the survey. Height values may be effortlessly obtained in areas of low textural variation, such as beaches and dunes. Wavelengths in the near infrared part of the spectrum (typically 900, 1,060 and 1,500 nm) are non-penetrative, so that pulses will be reflected from forest-stand foliage and other vegetation. However, the footprint has a certain extent, so when a laser pulse hits a tree crown part of the pulse will reflect on the leaves, but where there are holes in the canopy the pulse will continue to proceed. As a result a laser

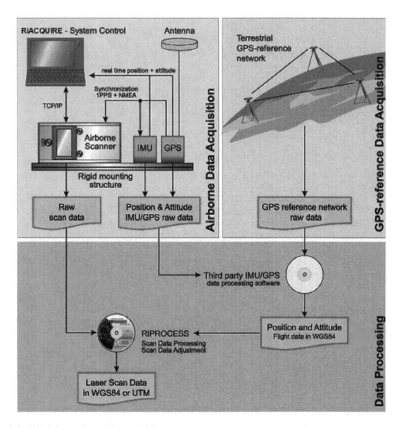

Fig. 8.5 Workflow of an airborne Lidar survey

pulse can partly penetrate in tree canopy to a level dependent on the density of structure and the angle of incident (Kraus and Pfeifer, 1998). The return signal will thus contain information from interaction with canopy structures at several depths. A part of the signal may even reach the bare ground, which is beneficial for the creation of DEMs (Ackermann, 1999). The last part of the return signal may thus represent distance to the ground and the first part canopy height (Fig. 8.6).

The beam divergence of the various commercial systems varies from 0.2 to 11 mrad. Beam divergence is an angular measure which determines the size of the footprint: the cross-sectional diameter at reflecting surface when the pulse hits an object or the ground (Fig. 8.7). Beam divergence is an important parameter because it determines the penetration level of the pulse in forests and other vegetated areas, the level of detectable detail, the sharpness with which outlines of buildings and other objects can be recorded, and the level of eye safety. For creating bare-ground digital elevation models (DEM) or determining tree height it is desirable that the pulse hits not only the trees but also the ground surface. The higher its energy, the greater the chance that the pulse passes through spaces between leaves to reach the ground. Small beam divergence warrants a high energy level present in the footprint

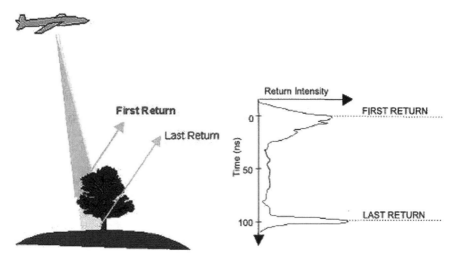

Fig. 8.6 First and last return

Fig. 8.7 Relationship between beam divergence, footprint and flying height

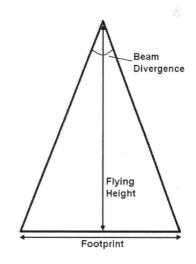

and is thus beneficial for forestry applications; however, there is also a raised probability of causing damage to sensitive tissues, such as the eye, if a person is hit by the pulse. When surveying urban areas for 3D modelling of buildings, bridges, channels and so on, the outlines of objects should be reconstructed as sharply as possible, requiring small footprint in tandem with high point density. In contrast, when small or thin objects such as electric power lines are being surveyed, the footprint should be large to increase the probability of hitting the object.

Most commercial systems produce a small footprint, typically 20 cm to 110 cm at 1 km flying height, but they also allow divergence tuning by the operator, introducing the advantage of adjusting footprint size to specific application, some mentioned above. Discrete switching is possible from one beam width to another, for example, from 0.3 to 0.8 mrad, or tuning may be effected on a continuous scale. Tuning requires placement of moving mechanical parts in the optical system, which may introduce a shift in optical axis and thus degradation of laser pointing accuracy. Planimetric error may be a few decimetres at an altitude of 1 km, and changes linearly with flying height. The relationship between beam divergence expressed in milliradians (mrad) and footprint size is quite simple (Lemmens, 2011a). At a flying height of 1 km, beam divergence of 0.2 mrad equals a footprint size of 0.2 m; footprint size of 0.8 mrad equals 0.8 m (Table 8.1). At 2 km the size of the footprint doubles as it varies linearly with flying height.

Some systems, in addition to the first and/or last part of the return signal, collect four or eight samples, or even the full waveform of the return pulse. Full-waveform digitisation of each backscattered pulse enables determination of vertical surface structure such as roughness, height and shape of objects, canopy density and height of trees, and reflectivity. Capturing the full waveform provides additional and more detailed information on the structure and the physical characteristics of the surfaces of objects (Mallet and Bretar, 2009). For example, scrub and other low vegetation may be differentiated from the ground surface and height measurements of canopy may be improved (Hug et al., 2004). Indeed, the capability of airborne Lidar systems of digitising the return signal in parts is especially appreciated by forestry practitioners as they can estimate vegetation structural variables from such signals (Chasmer et al., 2006). The waveform provides an indication of the top of the canopy, underlying layers and ground surface, allowing more accurate determination of ground surface DEM and canopy height (Fig. 8.8). Full-waveform Lidar is also investigated for the classification of tree species. Using the intensity of the return signal, its width and the number of targets popping up in the signal Heinzel and Koch (2011) could classify six tree species with an overall accuracy of 57% and four tree species with 78% accuracy. Constraining themselves to conifers and broadleaved trees a 91% accuracy could be achieved. In urban areas roofs can be distinguished from trees, cars from garages and so on. Systems using full waveform technology became operational in 1999, primarily for forest applications, and by 2004 they were implemented in commercial systems.

Table 8.1 Relationship between beam divergence and footprint size at flying altitudes 1 and 2 km	Beam divergence [mrad]	Footprint [m] at 1 km	Footprint [m] at 2 km
	0.1	0.1	0.2
	0.3	0.3	0.6
	0.8	0.8	1.6
	1	1	2
	3	3	6
	12	12	24

Fig. 8.8 Full waveform
digitisation enables increased
detail capture of objects in
trajectory of pulse

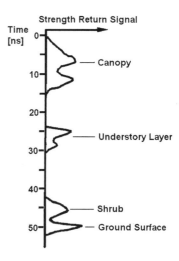

8.4 Multiple Pulses in Air

As stated earlier airborne Lidar has matured to an accurate technology for the highly
automated capturing of terrain data through 3D point clouds. 'Mature' means that
improvements are no longer founded on major technological breakthroughs, but
are incremental. From 2007 onwards the steps comprise full waveform digitisation,
multiple pulses in air, and increased accuracy resulting from enhanced GNSS positioning and inertial navigation attitude determination. Let us go back, for a moment,
to the AHN, the DEM of the Netherlands we started this chapter with. Upgrading
the AHN to one point per square foot (10 points/m^2) with an accuracy of 5 cm, both
RMSE and systematic error, is in progress and will result in AHN2. The centupling
of resolution and tripling of accuracy compared to the first release of the AHN in
a cost effective way has been enabled by the recent advances mentioned above,
most significantly multiple-pulses in air (MPiA) technology, which has become
commercially available by 2007.

MPiA allows the firing of the next laser pulse before the reflection of the previous pulse has been received (Lemmens, 2009b). The pulse rate of single-pulse
in air systems is determined by the ratio of the speed of light and two times
the flying height; e.g. at 1 km, the maximum pulse rate is 300,000 km/s divided
by 2 km; that is 150,000 pulses per second; at 2 km this number diminishes to
75,000. The basic advantage of MPiA becomes clear when surveying at higher
altitudes. It reduces acquisition time (and hence costs) and also reduces occlusion
by relief variations; the latter enables the reduction of the width of across overlap. Furthermore, air turbulence at higher altitudes is less severe than closer to the
ground, so the flight is more comfortable and the crew can stay in the air longer.
MPiA technology is beneficial for capturing data on large areas, such as creating
high accuracy nationwide DEMs, flood risk mapping in delta areas and mapping

of coasts. Are there any disadvantages? Higher altitudes will magnify the effects of errors in the angular measurements of the inertial navigation system resulting in larger error ellipses of the pulses hitting the ground. However, inertial navigation systems are becoming more accurate by the year. An overview of applying MPiA technology in practice for large area survey is given by Roth and Thompson (2008).

8.5 Data Handling

A Lidar sensor generates hundreds of millions or even billions of three-dimensional coordinates in a few hours. From this massive number of sampled points one wants to reconstruct surfaces and the structure of objects such as trees. Lidar scanning, due to characteristics of its acquisition, generates irregularly distributed points Therefore, unprocessed Lidar points are sometimes described as being 'unorganised'; they form clouds. How to convert the unorganised into the organised? The central issues when dealing with Lidar data concern archiving and management of the data, generation of output suitable for use within GIS and CAD software, filtering of data for reconstruction of bare-earth ground surfaces and features such as buildings and trees, and visualisation of the data. To carry out data handling and information extraction from Lidar point clouds, specialised software is on the market. For an overview of software available on the commercial market and specifications refer to Lemmens (2007b, 2010a).

8.5.1 Interpolation

The implementation of many mathematical operations, such as area and volume computations, requires data arranged on a grid of square cells. Therefore, the representation of elevation data within a GIS and CAD environment is carried out preferably in raster format, elevation values being stored in the cells of a regular planar raster. Airborne Lidar data are recorded as an irregular point-cloud; the spatial distribution of the points depends on scanning method, flight height and flight speed (Fig. 8.9).

The transformation of an irregular point-cloud into a regular raster – gridding – requires interpolation: computation of elevation values for non-sampled terrain points from sampled points. Reconstruction of the continuous surface requires definition of a function that passes through the sampled points. An infinite number of functions will fulfil this constraint and, unfortunately, no simple rule exists for determining which is best suited to a given data set. Additional conditions have thus to be defined, which has resulted in the development of a wealth of interpolation techniques. Some conditions are based on geostatistical concepts (kriging) and others on locality. The latter assumes a relationship between the elevation of each point

Fig. 8.9 Irregular
distribution of points in a raw
airborne Lidar data set. The
pattern indicates flight
direction from *left* to *right*.
Some objects, particularly
water, do not reflect sufficient
energy, visible as white
patches

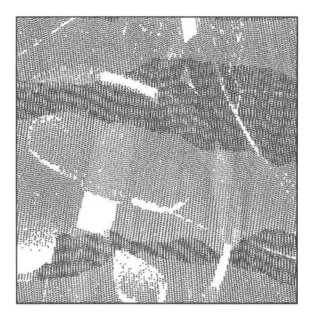

Fig. 8.10 Transferring the
data set of measured height
points into regular grid by
interpolation. All measured
points in the search area
contribute dependent on
distance

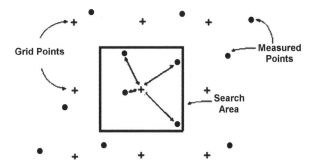

and other points in the vicinity, up to a certain distance away. One of the most simple and available methods is 'inverse distance-weighted interpolation' (Fig. 8.10). This method assumes an inverse relationship between elevations of neighbouring points and their distance: the greater the distance, the less the elevation of a sampled point will contribute to computation of the elevation of the non-sampled point. Selection of the method of interpolation is often based on experience, experiment and availability of algorithms in the GIS system.

The hundreds of millions of points generated by each Lidar survey cause computational complications during interpolation since all this data cannot be stored in the internal memory of even the most sophisticated computer. As a consequence the data have to remain on larger but significantly slower disks. Operational weakness arises from data swaps between disk and RAM rather than computation when pro-cessing such massive volumes of data. Many practical algorithms therefore perform 'segmentation', breaking down the point-cloud into a set of non-overlapping sub-clouds each containing a small number of sampled points. The points in each segment are then independently interpolated.

8.5.2 Ground Filtering

Ground filtering is the removal of unnecessary points, e.g. those reflected from veg-etation when creating a 3D city model, or from vegetation and buildings when creating a bare ground DEM. When creating such a DEM all objects on top of the bare ground have to be removed preferably automatically (Fig. 8.11). For this purpose procedures can be developed and implemented in existing GIS software packages (Brovelli et al., 2004). As ground filtering is the primary step for DEM generation, numerous algorithms have been developed to separate the ground from the objects on top of it. In the course of time, many different filtering methods have been developed, see, e.g., Vosselman (2000), Kraus and Pfeifer (2001), Raber et al. (2002), Zhang et al. (2003), Filin and Pfeifer (2006), Lloyd and Atkinson (2006), Silván-Cárdenas and Wang (2006), and Zheng et al. (2007). The resulting algo-rithms are based on a wide variety of mathematical modelling of the point cloud data. Kraus and Pfeifer (2001) fitted a low-degree polynomial through a set of neigh-bouring points. Next the distances of all the neighbouring points to this trend surface are computed. When the distances exceed a predefined threshold, the concerning points are removed from the data set. The method is developed to remove points reflected on trees in forests to obtain the bare ground and is not dedicated to remove buildings. For the purpose of removing sets of nearby off-the-ground points, such as those reflected from buildings and building roofs, Vosselman (2000) developed a

Fig. 8.11 The aim of many airborne Lidar surveys is to create nationwide, accurate and detailed DEMs of the bare-earth ground surface

slope-based filter that identifies ground data by comparing slopes between a point and its neighbours. A point is an off-the-ground point if the maximum value of all these slopes is larger than a predefined threshold. This filter yields satisfactory results when applied to rather flat urban areas but fails when applied to vegetated mountain areas with a large slope variation and to filter such areas, morphological filters have been developed (Zhang et al., 2003). Recognising that current methods can be grouped into two major categories (neighbourhood-based approaches and directional filtering) Meng et al. (2009) proposed a mixed algorithm to incorporate a two-dimensional neighbourhood in the direction of scanning to diminish the scan-direction sensitivity to errors. The predominant assumption underlying the design of all the filters is that the heights of nearby points are strongly correlated. So, when the height of one or more points strongly deviates from those of the surrounding points, this anomaly is attributed to the reflection of an above-ground object. Accordingly such points should be removed from the data set.

The wide variety of methods developed to remove points reflected on above-ground objects from the terrain is an apparent indication of the complexity of the filtering problem. The basic problem is that terrain characteristics can not be uniquely modelled. For example, when one wants to remove houses from the data set using a particular filter, it may occur that dune tops, which have the same morphological structure as houses, are removed. To decrease the chance on errors of commission (points are wrongly classified as non-ground points, so they are removed while they should be kept) or errors of omission (non-ground points are wrongly classified as ground points, so they are kept while they should have been removed), one can divide the area in homogeneous segments covered by the same terrain types, such as agricultural, forest or built-up. Such a segmentation of the data set can be done manually or by automatic means using statistical pattern recognition techniques. New users may find it difficult to select the proper filtering algorithm from the plethora available. A comprehensive categorisation of Lidar ground filtering algorithms for guiding users to select the optimal method for their specific applications is given by Meng et al. (2010).

8.5.3 Manual Editing

Data handling software should also allow onscreen zoom and rotation of the point-cloud, and navigating through it, as these are crucial for assessing data quality, planning and control of subsequent steps in processing, and editing for manual selection of individual points (Lemmens, 2010b). Basically, points may be represented as white dots on a dark background, but usually they are coloured according to intensity or textured using co-registered imagery. Presentation of results may require facilities such as hillshading, diffuse lightening, multi-layering, altitude colouration or automatic draping of aerospace imagery over the point-cloud. Good visualisation is also necessary for fitting through a group of selected point features such as lines, circles, planes, spheres, cylinders and cones. Selection facilities enable measurement of the distances and angles of lines connecting points. Some software dedicated to the processing of laser point clouds support point-cloud viewing in stereo using graphics hardware.

8.5.4 Deriving Parameters

DEMs belong to the main category of Lidar products, and processing software may enable generation of contour lines and cross-sections, and detection of break lines, usually based on a Triangulated Irregular Network (TIN) earlier created from the point-cloud. Such software may contain refined DEM modules such as calculation of lines of sight, slopes, volumes and simulation of floods or other hazards. The software may also allow automatic classification of groups of points into buildings, roads, trees or power lines.

The output of laser data handling software will often be imported into other software, usually a GIS. Since many data formats exist it is important that software supports a wide variety of import and export formats, both for importing data into the software itself and for importing the output into other systems.

8.6 Manufacturers

Manufacturers of airborne Lidar systems are coming from two opposite directions: system building versus service provision. System builders include Leica Geosystems and Optech Inc., but most conspicuous is Riegl. Until late 2006, the latter was a manufacturer for manufacturers only, offering 'laser sensors for airborne applications but not complete airborne Lidar systems'. But in 2007 Riegl released its full-fledged systems LMS-S560 and LMS-Q560. On the service provision side are Fugro, TopEye and Terrapoint. Fugro developed its own in-house system, Flimap, designed for corridor mapping using helicopters. When manufacturers cross the line, differences become more diffuse. German company TopoSys combines manufacturing with self-executed surveys and considers this dual role as a base for expertise. Product features are quite similar to those of the Riegl systems even in name: LMS-Q560 versus Harrier 56 and LMS-Q680 versus Harrier 68; indeed, the core of the TopoSys systems are Riegl sensors. This is also true for German firm IGI's airborne Lidar terrain mapping system LiteMapper. TopEye was initially both a system builder and a service provider. With its origins in Swedish company Saab, TopEye became part of the Blom Group – which uses the technology for geodata collection – in July 2005. Another Saab development is a system for hydrographic and topographic surveys. In 2002, Saab sold the rights for the system to three former employees who launched Airborne Hydrography. The company manufactures Dragon Eye topographic Lidar sensor and Hawk Eye II bathymetry and topography airborne laser system. The competition between airborne Lidar manufacturers mainly focuses on increasing the frequency with which pulses are emitted. The pulse repetition rate has increased from less than 50 kHz in 2001 to 400 Khz 10 years later (Lemmens, 2011b).

8.7 Applications

As is the case for terrestrial laser scanning (Chapter 6), the value of airborne Lidar data appears to full advantage when combined with other data sets such as terrestrial, aerial and satellite imagery, topographic data and terrestrial laser scan data. When fused with multispectral imagery Lidar data can be used to automatically detect residential buildings (Awrangjeb et al., 2010). Combined with aerial images Lidar DEMs have been exploited to quantify intertidal complex landforms volumes (Noernberg et al., 2010).

Airborne Lidar provides useful information for all stages of infrastructure works, such as corridor planning, environmental-impact simulation, optimal movement of earth works, determination of (rail)road deformation, detection of obstructions such as fallen trees after storms and dike improvement (Fig. 8.12). Lidar also provides essential information for the creation of 3D digital city models (Lemmens et al., 1997; Sohn and Dowman, 2007); and aerial monitoring of electricity power lines (McLaughlin, 2006; Jwa et al., 2009). The latter can be done from unmanned airborne platforms, such as helicopters, flying at a maximum height of 60 m and using a compact and lightweight scanner optimised for detecting low-cross-section targets including power line wires. Such a system has been developed by Riegl. Combining Lidar data with geo-referenced oblique images (see Section 7.9) enables

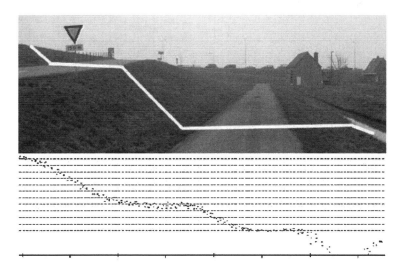

Fig. 8.12 Profile through an airborne Lidar DEM across a dike; the water in the ditch at the right did not reflect laser pulses

to create 3D city models in a highly automated manner. Other applications include flood-hazard zoning, river-flood modelling, assessment of post-disaster damage, tree species identification for natural resource management and monitoring activities (Kim et al., 2009) and mapping of leaf area index (Zhao and Popescu, 2009).

Lang and McCarty (2009) investigated the suitability of airborne Lidar for capturing forested wetlands, using the eastern shore of Maryland, USA, as study site and found that the potential of Lidar intensity data is strong for detecting inundation below the forest canopy. Jones et al. (2010) assessed the potentials of combining hyperspectral airborne imagery and small footprint airborne Lidar data for mapping tree species in coastal South-western Canada. Adding tree heights and volumetric canopy profile data derived from Lidar point clouds to the hyperspectral imagery improved accuracy from around 60 to 90% with 5 to 12%.

8.8 Concluding Remarks

Although airborne Lidar has become a standard approach for the generation of DEMs, the pathway from raw data as captured by the sensor to the final DEM is paved with challenges. The most critical and difficult step is the separation of points in point cloud, which refers to the bare ground from non-ground points (Liu, 2008). Fisher and Tate (2006) review sources and nature of errors in DEMs and derived models such as slope are reviewed by Fisher and Tate (2006). Lemmens (1997) provides a list of all error sources within airborne Lidar data, whether they were random or systematic and their approximate magnitude. The list is based on thorough theoretical considerations.

In November 2008 a book was published edited by Shan and Toth (2008), providing the first systematic, in-depth, introduction to the principles of capturing and processing Lidar data. The December 2006 issue of *Photogrammetric Engineering & Remote Sensing*, Journal of the American Society for Photogrammetry and Remote Sensing, is a special issue on the use of airborne Lidar and Terrestrial Laser Scanning for vegetation and habitat mapping; most articles focus on forestry applications.

References

Ackermann F (1999) Airborne laser scanning present status and future expectations. ISPRS J Photogramm Remote Sens 54:64–67

Awrangjeb M, Ravanbakhsh M, Fraser CS (2010) Automatic detection of residential buildings using LIDAR data and multispectral imagery. ISPRS J Photogramm Remote Sens 65(5): 457–467

Baltsavias EP (1999a) Airborne laser scanning: basic relations and formulas. ISPRS J Photogramm Remote Sens 54(2–3):199–214

Baltsavias EP (1999b) A comparison between photogrammetry and laser scanning. ISPRS J Photogramm Remote Sens 54(2–3):83–94

Brovelli MA, Cannata M, Longoni UM (2004) Lidar data filtering and DTM interpolation within GRASS. Trans GIS 8(2):155–174

Chasmer L, Hopkinson Ch, Treitz P (2006) Investigating laser pulse penetration through a conifer canopy by integrating airborne and terrestrial Lidar. Can J Remote Sens 32(2):116–125

Filin S, Pfeifer N (2006) Segmentation of airborne laser scanning data using a slope adaptive neighbourhood. ISPRS J Photogramm Remote Sens 60(2):71–80

Fisher PF, Tate NJ (2006) Causes and consequences of error in digital elevation models. Prog Phys Geogr 30(4):467–489

Heinzel J, Koch B (2011) Exploring full-waveform Lidar parameters for tree species classification. Int J Appl Earth Obs Geoinf 13(1):152–160

Hug C, Ullrich A, Grimm A (2004) LiteMapper-5600: a waveform digitizing Lidar terrain and vegetation mapping system. Int Arch Photogramm Remote Sens Spat Inf Sci XXXVI(Part 8/W2):24–29. Freiburg, Germany

Jones TG, Coops NC, Sharma T (2010) Assessing the utility of airborne hyperspectral and LiDAR data for species distribution mapping in the coastal Pacific Northwest. Can Remote Sens Environ 114(12):2841–2852

Jwa Y, Sohn G, Kim HB (2009) Automatic 3D powerline reconstruction using airborne Lidar data. Int Arch Photogramm Remote Sens Spat Inf Sci XXXVIII(Part 3/W8):105–110. Paris, France

Kim S, McGaughey RJ, Andersen H-E, Schreuder G (2009) Tree species differentiation using intensity data derived from leaf-on and leaf-off airborne laser scanner data. Remote Sens Environ 113(8):1575–1586

Kraus K, Pfeifer N (1998) Determination of terrain models in wooded areas with airborne laser scanner data. ISPRS J Photogramm Remote Sens 53:193–203

Kraus K, Pfeifer N (2001) Advanced DTM generation from Lidar data. Int Arch Photogramm Remote Sens XXXIV(3/W4):23–30. Annapolis, MD

Lang MW, McCarty GW (2009) Lidar intensity for improved detection of inundation below the forest canopy. Wetlands 29(4):1166–1178

Lemmens MJPM (1997) Accurate height information from airborne laser-altimetry. Proceedings of IGARSS'97, remote sensing: a scientific vision for sustainable development, Singapore, pp 423–426. ISBN 0-7803-3839-1

Lemmens MJPM (1999a) Uncertainty in automatically sampled Digital Elevation Models, chapter 47. In: Lowell K, Jaton A (eds) Spatial accuracy assessment: land information uncertainty in natural resourses. Ann Harbor Press, Chelsea, MI, pp 399–407. 1 ISBN 1-57504-119-7

Lemmens MJPM (1999b) Quality description problems of blindly sampled DEMs. In: Shi W, Goodchild MF, Fischer PF (eds) Proceedings of the international symposium on spatial data quality'99, Hongkong, pp 210–218

Lemmens MJPM, 2001, Height Information from laser-altimetry for urban areas. GIS Development, The Asian GIS Portal, Map India, 2001, pp 1–5

Lemmens M (2007a) Airborne Lidar sensors. GIM Int 21(2):24–27

Lemmens M (2007b) Airborne Lidar processing software. GIM Int 21(2):52–55

Lemmens M (2009a) Airborne Lidar sensors. GIM Int 23(2):16–19

Lemmens M (2009b) Multiple-pulses in air. GIM Int 23(2):57

Lemmens M (2010a) Airborne Lidar processing software. GIM Int 24(2):14–15

Lemmens M (2010b) ALPS. GIM Int 24(2):57

Lemmens M (2011a) Airborne Lidar sensors: status and development. GIM Int 25(2):33–39

Lemmens M (2011b) Airborne Lidar Sensors. www.gim-international.com/productsurvey/

Lemmens M, Deijkers H, Looman P (1997) Building detection by fusing airborne laser-alimter DEMS and 2D digital maps. Int Arch Photogramm Remote Sens 32(Part 3–4/W2):42–49

Lemmens M, Lohani B (2001) Geo-Information from Lidar: how India may benefit from airborne laser-altimetry. GIM Int 15(7):30–33

Liu X (2008) Airborne Lidar for DEM generation: some critical issues. Prog Phys Geogr 32(1):31–49

Lloyd CD, Atkinson PM (2006) Deriving ground surface digital elevation models from Lidar data with geostatics. Int J Geogr Inf Sci 20:535–563

Mallet C, Bretar F (2009) Full-waveform topographic lidar: state-of-the-art. ISPRS J Photogramm Remote Sens 64(1):1–16

McLaughlin RA (2006) Extracting transmission lines from airborne Lidara data. IEEE Geosci Remote Sens Lett 3(2):222–226

Meng X, Currit N, Zhao K (2010) Ground filtering algorithms for airborne Lidar data: a review of critical issues. Remote Sens 2(3):833–860

Meng X, Wang L, Silván-Cárdenas JL, Currit N (2009) A multi-directional ground filtering algorithm for airborne Lidar. ISPRS J Photogramm Remote Sens 64(1):117–124

Noernberg MA, Fournier J, Dubois S, Populus J (2010) Using airborne laser altimetry to estimate Sabellaria alveolata (Polychaeta: Sabellariidae) reefs volume in tidal flat environments. Estuar Coast Shelf Sci 90(2):93–102

Petersen YM, Burman Rost H (2011) Swedish Lidar project: new nationwide elevation model. GIM Int 25(2):20–23

Raber GT, Jensen JR, Schill SR, Schuckman K (2002) Creation of digital terrain models using an adaptive Lidar vegetation point removal process. Photogramm Eng Remote Sens 68:1307–1316

Romano ME (2004) Innovation in Lidar processing technology. Photogramm Eng Remote Sens 70:1202–1206

Roth RB, Thompson J (2008) Practical application of multiple pulse in air (MPiA) Lidar in large area surveys. Int Arch Photogramm Remote Sens Spat Inf Sci XXXVII(Part B1):183–188. Beijing

Shan J, Toth ChK (eds) (2008) Topographic laser ranging and scanning: principles and processing. Taylor Francis, Boca Raton, London, New York, 608 p. ISBN-13 9781420051421

Silván-Cárdenas JL, Wang L (2006) A multi-resolution approach for filtering LiDAR altimetry data. ISPRS J Photogramm Remote Sens 61(1):11–22

Skaloud J, Schaer Ph, Stebler Y, Tomé P (2010) Real-time registration of airborne laser data with sub-decimeter accuracy. ISPRS J Photogramm Remote Sens 65(2):208–217

Sohn G, Dowman I (2007) Data fusion of high-resolution satellite imagery and Lidar data for automatic building extraction. ISPRS J Photogramm Remote Sens 62(1):43–63

Vosselman G (2000) Slope based filtering of Laser altimetry data. Int Arch Photogramm Remote Sens Spat Inf Sci XXXIII:935–942

Wehr A, Lohr U (1999) Airborne laser scanning: an introduction and overview. ISPRS J Photogramm Remote Sens 54(2–3):68–82

Zhang K, Chen S-C, Whitman D, Shyu M-L, Yan J, Zhang C (2003) A progressive morphological filter for removing measurements from airborne Lidar data. IEEE Trans Geosci Remote Sens 41(4):872–882

Zhao K, Popescu S (2009) Lidar-based mapping of leaf area index and its comparison with satellite Globcarbon Lai products in a temperate forest of the southern USA. Remote Sens Environ 113(8):1628–1645

Zheng S, Shi W, Liu J, Zhu G (2007) Facet-based airborne light detection and ranging data filtering method. Opt Eng 46(6):066202-1–066202-15. doi:10.1117/1.2747232

Chapter 9
Earth Observation from Space

Permanent observation of the Earth from space started in the early 1970s. It is a method of collecting synoptic imagery of (nearly) the whole globe. During the first 20 years the emphasis was on development of the technology and strategic applications in a strong national context. In the late 1980s, a process of change of mindset started and Earth observation from space gradually moved away from governmental umbrellas to commercialisation and privatisation. Since the turn of the millennium many Earth observation satellites equipped with advanced imaging sensors have been launched. Satellite images with high spatial resolution provide an up-to-date and cost-effective means of producing image maps, derived topographic maps and cadastral maps for all areas of the world. The ability to extract from 5-m to 50-cm imagery a wide variety of topographic data and to locate features at an absolute accuracy of up to 1 m or even better provides an unprecedented opportunity for the cost-effective production of accurate maps of areas ranging from small cities to entire countries. Images with a Ground Sample Distance (GSD) of 1 m or better enable to extract buildings and roads making them suitable for the production of scale 1:10,000 topographic maps. This chapter considers the status of those satellites and sensors suited for geomatics purposes and treats features of the captured imagery. The considerations are followed by a view on the future.

9.1 Remote Sensing

Earth observation from space is based on the combined use of space technology and sensor technology. The capturing of earth-related features from a distance is called remote sensing. Rather than being a discipline, remote sensing is a conglomerate of measuring and processing techniques for the collection and analysis of geo-information by recording reflected or emitted Electromagnetic (EM) energy by instruments mounted on terrestrial, airborne or spaceborne platforms. The measurements may not only be represented as images but also as temperature fields or wind velocities. Most of the ten sensors mounted on ESA's Envisat satellite – launched early 2002 – are aimed at the scientific objective of arriving at a better understanding of the sea and the atmosphere. For geomatics applications remote

M. Lemmens, *Geo-information*, Geotechnologies and the Environment 5, DOI 10.1007/978-94-007-1667-4_9, © Springer Science+Business Media B.V. 2011

sensing techniques, which capture images from which topographic maps, can be created and DEMs are of particular interest.

Remote sensing instruments not only record reflected sunlight in the blue, green and red part of the EM spectrum as the human eye does but these sensors also go beyond what the human eye sees; they can "see" what remains hidden for the human eye and record EM energy emitted by the objects themselves. These are thermal bands that human cannot see but feel as warmth or heat. Other instruments generate EM energy themselves and record EM energy that is scattered back in the direction of the instrument. The instruments that record reflected sunlight – passive sensors – produce visible, near-infrared and thermal images. In contrast, active sensors emit energy in the visible and near-infrared part of the EM spectrum, which are generated by lasers or microwaves produced by Radar systems. Active sensors emitting visible and near-infrared energy are Lidar systems used to create highly automatically three-dimensional representations of objects and the surface of the Earth (see Chapter 8). Microwaves can be used to produce images as well as DEMs.

Active systems are highly insensitive to sunlight conditions; they can operate day and night. Radar has an additional advantage, which makes its operation highly weather independent; the wavelengths of microwaves range from the centimetre level to the metre level. That means microwaves are much larger than the water particles present in the air so that the waves are not or just little affected by them. They "wave" around the particle and the larger the wavelength the less sensitive microwaves are for water particles. As a result Radar can see through clouds, a beneficial property especially in tropical areas, where the optical view from space is (almost) permanently impeded. The larger wavelengths can even penetrate dense vegetation cover and upper soil layers.

9.2 Earth Observation Systems

The first permanent earth observation satellite, initially called Earth Resources Technology Satellite 1 but today known as Landsat 1, was launched in July 1972 by NASA, the National Aeronautics and Space Administration of the USA. With Ground Sampled Distance (GSD) of 79 m the Landsat Multispectral Scanner images provided an astonishingly comprehensive and panoramic view of areas never before mapped. GSD of the US Landsat family has subsequently been improved, first to 30 m with the launch of Landsat 3 and then to 15 m with Landsat 7 ETM. These are moderate GSDs and they do not allow seeing individual buildings but large man-made objects such as highways can be identified. In 1986 the French/European SPOT (Satellite Pour l'Observation de la Terre) satellite was put into orbit.

9.2.1 Overview of Systems

In the meantime earth-observation satellites are constructed and launched at conveyor-belt speed and the imagery is a commodity like any other. The Ikonos-2 satellite (Ikonos-1 was launched first but never operational) was launched in

September 1999 and has been delivering commercial imagery since early 2000. QuickBird was launched on 18 October 2001 and the scene sizes delivered as standard are 16.5 km × 16.5 km and 16.5 km × 165 km. The SPOT 5 satellite was launched on 4 May 2002 and orbits at an altitude of 810 km; scene sizes are typically 60 km by 60 km. OrbView-3 is in operation since 2004 and offers images of 1 m panchromatic and 4 m multispectral GSD, but in contrast to the other satellites, the panchromatic and multispectral bands cannot be recorded simultaneously.

On 27 October 2005, a Kosmos-3 M booster blasted off from Russia's Plesetsk Cosmodrome carrying, among others, two earth-observation satellites: UK's TopSat (GSD 2.5 m) and China's Beijng-1, carrying black-and-white (panchromatic) and colour cameras with spatial resolutions of 4 and 30 m, respectively (Lemmens, 2007). Daichi, Japan's Advanced Land Observing Satellite (ALOS), has been orbiting since 24 January 2006. On 25 April 2006, EROS B (Earth Resources Observation System) was launched, the successor of EROS A, launched on 5 December 2000. The scene width is typically 7 km. Kompsat 2, also referred to as Arirang-2 and developed by KARI (Korea Aerospace Research Institute) to continue the observation program of the Kompsat-1 mission, was launched on 28 July 2006. The Indian Remote Sensing Satellite IRS Cartosat-2 was successfully launched into polar orbit on 10 January 2007. DigitalGlobe's WorldView-1, blasted beyond Earth's atmosphere on Tuesday 18 September 2007, was the thirteenth earth-observation satellite launched in 2007; an unprecedented year, the number of orbiting earth-observation satellites reaching 25, the largest annual total ever. On Saturday 6 September 2008, GeoEye-1 was launched. To date, over 30 mid- and high-resolution, optical land-imaging satellites and five radar systems are in orbit, involving 17 countries. If all plans become reality, the number of mid- and high-resolution satellites will double over the coming years and orbiting radar systems will triple, involving 24 countries. Up to 2015, nearly half of launches will be on behalf of Asian Countries; fewer than 10% will be US-based.

Together with SPOT 5, Formosat from SPOT Image, Ikonos and Quickbird the above satellites produce very-high-resolution (VHR) optical images: GSD better than about 5 m (Table 9.1). In the panchromatic (black and white) mode the GSD of SPOT 5 is 5 m, of both Ikonos and Kompsat 2 the GSD is 1 m, Quickbird's

Table 9.1 Main characteristics of some of high-resolution optical satellite imagery

	Multispectral GSD (m)	Panchromatic GSD (m)	Frame size (km)	Intensity range (bits)
Worldview 2	1.85	0.46	16.4	11
GeoEye-1	1.65	0.41 (0.50)	15.2	11
QuickBird	2.44	0.61	16.5	11
Ikonos	4	1	11	11
Orbview 3	4	1	13.5	11
Eros B	n/a	1.8	7	10
Eros A	n/a	0.7	14	10
Spot 5	10	2.5	60	8

Source: Lemmens (2008)

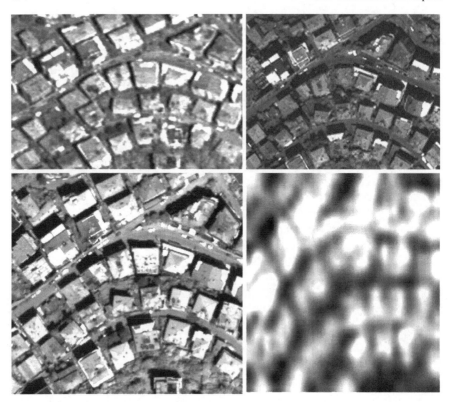

Fig. 9.1 Information content of satellite images depends mainly on ground sample distance (GSD). All images cover the area of Zonguldak, Turkey: (from *left* to *right* and *top* to *bottom*) OrbView-3 (GSD: 1.07 m), Ikonos (GSD: 1.00 m), QuickBird (GSD: 0.62 m) and Spot 5 (GSD 5 m)

GSD is 0.60 m, Daichi's PRISM 2.5 m and GeoEye1 has a GSD of 0.41 m. The GSD of the panchromatic band is typically four times better than that of the multi-spectral bands; for example, multispectral bands of QuickBird have a GSD of 4 m, Formosat-2 launched 20 May 2004 has a GSD of 2 m in the panchromatic mode and 8 m in the multispectral mode and the multispectral images of GeoEye-1 have a GSD of 1.65 m. All these satellites have a global coverage. They are polar-orbiting (inclination around 98°) and sun-synchronous, passing over a given area at about 10.30 a.m. local time. Figure 9.1 provides an indication of the information contents of the panchromatic bands of OrbView-3 (GSD: 1.07 m), Ikonos (GSD: 1.00 m), QuickBird (GSD: 0.62 m) and Spot 5 (GSD 5 m).

9.2.2 GeoEye-1 and Ikonos

GeoEye-1 orbits at 681 km above the surface of the Earth, an altitude that allows 15 orbits per day. The nominal swath width is 15.2 km at nadir. But the sensor can be rotated forward, backward or sideways, allowing it not only to record single images of multiple targets during a single pass or multiple images of the same target to create a stereo picture, but also revisit times up to 2 days. In panchromatic mode an area measuring 700,000 km^2, which is more than the size of the Iberian Peninsula or the Ukraine, can be recorded in a single day; in multispectral mode this is 350,000 km^2. An accuracy of 3 m can be achieved without using ground-control points. Together, Ikonos and GeoEye-1, both satellites being managed by GeoEye, can capture almost one million km^2 per day. The GSD of 41 cm is resampled to 50 cm because NOAA licensing restrictions require doing so before making the imagery available for sale to commercial customers. But over time such government limitations are being lifted or modified. For example, in June 2007, NOAA removed the 24-h hold rule for imagery collected at less than 0.82-m resolution. Also, the US Government is currently reviewing GSD licensing restrictions and will make some new determinations next year. The study, requested by Congress, started in October 2008. GeoEye 1 produces the world's highest resolution satellite imagery commercially available. Figure 9.2 shows the first image of GeoEye1 released on 8 October 2008. In the meantime GeoEye company is building GeoEye-2. Weighing over 2,000 kilos and scheduled for launch in late 2012, this satellite will capture the Earth's surface with a GSD of 25 cm at nadir, going down to 36 cm off-nadir. But as with GeoEye-1, US government restrictions enforce re-sampling to 50 cm.

Fig. 9.2 Detail of the panchromatic band of the first image of GeoEye1, Kutztown University PA, Pennsylvania

9.2.3 WorldView-1 and 2

WorldView-1 provides panchromatic imagery with GSD of 50 cm at nadir and dynamic range of 11 bits per pixel. The swath width at nadir is 17.6 km and one day of data acquisition may result in up to 750,000 km² being captured. During a single pass, contiguous areas of 60 km by 110 km can be covered in mono and 30 × 110 km in stereo. The platform is so stable and the positioning sensors accurate enough for achieving an accuracy of 3–7.6 m without using geometric information in the form of DEMs and ground control points, 2 m with using these geometric information sources. The same area on earth can be captured within just 6 days of a previous visit, so that if a GSD of 1 m suffices, revisit frequency may rise to 1.7 days. WorldView-2, launched 8 October 2009 and full operational since 4 January 2010, orbits at an altitude of 770 km and has a data capturing capacity of 975,000 km² per day of panchromatic images with a GSD of 50 cm and a GSD of 1.8 m for multispectral images. The average revisit time is 1 day.

The sensors of high-resolution satellites, such as QuickBird, typically capture four multispectral bands: blue, green, red and near infrared (NIR), but WorldView-2 adds another four bands to this quadruple (Fig. 9.3). The water and chlorophyll penetrating features of the coastal band (400–450 nm) facilitates bathymetric and vegetation studies. The rate of atmospheric scattering depends on wavelength of electromagnetic signals; the smaller the wavelength the more a signal will be scattered when passing the atmosphere. Hence, the coastal band is suited for investigating atmospheric correction methods. The yellow band (585–625 nm) aids vegetation studies and colour corrections for visual interpretation of the images. The additional near infrared band (NIR2) overlaps partly NIR1 but covers longer wavelengths for the major part and hence is less affected by atmospheric scattering. NIR2 allows discrimination between vegetation types such as irrigated (parks)

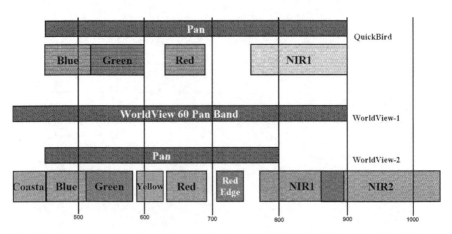

Fig. 9.3 Comparison of spectral characteristics from QuickBird, WorldView-1 and WorldView-2 imagery

and non-irrigated grasslands. The band covering the wavelengths between 705 and 745 nm lies at the edge of the red band and NIR band and supports the analysis of plant health. Without using DEMs and ground control points, an accuracy of 6.5 m can be achieved.

9.2.4 Cartosat

The Indian remote sensing satellites Cartosat-1 and Cartosat-2 have been designed to provide high-resolution along-track stereo imagery for mapping purposes (Krishnaswamy, 2002; Radhadevi et al., 2009). The sun synchronous Cartosat-1, orbiting at an altitude of 618 km, carries two panchromatic optical linear array sensors (12,000 pixels each), which are +26° and -5° tilted along track with respect to nadir. This configuration of the two cameras provides along-track stereo images. The swath width is 27 km, nominal GSD is 2.5 m and the radiometric resolution is 10 bits. The nominal revisit time is 11 days, but using the off-nadir across track steering facility the revisit time can be reduced to 5 days.

Cartosat-2, launched in 2007, is also a sun synchronous satellite and orbits at an altitude of 630 km. The platform carries one panchromatic optical linear array sensor (12,000 pixels), which captures images with a GSD of 1 m. The swath width is 9.6 km; however, the sensor can be pointed in any direction enabling the recording of four swaths in one overpass; in this way the nominal image width (9.6 km) can be extended to 38 km. The pointing facility also allows to generate three views on the same area – nadir, forward and backward – in one overpass. The resulting three images form a stereo triplet.

9.2.5 Daichi

Japanese satellite Daichi (ALOS) has on board three sensor systems (Lemmens, 2006b):

– Panchromatic Remote-sensing Instrument for Stereo Mapping (PRISM)
– Advanced Visible and Near Infrared Radiometer
– Array type L-band Synthetic Aperture Radar (PALSAR).

The PRISM looks simultaneously forward (swath width 70 km), nadir and backward (swath width 35 km), which allows for stereo mapping and creation of digital elevation models. Because there is virtually no time difference between the recordings of the overlapping images (stereo-pairs), three-dimensional feature extraction and DEM generation can be carried out in a reliable way. The in-track distance between the two images of the same terrain point acquired by the forward-looking sensor and the backward-looking sensor is around 700 km, resulting in a favourable base-to-height ratio of 1.0. AVNIR-2 is the successor to AVNIR, a sensor onboard the Advanced Earth Observing Satellite (ADEOS) launched in August 1996. The PALSAR is a SAR sensor (L-band), which allows for day-and-night and all-weather land observation. A pilot project, carried out by Amhar and Mulyana (2009), aimed at exploring the technological and economical aspects of using ALOS imagery for applications in Indonesia. Their study demonstrated that the planimetric accuracy of images captured by the PRISM sensor system is better than 5 m, which is sufficient for mapping at scale 1:25,000.

9.2.6 EROS

EROS B is an Israeli commercial/military satellite that captures scenes in panchromatic mode only. As with EROS A, EROS B also is a lightweight satellite; its mass is just about 20% of the regular Earth observations satellites. For example, the weight of GeoEye-1 is 1,955 kg while EROS B mass does not exceed 290 kg. The SPOT 5 satellite has a weight of 3,030 kg. With a GSD of 70 cm the satellite captures scenes of 7 km width and a length up to 150 km. The pointing facility creates oblique viewing capability enabling the satellite to view virtually any area on the Earth every 3–4 days.

9.2.7 Radar Satellites

In 2007 two high-resolution Synthetic Aperture Radar (SAR) satellites were launched both from Russia's Baikonur Cosmodrome in Kazakhstan: German TerraSAR-X and Canada's Radarsat-2; the first on 15 June and Radarsat-2 on 14 December.

Radar makes use of microwaves of which the wavelengths range from 8 mm to 30 cm. These wavelengths can penetrate clouds very well and are largely insensitive to haze and other weather conditions, which cause scattering of optical wavelengths (Hanssen, 2001). The all-weather capacity has been the main reason to use radar since the late 1960s for mapping tropical regions. Although capturing of radar images is daylight- and (nearly) weather-independent, their radiometric quality is lower compared to optical images because SAR images are affected by speckle noise, radar shadow and layover. The GSD of imaging radar is directly proportional to the aperture of the antenna. The aperture is proportional to the ratio of the wavelength of the radar signal and the length of the antenna. In other words, the resolution is proportional to the wavelength but inversely proportional to the length of the antenna. Improvement of the resolution thus requires either the use of very small wavelengths or the use of a very long antenna. Lessening of wavelength faces absorption problems because the atmosphere blocks very short microwaves. Extension of the antenna meets physical and operational limitations. The solution has been found in the 1960s and reads: make use of the movement of the platform. The signals hitting the earth in front of a moving platform will be compressed, whilst the signals hitting the ground back of the platform will be elongated. This effect on the frequency of signals, which are transmitted from a moving platform, is already known for a very long time and has been described by Doppler in 1842. By registering along with signal strength and travel time the Doppler frequencies of the backscattered microwave signals the GSD and thus the aperture is considerably augmented in a synthetic way and that is why this technology is called Synthetic Aperture Radar, abbreviated as SAR. Since the velocity of a platform is just in the order of 1 ppm (part per million) of the velocity of light, the associated processing steps will be highly complicated demanding high computational capacity. Consequently, the maturing of Earth observation with radar has gone along a rather flat time curve.

Since the early 1990s numerous software packages for processing SAR data have been developed. In particular, Canada and Europe have carried out intensive radar research. However, most packages are products of stand-alone efforts from research institutes and are a result of lack of commercially available software. Today software, developed in the past in research environments, is increasingly adapted to market requirements as modules within commercial software like PCI's Geomatica products, Leica Geosystems' Erdas Imagine and within Intermap's STAR production chain. These software developments together with the rapidly growth in electronic computation power, which make carrying out complex computations cheap and fast, are continuously causing a sharp increase of production and use of high-quality SAR products.

Radarsat-2 will be able to produce images with a resolution of 3 m; three times better compared to its precursor Radarsat-2. TerraSAR-X acquires data in SpotLight, StripMap and ScanSAR modes with single-, dual- or full polarisation (Table 9.2, Fig. 9.4) (Weber and Koudogbo, 2009). With a bandwidth of 300 MHz the *SpotLight Mode* achieves 1-m resolution across flight direction. In flight direction, the radar beam can be steered like a spotlight, illuminating a ground scene of 5 km × 10 km, achieving 1-m resolution; an area of 10 km × 10 km can be covered in 2-m resolution. For acquisitions in *StripMap Mode*, the ground swath is illuminated with a continuous sequence of pulses, while the antenna beam is fixed in both looking directions, resulting in an image strip of 30 km by 50 km (standard scene size, length extendable to 1,650 km) with a continuous image quality (in flight direction) at a resolution of 3 m. In *ScanSAR Mode*, areas of 100 km × 150 km (standard scene size, length extendable to 1,650 km) are covered in 18-m resolution. The swath width of 100 km is achieved by scanning four adjacent ground subswaths with quasi-simultaneous beams, each with a different incidence angle. The Basic Image products correspond to the Committee on Earth Observation Satellites (CEOS) Level 1b quality. The highest quality is achieved by compensating for distortions using a standard DEM but also customer delivered DEMs can be used to create improved orthorectified images. Merging orthorectified images acquired in ascending and descending orbits enables to replace no-information patches in the one image, due to, e.g. foreshortening, radar layover and shadows by information patches in the other image, especially useful for steep mountainous areas. The data form a basis for services including client-specific image interpretation, topographic

Table 9.2 Overview of TerraSAR-X imaging modes

	SpotLight mode		StripMap mode		ScanSAR mode
	single: HH or VV	Dual: HH/VV	single: HH or VV	dual: HH/VV, HH/HV or VV/VH	HH or VV
Polarisations					
Scene dimensions	10 km × 10 km (SL) 10 km × 5 km (HS)	10 km × 10 km (SL) 10 km × 5 km (HS)	50 km × 30 km	50 km × 15 km	150 km × 100 km
Full performance range	20–55°	20–55°	20–45°	20–45°	20–45°
Azimuth resolution	1.7 m (SL) 1.1 m (HS)	3.4 m (SL) 2.2 m (HS)	3.3 m	6.6 m	18.5 m
Ground range resolution (@ incidence angle)	1.48–3.49 m (@55°…20°) HS with 300 MHz 0.74–1.77 m (@55°…20°)	1.70–3.49 m (@55°…20°)	1.70–3.49 m (@45°…20°)	1.70–3.49 m (@45°…20°)	1.70–3.49 m (@45°…20°)

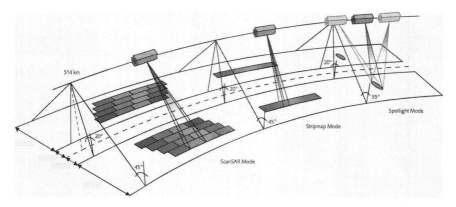

Fig. 9.4 Scheme of TerraSAR-X Imaging Modes

maps, geo-spatial databases, land cover assessments, terrain analysis and monitoring services. Users also include insurance companies.

June 2010 TerraSAR-X has been accompanied by an almost identical satellite together constituting the TerraSAR-X/TanDEM-X constellation. Orbiting only a few hundred metres apart the twin records data synchronously in the StripMap mode. The resulting homogeneous DEM has a GSD of 12×12 m^2, a relative vertical accuracy of 2 m and an absolute vertical accuracy of 10 m (Fig. 9.5). The entire land mass of the globe, 150 million km^2, will be covered by 2014.

The application potential of radar reaches further than just mapping and cartography, and concerns a wide variety of geo-management activities, including management of disasters like floods and oil spills, determination of crop type and crop yield and forest monitoring. For example, TerraSAR-X and the French optical satellite SPOT 5 played an important role in the response stage of Sichuan Earthquake, which shook the steep eastern margin of the Tibetan plateau in Sichuan Province, China, on 12 May 2008 (see Chapter 14). In addition, radar is increasingly

Fig. 9.5 First DEM created by the TerraSAR-X/TanDEM-X tandem

being used for the generation of DEMs. In particular, interferometric synthetic aperture radar – in the North Americas abbreviated to IFSAR, in Europe to INSAR – has proven to enable generation of high-quality DEMs and detection of elevation changes over time of dikes, mining areas, buildings and other objects at a variety of scales (Yue et al., 2005). INSAR is not a competitor to photogrammetry or other conventional techniques, or to novel techniques such as Lidar, but a complementor to these techniques. By combining data stemming from different geo-data acquisition systems, the strengths of each can be used optimally whilst weaknesses of the individual systems can be balanced out.

Inventories of the main characteristics of high-resolution satellite imagery acquired by passive optical and active radar sensors suited for mapping purposes are given in Lemmens (2006a, 2008). An inventory of geo-referenced data products derived from satellite imagery and deliverable off-the-shelf is given in Lemmens (2010a).

9.3 Processing of Satellite Images

9.3.1 Image Enhancement

Image enhancement involves improvement of the visibility of image features without the use of external or additional information. Virtually every digital image processing package, even the most humble and free downloadable one, is provided with image enhancement utilities, including the following:

– Contrast manipulation, such as stretching and level slicing. Level slicing is also useful for visualisation of DEMs
– Manipulation of differing spectral bands, such as band rationing
– Filtering operations based on convolution for noise reduction (smoothing), sharpening and edge enhancement
– Non-linear smoothing filters

Remote sensing imagery often consists of multiple spectral bands. The reflectance values in these bands are frequently highly correlated (Jensen, 2007). To reduce the number of bands without losing information, decorrelation is carried out by applying Principal Component Transformation (PCT). PCT results in a reduced set of modified images, which are suited as input for multispectral classification. However, the visual appearance of PCT images looks odd, which makes them inappropriate as a source for image map creation.

9.3.2 Pan-sharpening

The level of detail visible in an image depends, among other features, on its GSD. Obviously, for mapping purposes it is appropriate to use the highest GSD and that is in the panchromatic band. In most cases this black-and-white band not only covers the visible part of the electromagnetic spectrum (0.4–0.7 µm) but also stretches beyond the red band into the near-infrared increasing the visibility of features. Colour is an important visual clue for detecting and identifying features in images, although mapping with pan-sharpened images simplifies identification of man-made objects, but does not increase the number of objects that can be recognised compared to panchromatic images (Topan and Jacobson, 2006). To optimise level of detail and level of detection and identification, the panchromatic band is usually fused with three of the multispectral bands captured at lower GSD so that a true or false-colour image is created with the geometric fidelity of the panchromatic band. Either the user can do the job of fusion himself or the supplier can supply the imagery in pan-sharpened mode. Pan-sharpened images are not only suited for mapping purposes, such as creating or updating topographic maps, but they can also be used as a background in GIS applications. With a resolution of 1 m and better, buildings, roads, bridges and other detailed infrastructure are well identifiable. The information content of OrbView-3, Ikonos, QuickBird and GeoEye images are well suited for creation of scale 1:10,000 topographic maps. Using a rigorous geometric correction model and ground control points, precision can be achieved which meets scale 1:5,000, and better, mapping requirements (Volpe and Rossi, 2005).

9.3.3 Radiometric Rectification

The quality of satellite images does not only depend on the quality of the raw imagery but also on the level of post-processing. Remote sensing instruments are sensitive optical–mechanical–electronic constructions operating in a physical environment. The process of sensor construction inevitably results in its final radiometric and geometric characteristics deviating from the descriptive model, determined in the design stage. All providers carry out corrections on radiometry and geometry distortions introduced by the sensor system itself prior to delivery of the images to customers. Radiometry, here, refers to the measurement of EM energy by the sensor. The intensity range between the minimum amount of EM energy and the maximum, detectable by the sensor, is determined by the number of bits available to store the intensity values. When eight bits are available, the range is divided into 256 ($=2^8$) intensity values. Today's satellite sensors usually have a radiometric resolution of 11 bits, with which 2,048 grey values can be represented. However, the grey values within one scene will not cover the whole range and by applying contrast manipulation tools dark areas, primarily caused by shadow may be made brighter so that details become better visible.

During operation the instrument may gradually drift away from the design characteristics. When information, usually obtained by calibration, is available about the type and size of deviations the raw imagery can be restored to as close an approximation of the original scene as is possible. An example of a source causing deviations is irregular sensitivity of the sensor elements in the scanning devices to the EM energy of the band they are supposed to record. The deviations are visible as striping in the image. Even line drop may occur when one or more sensor elements are defective. Another example is blur, caused by motion of the platform or imperfections in the lens system. The medium through which the EM energy travels is not a vacuum. The particles present in the atmosphere may cause scattering visible in the image as a veil layer. In addition, radar imagery may be corrupted by speckle noise.

The need to correct for and the level at which depend on the application. Such deviations do matter for applications involving water quality, vegetation health, biomass and other subtle ecological phenomena. But also for geomatics applications corrections are often necessary, for example, when image map production requires a mosaic of images recorded at different times, sun elevation correction becomes necessary. For hilly terrain this requires the use of a DEM.

9.3.4 Geometric Rectification

The geometric properties of raw images deviate essentially from the geometric integrity of maps. Deviations are caused, among other things, by sensor projection characteristics, internal deficiencies such as lens distortion and non-linear scanning speed. External factors include varieties in the attitude and position of the sensor, terrain relief, earth rotation and curvature, and atmospheric refraction. Hence, the geometric quality depends largely upon the use of additional information, in particular Ground Control Points (GCP) and DEMs. Most providers deliver products on several levels of accuracy: raw imagery, geo-referenced imagery (using GCP) and orthorectified imagery, using DEM. Of course, the quality of used GCPs and DEM will affect the final quality of the satellite imagery and products derived from them. One may choose from several levels of processing depending on need and budget; as may be expected, the more advanced the higher the price. Arranged from the least to the most advanced correction level, geometric processing steps include the following:

– Geo-referenced by using information about the flight dynamics of the space-craft (emphemeris, attitude data and so on). This level is especially suited for the experienced customer, who wants to control the entire processing himself
– Corrected for distortions introduced by the rotation and curvature of the earth
– Transferred to a predefined cartographic projection system, such as UTM or conformal Lambert
– Corrected for relief distortions by using a DEM, resulting in ortho-images. This level results in a cartographic product with high geometry fidelity, being ready for use for mapping

When processing the panchromatic and pan-sharpened SPOT 5 images in high-precision mode using GCP, a root mean square error (RMSE) precision can be achieved of 5 m while for Ikonos-2 this measure is about 2 m. With QuickBird data and using commercial software it is possible to achieve accuracies of the order of 1 m RMSE, sufficient for topographic mapping at scales of 1:10,000 and better. Once the imagery has been orthorectified, consumer-level image processing software can enhance the image for feature extraction using contrast stretching and sharpening functions. From GeoEye-1 images natural and man-made features can be mapped in stereo to within 3 m of their actual locations on the surface of the Earth, without the use of GCPs.

9.4 Stereo Images

In addition to planimetric information, overlapping (stereo) images provide height information of objects and extracting the third dimension from stereo images is an established photogrammetric technique. Most Earth observation systems can create stereo-images and these can be acquired along-track and across-track. The first are generated while the satellite is on the same orbital pass. So the image pair is recorded nearly instantaneously, with typical time differences in the order of seconds. Across-track images are created from overlapping images recorded from two neighbouring orbits. Accordingly, across-track stereo pairs cannot be acquired near simultaneously; the time difference is at best several days but may be also several months as a result of the presence of clouds or limitations induced by the operational schedule. The extraction of three-dimensional information from near simultaneously recorded image pairs is substantially easier than from image pairs characterised by a time shift because growing season and man-induced changes may have substantially modified appearance of the surface of the Earth between the two recordings. Stereo-images are an often used source to create DEMs semi-automatically by applying image-matching techniques. Very-high-resolution satellite stereo-images are a feasible substitute for aerial images or airborne Lidar data for obtaining heights of buildings and other objects using image matching techniques (Alobeid et al., 2009).This is especially true for countries where airborne datasets are unavailable, too expensive or classified.

9.5 Information Extraction

The information content of satellite imagery is primarily depended upon four factors:

– Ground Sample Distance (GSD) also called spatial resolution: the pixel size expressed in ground units (see Fig. 9.6)
– Number of spectral bands, also called spectral resolution

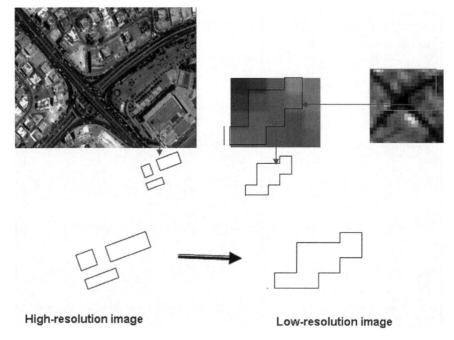

High-resolution image   **Low-resolution image**

Fig. 9.6 Swimming pool extracted from an IKONOS image (GSD 1 m) (*left*) and from a Landsat image (GSD 30 m); images taken in the same year

– How often the same area is recorded by the satellite, also called temporal resolution
– Intensity range also called radiometric resolution: basically the number of bits used to store the intensity values of a spectral band

Figure 9.6 illustrates the differences in information content of an IKONOS image (GSD 1 m) (*left*) and a Landsat image (GSD 30 m); in the 1 m GSD image three pools can be identified in the other only one (Rawashdeh, 2006). The level of detail visible in a satellite image is important for mapping purposes. Mixed pixels (mixels) cause degradation in shape and in number of recognisable features. Course GSDs also introduce overestimation of the area size.

Any information extraction process requires that knowledge be available about the road leading from the available data to the requested information. Within the framework of satellite remote sensing, this road is often complicated. Suppose that, for planning purposes, the administration of a metropolis in a developing country wants to gain insight into the population density of an informal settlement. There are many ways to obtain this information, the one way being more expensive than the other. Efficiency and effectiveness counts and therefore what is required is the best possible estimate for the least possible effort and cost. Satellite imagery provides one suitable alternative although it will, of course, be impossible to count individuals from images taken from space. But by using an indirect measure of the number of dwellings, or more generally housing density, a sophisticated guess may be arrived at. However and in turn, housing density is very hard to measure in a direct sense in satellite images. Again, an indirect measure is needed. For this, the sealed surface can be used: the result of all types of construction activities, including buildings, roads and parking places.

Another example to demonstrate the complexity of information extraction from satellite imagery is the determination of areas of informal settlement within and at the fringe of an urban area extending over thousands of square kilometres. The outlining of such areas would seem at first glance quite simple. Satellite images with a GSD in the metre or sub-metre range covering the entire city can be purchased. Next built-up areas are manually delineated by heads-up digitisation. In addition, the boundaries of the built-up areas, which are legally established, are taken from town planning administration and superimposed on the image. Subsequently, some appropriate spatial computations are straightforwardly carried out using a GIS. However, it is often not simple like that because in many developing countries local administration authorise are not so well organised that town planning information is available on the computer hard disk within a reasonable time and in the right format.

One special and advanced form of information extraction from satellite images is Multispectral Classification (MSC). Without the presence of MSC modules, no image processing package may be prefixed 'remote sensing'. Indeed, one of the most central routines in a remote sensing package is classification. MSC is primarily based on the use of pattern recognition methods. The individual pixels are treated as separate objects without exploring their spatial relationships with adjacent pixels. Unsupervised MSC and supervised MSC are the main approaches. The underlying methods used in each of them may widely vary in principle and scope. In a supervised MSC, the characteristics of the scene-related classes are known in advance. This knowledge is acquired by collecting data about what is actually present on the ground for a limited number of areas – training samples. The multispectral characteristics of each pixel are compared to those of the training samples, using some measure of similarity. To delineate the training samples, heads-up digitisation of polygons in the image is important. From an operational point of view it is also important when the polygons and their classes can be incorporated from a GIS database. The pixel is assigned to the class to which the grey values of the multispectral bands exhibit the closest resemblance according to the similarity measure. In the course of time many multispectral classification methods have been developed to decide to which class each of the pixels of the image should be assigned. Basically, the comparison can be done on the statistical means of spectral values, their dispersion, or both, resulting in such well-known classifiers as nearest-neighbour, parallelepiped (box) or maximum likelihood classification (Schowengerdt, 1997; Curran, 1993; Lillesand et al., 2004; Jensen, 2005). In unsupervised classification, the pixels are grouped, prior to class-assignment, in clusters having similar reflectance characteristics. The analyst then assigns these scene-related classes to these clusters. There are many methods developed for the clustering of pixels.

Treating individual pixels as separate objects is a rather artificial decomposition of reality and one which causes several problems including

– sensitivity to noise
– clash between spectral classes and scene-related classes
– possibility of one class consisting of several subclasses with incongruent spectral characteristics
– inability to distinguish between different classes with the same spectral characteristics

As a result, the accuracy of per pixel classifiers often stalls at the 80% level or even lower. Use of *contextual information* that is information from neighbouring pixels, usually in the form of texture measures (Haralick et al., 1973), as an additional feature assigned to the pixel may improve the classification result. Typically texture measures are derived from the same multispectral images as is used in the multispectral classification process. However, often this approach fails to result in yielding enhanced classification results for optical satellite imagery of urban scenes where residential and central city areas with tall office buildings can cause heavy spectral confusion. This is because the texture value of a pixel requires the involvement of adjacent pixels and texture filters applied to the same image as used for multispectral classification are unable to discriminate between the pixels of the object itself and those of adjacent objects. As are result, the texture value of pixels lying near an object boundary are contaminated by the texture of adjacent objects. A first remedy here is to use a priori delineation of objects boundaries. When the filter collides with a boundary, pixels on the other side of the boundary are not used in the computation process of the texture values. This option is, however, impractical because no techniques that are satisfactory to automate the delineation of object boundaries have yet been developed despite the abundance of research that has been carried out in the field of edge detection and boundary delineation (see, e.g. Lemmens et al., 1990; Lemmens and Wicherson, 1992; Lemmens, 1996a, b and the references given there). Manual delineation is very labour intensive and nullifies the advantages of multispectral classification. Another option developed by Hornstra et al. (1999) is to compute the intra-pixel reflectance variability. The rationale underpinning this approach is that for each pixel in a multispectral image the intra-pixel reflectance variability is computed using panchromatic images, which usually do have a higher spatial resolution (GSD) than their multispectral counterpart. For example, the GSD of the panchromatic band of Ikonos imagery is 1 m and the GSD of multispectral bands 4 m; for QuickBird these values are 60 cm and 2.4 m, respectively. So, in these images 4 × 4 panchromatic pixels cover one multispectral pixel. The patch of 4 × 4 pixels is sufficient to compute an uncontaminated texture measure. In this way, a texture measure can be assigned to each pixel whilst the boundary effect is abandoned.

Another type of information that can be used to improve the classification result is incorporating *ancillary (or external) data*, such as outdated land use data as a priori probabilities. For the incorporation of a priori probabilities a Bayesian classification should be applied, which is an advanced form of maximum likelihood classification. The classification result can also be improved by grouping neighbouring pixels together in regions on the basis of reflectance or texture values. This way of grouping pixels together is called segmentation. Scientific research and practical experiences demonstrate that segmentation significantly improves classification results (Gorte, 1998).

Another method, and one gaining increasing in practitioner interest, classifies not single pixels but groups neighbouring pixels together in regions on the basis of spectral and/or texture values prior to starting the classification process (Blaschke, 2010). Object-based image analysis starts with the segmentation of the image into meaningful objects. This is a key step as all later stages in the classification process depend on the segmentation result. One of the earlier approaches of object-based classification avoids the segmentation step by grouping the pixels on basis of the boundaries of objects stored in a geo-database (Lemmens and Verheij, 1988; Walter, 2004). The training samples are automatically derived from the geo-database and next. When no a prior information on the boundaries of objects is available in the geo-database or if this information is not fit for purpose the image is divided into non-overlapping segments on basis of similarity of grey values or shape using digital image processing methods based on region growing. After this initial segmentation stage, statistical measures such as mean and variance of spectral and/or texture values, and shape and size parameters, are calculated for each region (Lemmens, 2010c). Adjacent regions may be joined on the basis of a priori defined thresholds. This refinement stage enables reduction in fragmentation. Boundaries may thus also be smoothed, and buildings squared. The region is then assigned to the class to which it exhibits the closest resemblance. Since there is a high probability the resulting regions in the image correspond with objects in reality, the method has been termed *object-based image analysis*, usually abbreviated to OBIA, or more specifically GeOBIA. It is a more natural method than pixel-based classification, and accuracy is comparable with human visual interpretation, even for such fragmented scenes as urban areas (Jalan and Sokhi, 2010). Due to the significant presence of (high-rise) buildings and other constructions the results of applying MSC to high-resolution imagery of urban areas are affected by shadows. GeOBIA is able to improve the classification of shadowed areas (Zhou et al., 2009). The existing boundaries as stored in the geo-data can also be used to trigger the segmentation of the image, an approach particularly feasible for (semi-)automatically updating geo-databases of urban areas (Bouziani et al., 2010).

It often becomes necessary to enhance the image products of the above processing steps with cartographic features such as geographical names and coordinate grids. For that map editing utilities are necessary. Since mapmaking is as much an art as is the processing of remote sensing images, the analyst may benefit a lot from cartographic expert knowledge embedded in the software. A comprehensive introduction and good background how to interpret and classify remote sensing images is given in Lillesand et al. (2004).

An inventory listing the main specifications of remote sensing image processing software available at the market place is given in Lemmens (2010b).

9.6 Change Detection

To follow dune erosion, soil degradation, fragmentation of mangroves, deforestation, expansion of urban conglomerates and many other environmental processes geo-data captured at several time stamps (multi-temporal data) should be available (see, e.g. Lunetta and Elvidge, 1999). For example, Coppin and Bauer (1996) used Landsat TM imagery from three consecutive years – 1984, 1986 and 1990 – captured in the July–August summer period to detect changes in forest ecosystems. Dagar (2010) used topographical and geophysical maps along with Landsat TM and Spot-5 imagery captured at two time stamps, 1990 and 2000, respectively, to delineate environmentally vulnerable zones and sites suitability for urban and industrial development, wildlife and forest conservation and mineral exploration in western Rwanda. However, change detection is not limited to the use of maps and satellite imagery. For example, Murakami et al. (1999) used multi-temporal DEMs yielded from airborne Lidar data to detect changes of buildings. Change detection is the process of identifying and quantifying differences in the appearance of an object or phenomenon by observing it at a different time stamps using multitemporal data sets (Singh, 1989). When using satellite images the primary assumption is that changes in land cover result in changes in grey values in the diverse spectral bands and these differences should be large enough to distinguish them from differences introduced by atmospheric conditions, sun angle and soil moisture.

Once the problem area has been defined the process of change detection can be subdivided into four general steps (Jensen, 2005):

- Identification of the environmental parameters, which should be quantified and followed over time in order to arrive at sufficient information suited for tackling the issue under investigation
- Identification of the suitable remotely sensed geo-data sets in conjunction with other data sources from which the above parameters can be derived
- Applying the proper interpretation and analysis methods and techniques to the selected geo-data sets
- Translating the results of the previous step to how the study area did evolve and will continue to evolve in the future with respect to the parameters studied

Let us demonstrate the process by taking as an example change detection of land use in Beijing, the capital of China. Huang et al. (2008) wanted to know how Beijing expanded over the 1984–2005 period and the relationship of land use change with Gross Domestic Product (GDP), population density and other socio-economic factors. They identified six land use classes (urban; horticultural; cultivated; grassland and forest; waste land; and water bodies, such as rivers, lakes and reservoirs) as relevant environmental parameters to be determined. Lands TM images (GSD 30 m) recorded in 1984, 1988, 1991, 1994, 1997 and 2001 and SPOT 4 images (GSD 10 m) recorded in 1998 and 1999 and four SPOT 4 images recorded in 2005 were acquired. Existing maps used included neighbourhood maps and road network maps both compiled in the year 2000 at a scale of 1:10,000. Socio-economic data were obtained from Beijing's statistical yearbooks covering the period 1978–2005. To determine the six land use classes, maximum likelihood classification was employed on the satellite imagery once they were co-registered in a preprocessing stage. Using ground truth, confusion matrices were determined and these showed overall accuracy percentages of the classification process between 80 and 90 for the seven time stamps. Regression analysis and spatio-temporal statistical techniques were used to determine the relationship between urban area in a certain year at one hand and GDP and population in the same year at the other hand. From the seven land use maps derived from the multi-temporal satellite imagery the scientists estimated that the urban land area of Beijing will be nearly 4,200 km^2 in the year 2020 while inhabited by 18 million people. In 1984 these values were 1,241 km^2 and one million, respectively, which had been increased to 2,900 km^2 and 15 million in the year 2005. Change detection by regression analysis is still subject of research, for example, Dianat and Kasaei (2009) improved the method through incorporating spatial relations among neighbouring pixels.

9.7 Applications

Sandau et al. (2010) give for eight application areas an overview of the requirements with respect to GSD and temporal resolution (Fig. 9.7). GSD and accuracy are so high that high-resolution satellite imagery could form the geometric and topographic foundation for the spatial data infrastructure in many countries, especially those lacking up-to-date, accurate and detailed topographic maps, such as China, India and

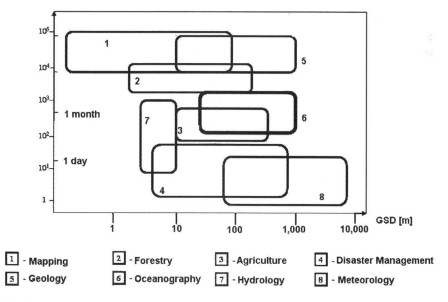

Fig. 9.7 Earth observation requirements with respect to GSD (horizontal axis) and temporal resolution (vertical axis) for eight application areas, ranging from Mapping to Meteorology

many countries in sub-Saharan countries. Satellite images also provide keen means
for establishing, extending and updating Land Information Systems especially for
rural areas in developing countries. In many parts of the world human activity is so
intensive that 'the present moment' lasts no longer than the time needed to pile up
bricks, anchor concrete and pave a surface with asphalt. Indeed, within a few years
of imagery having been acquired it has lost virtually all its value for planning and
monitoring purposes.

Only for applications involving change detection, historical images keep their
worth. A non-limitative list of applications of high-resolutions satellite images is as
follows

– Topographic maps updating up to the scale of 1:10,000
– Creation of DEMs
– Cadastral mapping
– Monitoring of agricultural land use, crop rotation and land degradation
– Supporting rapid response to earthquakes, floods and other disasters
– Change detection of the natural environment (vegetation, soil)
– Generation of thematic maps for environmental monitoring.

Fraser et al. (2008a, b) used stereo Ikonos, Quickbird, SPOT5 and ALOS PRISM
imagery to map the small Himalayan Kingdom Bhutan. All these high-resolution
satellite images yielded ground point positioning to 1-pixel accuracy. DEMs were
created from SPOT5 and PRIMS imagery (Fig. 9.8) resulting in a height precision of
5–7 m root mean square error (RMSE) except in areas with very steep slopes, where
the area-based matching techniques often failed to produce a good match between

Fig. 9.8 DEM covering 120 km^2 of the Paro Valley, Bhutan, generated from Ikonos stereo
imagery: elevation range, 2,200–3,700 m

Fig. 9.9 Digital cadastral data overlaid onto an Ikonos ortho-image

corresponding points. Next, using the DEMs, ortho-images were produced. Three-dimensional coordinates of points can be extracted from combining ortho-images with the corresponding DEM; the planar coordinates from the ortho-images and the height component from the DEM. This way of mapping, called monoplotting, was used for extracting cadastral boundaries. Figure 9.9 shows cadastral parcel boundaries superimposed on an Ikonos ortho-image. Differences between boundary data and the current situation on the ground are readily apparent. In addition, the ability to reliably position boundary points in high-resolution satellite images is beneficial in the resolution of land disputes.

Topan et al. (2009) investigated panchromatic and pan-sharpened Ikonos and Quickbird images as well as panchromatic OrbView-3 images on their suitability for manual and automatic feature extraction. The pilot study, covering a mountainous urban area in the Western Black Sea area of Turkey, revealed that the geometric and radiometric information of the above very-high-resolution satellite images is appropriate for manual feature extraction for topographic mapping purposes, scale 1:10,000. Baghdadi et al. (2009) used multi-temporal TerraSAR-X, ASAR/ENVISAT and PALSAR SAR data for studying how the different radar parameters (wavelength, incidence angles and polarisation) could provide information for monitoring sugarcane harvest. Vogelmann et al. (2009) used a time series (1988–2006) of eight Landsat TM images capturing the San Pedro Parks Wilderness area, New Mexico, USA, during autumn, to detect long-term, short-term and abrupt changes in land cover.

9.8 Commercialisation

The 1990s saw a world wide trend of governments privatising many of their services and facilities. In the trace of this wave, the US government issued licences for permanent Earth observation to private companies, resulting in the founding of Worldview, Space Imaging and OrbImage. In 1993 the first license allowing a private firm to build and operate a high-resolution satellite for commercial sale was granted to WorldView, which became EarthWatch by a merger in 1995 and was renamed to DigitalGlobe in 2002. DigitalGlobe operates QuickBird, WorldView-1 and WorldView-2. Orbimage acquired Space Imaging in January 2006. The name of the newly formed company is GeoEye and it flies the high-resolution Ikonos, Orbview-3 and Orbview-2 satellites and GeoEye-1. GeoEye-2 is scheduled for launch in spring 2013.

References

Alobeid A, Jacobsen K, Heipke C (2009) Building height estimation in urban areas from very high resolution satellite stereo images. ISPRS Hannover Workshop 2009 High-Resolution Earth Imaging for Geospatial Information. Hannover

Amhar F, Mulyana AK (2009) Mapping and map updating with ALOS data – the Indonesia experience. ISPRS Hannover Workshop 2009 High-Resolution Earth Imaging for Geospatial Information. Hannover

Baghdadi N, Boyer N, Todoroff P, El Hajj M, Bégué A (2009) Potential of SAR sensors TerraSAR-X, ASAR/ENVISAT and PALSAR/ALOS for monitoring sugarcane crops on Reunion Island. Remote Sens Environ 113(8):1724–1738

Blaschke T (2010) Object based image analysis for remote sensing. ISPRS J Photogramm Remote Sens 65(1):2–16

Bouziani M, Goïta K, He D-Ch (2010) Automatic change detection of buildings in urban environment from very high spatial resolution images using existing geodatabase and prior knowledge. ISPRS J Photogramm Remote Sens 65(1):143–153

Coppin PR, Bauer ME (1996) Change detection in forest ecosystems with remote sensing digital imagery. Remote Sens Rev 13:207–234

Curran P (1993) Principles of remote sensing. Wiley, New York, NY

Dagar P (2010) Environmental sensitivity and suitability: country-level planning for Rwanda. GIM Int 24(4):14–17

Dianat R, Kasaei S (2009) Change detection in remote sensing images using modified polynomial regression and spatial multivariate alteration detection. J Appl Remote Sens 3(1):1–12

Fraser C, Tshering D, Gruen A (2008a) High-resolution satellite imagery for spatial information generation in Bhutan. Int Arch Photogramm Remote Sens Spat Inf Sci XXXVII(Part B6a): 197–200

Fraser C, Tshering D, Gruen A (2008b) Satellite mapping in Bhutan: high-resolution imagery in generating spatial information. GIM Int 22(5):18–21

Gorte B (1998) Probabilistic segmentation of remotely sensed images. PhD thesis, Wageningen Agricultural University, ITC Enschede, The Netherlands, ITC Publication 63. ISBN: 90-6164-157-8

Hanssen RF (2001) Radar interferometry: data interpretation and error analysis. Kluwer, Dordrecht, The Netherlands

Haralick RM, Shanmugam K, Dinstein I (1973) Textural features for image classification. IEEE Trans Syst Man Cybern SMC-3(6):610–621

Hornstra TJ, Lemmens MJPM, Wright GL (1999) Incorporating intra-pixel reflectance variability in the multispectral classification process of high-resolution satellite imagery of urbanised areas. Cartography 28(2):1–9

Huang W, Liu H, Luan Q, Jiang Q, Liu J, Liu H (2008) Detection and prediction of land use change in Beijing based on remote sensing and GIS. Int Arch Photogramm Remote Sens Spat Inf Sci XXXVII(Part B6b):75–82

Jalan S, Sokhi BS (2010) Mapping urban fringes, object based image analysis a promising technique. GIM Int 24(6):13–16

Jensen JR (2005) Introductory digital image processing, 3rd edn. Prentice Hall, Upper Saddle River, NJ

Jensen JR (2007) Remote sensing of the environment: an earth resource perspective, 2nd edn. Prentice Hall, Upper Saddle River, NJ

Krishnaswamy M (2002) Sensors and platforms for high resolution imaging for large scale mapping applications – Indian Scenario. Indian Cartogra DAPI-01:1–6

Lemmens M (2006a) First images from Japan's Daichi: advanced land observing satellite. GIM Int 20(5):39–41

Lemmens M (2006b) High-resolution satellite imagery. GIM Int 20(4):36–39

Lemmens M (2007) Earth Observation Business. GIM Int 21(7):11

Lemmens M (2008) High-resolution Satellite Imagery. GIM Int 22(7):18–21

Lemmens M (2010a) Georeferenced data products. GIM Int 24(4):23–25

Lemmens M (2010b) Remote sensing image processing software. GIM Int 24(6):38–39

Lemmens M (2010c) OBIA. GIM Int 24(6):57

Lemmens MJPM (1996a) A survey on boundary delineation methods. Int Arch Photogramm Remote Sens XXXI(Part B3):435–441

Lemmens MJPM (1996b) Structure based edge detection: delineation of boundaries in aerial and space images, Ph.D. thesis, Delft University of Technology, ISBN 90-407-1366-9

Lemmens MJPM, Verheij KF (1988) Automatically GIS updating from classified satellite images using GIS knowledge. Int Arch Photogramm Remote Sens XXVII(Part B4):198–206

Lemmens MJPM, Wicherson RJ (1992) Edge-based region growing. Int Arch Photogramm Remote Sens 29(Part B3):793–801

Lemmens MJPM, Meijers PJM, Verwaal RG (1990) GIS-guided linear feature extraction. Int Arch Photogramm Remote Sens 28(Part B3):395–404

Lillesand ThM, Kiefer RW, Chipman JW (2004) Remote sensing and image interpretation, 5th edn. Wiley, New York, NY

Lunetta RS, Elvidge CD (eds) (1999) Remote sensing change detection: environmental monitoring methods and applications. Taylor & Francis, London. ISBN 0-7484-0861-4

Murakami H, Nakagawa K, Hasegawa H, Shibata T, Iwanami E (1999) Change detection of buildings using an airborne laser scanner. ISPRS J Photogramm Remote Sens 54(2–3): 148–152

Radhadevi PV, Nagasubramanian V, Mahapatra A, Solanki SS, Sumanth K, Varadan G (2009) Potential of high-resolution Indian remote sensing satellite imagery for large scale mapping. ISPRS Hannover Workshop 2009, High-Resolution Earth Imaging for Geospatial Information. Hannover

Rawashdeh SA (2006) Spatial data quality degradation: real-life examples from jordan. GIM Int 20(11):13–15

Sandau R, Briess K, D'Errico M (2010) Small satellites for global coverage: potential and limits. ISPRS J Photogramm Remote Sens 65(6):492–504

Schowengerdt RA (1997) Remote sensing: models and methods for image processing, 2nd edn. Academic, San Diego, CA

Singh A (1989) Digital change detection techniques using remotely sensed data. Int J Remote Sens 10(6):989–1003

Topan H, Jacobson K (2006) Mapping with OrbView-3 images: information content of high-resolution satellite image. GIM Int 20(12):14–17

Topan H, Oruc M, Jacobsen K (2009) Potential of manual and automatic feature extraction from high resolution space images in mountainous urban areas. ISPRS Hannover Workshop 2009 High-Resolution Earth Imaging for Geospatial Information. Hannover

Vogelmann JE, Tolk B, Zhu Z (2009) Monitoring forest changes in the southwestern United States using multitemporal Landsat data. Remote Sens Environ 113(8):1739–1748

Volpe F, Rossi L (2005) Mapping towns from QuickBird imagery: sub-metre resolution and high positioning accuracy. GIM Int 19(5):13–15

Walter V (2004) Object-based classification of remote sensing data for change detection. ISPRS J Photogramm Remote Sens 58(3–4):225–238

Weber M, Koudogbo FN (2009) TerraSAR-X – 1m spaceborne radar: use, features, products and upcoming Tandem-X mission. GIM Int 19(5):12–15

Yue H, Hanssen R, van Leijen F, Marinkovic P, Ketelaar G (2005) Land subsidence monitoring in city area by time series interferometric SAR data. Geoscience and Remote Sensing Symposium, Proceedings, IGARSS '05, IEEE International, Seoul Korea, pp 4590–4592

Zhou W, Huang G, Troy A, Cadenasso ML (2009) Object-based land cover classification of shaded areas in high spatial resolution imagery of urban areas: a comparison study. Remote Sens Environ 113(8):1769–1777

Chapter 10
Modelling and Exchanging Geo-information

All sorts of businesses wishing to sell products and services via user-friendly web pages are pushing rapid advances in web development technology. As a result, since the turn of the millennium this has become one of the fastest growing industries globally and important offspring of the worldwide desire to do business via the internet. This chapter starts with treating Unified Modelling Language (UML), a process and tool-independent modelling syntax for building software systems, but also usable for modelling processes involving geo-information such as transferring real estate from seller to buyer. Next we consider Open Source Software and languages to disseminate geo-data over the web.

10.1 Unified Modelling Language

Science and Engineering is all about developing solutions by reducing complex reality to tractable bits and pieces. At the heart of problem solving lies modelling; the essential aspects of the domain are isolated from the rest of the complicated real world. In course of time many tools have been developed to assist the researcher, developer and designer in modelling real-world processes. How to warrant that during design, development, implementation and maintenance of automated processes, business analysts, designers and programmers all speak a common vocabulary and to avoid Babel, a confusion of tongues? In the past decade UML (Unified Modelling Language) has emerged to act as such a universal language. The basics of UML and how to use it have been documented in Booch et al. (2005). A brief guide to UML is given by Fowler and Scott (2000).

UML has been accepted as an industry standard by the Object Management Group (OMG), an international, open membership, not-for-profit computer industry consortium in 1997 and it is OMG's most used specification. OMG defines UML as (www.uml.org):

> UML is a graphical language for visualizing, specifying, constructing, and documenting the artefacts of a software-intensive system. The UML offers a standard way to write a system's blueprints, including conceptual things such as business processes and system functions as well as concrete things such as programming language statements, database schemas, and reusable software components.

M. Lemmens, *Geo-information*, Geotechnologies and the Environment 5,
DOI 10.1007/978-94-007-1667-4_10, © Springer Science+Business Media B.V. 2011

The above definition implies that UML has been developed first and foremost as a process and tool-independent modelling syntax for building software systems. However, UML can be very well used for various other application areas. Within the framework of geo-information technology UML has been extensively employed to model storage and dissemination of geo-information. Aydinoglu et al. (2009) developed a geographic data sharing model for Turkey, which is fully described with UML class diagrams. FIG's Land Administration Domain Model (LADM) originally developed by Lemmen and van Oosterom (2002) uses UML as a model for building land administration systems, i.e. systems concerning (property) rights on real estate and land parcels, and for developing a basic land administration terminology, which is accepted and employed worldwide. Land Administration is a subject comprehensively treated in Chapter 15. The LADM distinguishes five basic components:

1. Persons and organisations, called parties
2. Rights, restrictions and responsibilities, called RRR
3. Parcels, buildings an construction works, called spatial units
4. Surveying
5. Geometry and mapping

Since its inception the model has been further developed and documented by numerous papers co-authored by the founding fathers Lemmen and van Oosterom, see, e.g. Lemmen and van Oosterom, 2003, 2006; Hespanha et al., 2008; Lemmen et al., 2009, 2010a, b. In November 2009 the 29th Plenary Meeting of ISO TC 211 held in Quebec City decided to forward the ISO 19152 Land Administration Domain Model as a Draft International Standard (DIS). ISO is the International Organization for Standardization joining 161 national standards institutes (www.iso.org). Standards are developed by Technical Committees (TC) and the average time to develop an ISO standard is nearly 3 years. TC 211 develops standards in the field of geographic information and geomatics (Kresse and Fadaie, 2004). Once ISO has accepted LADM as a standard (ISO 19515), which is expected to be decided in course of the year 2011, LADM will strengthen the relationship between Cadastre and other public registers and will stimulate the standardised registration of cadastral information worldwide (Uitermark, 2010).

The aim of this chapter is not to scrutinise LADM, for an in-depth explanation we refer to the extensive literature, but to provide a brief introduction to UML and we do this by using the land administration domain as an example. When one wants to (partially) convert a process (dynamic system) taking place in the land administration domain – such as registration of a transfer act within a land administration system (LAS) – into a software system, the process may be approached from several directions (Lemmens, 2005):

– What is the user's view on the process?
– What are the objects in the system and how do they interrelate? (Static system view)

- Who are the actors and what activities do they perform?
- At what point in time do actors come into action, in what sequence and for how long?
- How do the actors collaborate?
- Which software components are involved in running the process on a hardware system?

UML has been developed to describe the different views on a domain process in graphical notations in the form of diagrams. Nine modelling diagrams are distinguished:

1. Use case diagram
2. Class (package) diagram
3. Object diagram
4. Sequence diagram
5. Collaboration diagram
6. Statechart diagram
7. Activity diagram
8. Component diagram
9. Deployment diagram

In the sequel the above nine modelling diagrams will be further detailed, and most will be exemplified by diagrams created by Šumrada (2005), which describe the process of transferring a spatial unit (parcel) from seller to buyer in the context of land administration.

Use cases aim at obtaining system requirements from a user's view. The *use case diagram* is a collection of use cases, users of the system (actors) and their messages (Fig. 10.1). Use cases are represented by ovals, actors by stick figures and communications by lines that link actors to use cases. The *class diagram* describes the static structure of a system by displaying classes and the relationships among them. It is the backbone of UML and is essentially the communication language for designers. Classes are represented by rectangles divided into three parts: class name, attributes and operations (Fig. 10.2).

Lines represent relations between classes. Related classes can be grouped into packages so that overview is kept. A *Package diagram* organises elements of a system into related groups to minimise dependencies between packages. Packages are represented as folders (rectangles with small tabs at the top). The package name is on the tab or inside the rectangle while arrows represent dependencies (Fig. 10.3). Objects are the basic elements of a system; they interact by sending each other messages. *Object diagrams* describe the static structure of a system at a certain instant. Class and object diagrams describe the static part of the process; interaction diagrams are dynamic: they describe how objects work together over time.

A *sequence diagram* (Fig. 10.4) is an interaction diagram, which specifies which messages are sent when. These are arranged according to time: time progresses downward along the vertical axis. The objects involved are listed in rectangles along

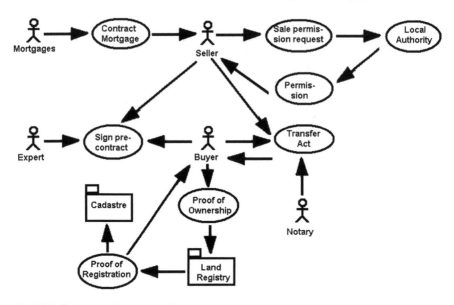

Fig. 10.1 Fragment of a use case diagram

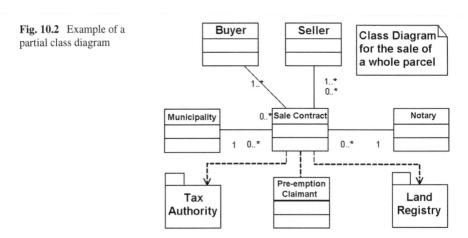

Fig. 10.2 Example of a partial class diagram

the horizontal axis at the top, from left to right according to their place in the message sequence. The lifeline is the time that an object exists and is represented by a vertical dotted line. Arrows represent messages. The length of the activation bar represents the duration of execution of the messages. In Figure 10.4, seller, buyer, sale contract and notary are all classes, which start sending messages at the same time, but the first message sent by the buyer takes longer than the first message of the others. The object rectangles are labelled either by class name, object name or both; class names are preceded by colons (:).

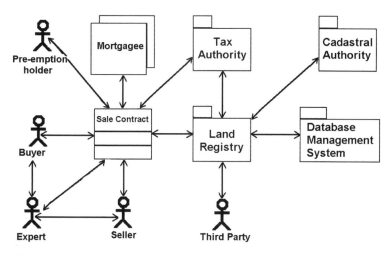

Fig. 10.3 Package diagram

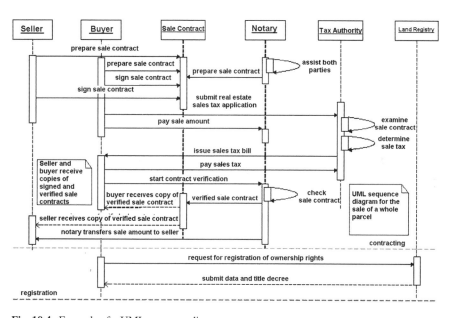

Fig. 10.4 Example of a UML sequence diagram

Collaboration diagrams are also interaction diagrams: they communicate the same information as sequence diagrams; however, the focus is now on roles of objects rather then on the dispatch times of messages. Since time is not represented, the messages are numbered to denote sending order.

Objects may be in different states at differing times, the state depending on current activity or condition. A *statechart diagram* shows the possible states and

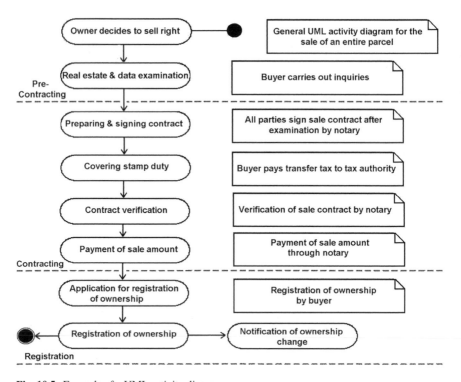

Fig. 10.5 Example of a UML activity diagram

the transactions that cause a change in state. States are represented by rounded rectangles while transactions are arrows connecting state. Events or conditions that trigger transitions are written beside the arrows. While a statechart diagram describes an object undergoing a process, an *activity diagram* focuses on the flow of activities involved in a single process (Fig. 10.5). The activity diagram, basically a general-purpose flowchart, shows how these activities depend upon one another.

Component diagrams are the software analogues of class diagram; they show the types of software components, their interfaces and dependencies. *Deployment diagrams* represent the physical configurations of software and hardware, including nodes, links and dependencies. Sequence diagrams enable modelling of time within the process, while activity diagrams show actions carried out by objects and components. Just as the 'use case diagram' comes closest to the real-world process, component and deployment diagrams are nearest to the software system: component diagrams show the types of software elements, their interfaces and dependencies, while deployment diagrams represent the physical configurations of software and hardware, including nodes, links and dependencies. These diagrams are mainly of interest to the IT tough guys – software programmers and hardware assemblers who keep the system running, possibly Web, but at least server-based.

A UML user will not commonly employ the entire set of above diagrams but will make a selection according to needs and the role he plays in the modelling train moving from real-world process to software system. For example, a designer may restrict himself mainly to the use of class diagrams because these form a pivotal link between the real-world domain process and its transformation into a software system running on hardware. The analyst, concerned with understanding what is actually happening within the domain process, will generally prefer modelling the problem along the lines, ovals and stick figures used in case diagrams.

Within the domain of geo-information technology UML is increasingly gaining momentum because it supports the interoperability of geo-information. This is important for dissemination of geo-information within the framework of establishing a National Geo-information Infrastructure. Today most standards such OGC, ISO TC211 and CEN are even written in UML. So the system developer who represents his design in accordance with UML syntax is developing along standardised lines, which is in itself a sound selling point.

10.2 Open Source Software

In the geomatics domain, open-source software is finding its way into more and more products and applications (Lemmens, 2007). For example, The Institut Geographique National (IGN) uses open source software to store the BDUni topographic base-map for France. Open-source provides flexible analytical tools of the sort that used to be done exclusively with Arc/Info and other workstation tools, allowing overlay, spatial joining and spatial summaries large and small. How did open-source become so popular, also in the geomatics domain? This section provides an overview of the developments of open source software, particularly focussed on geo-information technology based on a variety of sources including Lerner et al. (2006), Letellier (2008), Opensourcegis.org, Osterloh and Rota (2007), Rossi (2006), Schiff (2002), Schwarz and Takhteyev (2010), and en.wikipedia.org.

As the internet bubble approached bursting point at the start of the present century, voices arose against manufacturers who entrenched their software products within monopolistic bastions; in particular the Microsoft (quasi) monopoly became the target of verbal attack and lawsuits. At the turn of the millennium Microsoft controlled over 80% of the market in operating systems and 90% of the business-applications market. The company denied anti-competitiveness, arguing that the software market had benefited greatly from its efforts leading to standardisation and relatively low-cost software for the millions. Government institutions have been put in place to inhibit monopolisation and resistance also emanates from the user, particularly when it dawns on him that there is only one game in town and the manufacturer has taken him hostage. As a result, the general public revealed a broad interest in open-source software. Press coverage of the phenomenon was dominated by bewilderment at the emergence of such a thing as 'freely distributed' software, and software engineer Linus Torvalds, originator of Linux, became an icon. The

Linux operating system, emerging around since 1991, was an unparalleled success thanks to its openness. Users, themselves developers, fix bugs and adapt the system to new innovations. Improvements are made public and rapidly assimilated into the next official release of Linux Kernel. Non-technical users benefit by participating in the many internet discussion forums where answers may be got to questions on use and possible malfunctioning.

In the realm of geo-information technology free software was already available way back in the 1980s. A free GIS software package developed around 1980 and still used today is the Geographic Resources Analysis Support System, better known as GRASS (grass.fbk.eu). At that time there was no concept of free and open source software, but GRASS developed by the US Army, demonstrated that spatial analysis is first and foremost a free enterprise (Pradeepkumar and Radhakrishnan, 2008). Today this free GIS system is not only used by academics and scientists but also by many governmental agencies all over the world as well as commercial companies including environmental consultancy. The applications include geospatial data management and analysis, image processing, graphics/maps production, spatial modelling, and visualisation. GRASS is an official project of the Open Source Geospatial Foundation.

Most commercial software is made available in executable code consisting of ones and zeroes, the only language computers understand. By contrast, open-source programs always comprise the source code from which the executable code is compiled. Written in C++, Java, and the like, the code is accessible and comprehensible to any programmer. And the users have the legal right to alter and redistribute it. They are empowered to improve the software in all kinds of ways themselves, and not a vendor. 'All users may not make use of their freedoms under open-source, but all have the option', Paul Ramsey, Director, Open Source Geospatial Foundation (OSGeo) communicated to us (Lemmens, 2008c). Although the press, around 2000, reported on open-source software as if it was something astoundingly new, free exchange of source codes is as old as the place of computers as commodities in universities and research institutes. Indeed, from the very outset it was the policy of both academic institutions and commercial research centres that programmers unrestrictedly distribute and share the products of their intellectual efforts. Just as scientists publish their research results, software was considered a research product free for everybody, and by giving it away the scientific programming community hoped that others would use and improve on it. These researchers even considered the free distribution of software a prerequisite for cumulative furthering. Treating software as a tradable asset was far from anyone's mind and the rapid sharing of technologies led to huge progress, the sweet fruits of which we are all enjoying today.

After 25 years of proprietary software giving away software for free is again increasing in popularity. Why should somebody freely give away that for which he has laboured long and hard? 'Free' is here not synonymous with 'gratis' or 'without obligation'. Open-source software is distributed within the framework of conventional copyright law, in which ownership is asserted and exercised. Use requires a licence providing a number of permits and constraints. The licence may come gratis or be paid for by fee; 'free' here means freedom of use, not price. Why give software away? For the individual developer there is the excitement of mind-share: working with people around the globe who appreciate his or her software. Throwing software into the open-source ring also brings with it the possibility of fame, and maybe even fortune: a well-paid job in Silicon Valley. There is also the ideological motive conveyed by the free-software movement active since 1984 and driven by Richard Stallman, instigator of the GNU project. In proclaiming the essential contribution made to the development of computer science by freely available code, one is also recognising the need for measures to prevent pirates from making a profitable business from other people's efforts. The answer to this threat was the GNU General Public License (GPL), which allows anyone at will to copy and distribute software licensed under the GPL, provided that they do not restrain others from doing the same, either by charging or restricting through further licensing. The GPL also requires works derived from work licensed under the GPL too to be licensed under it.

Open-source software enables small, innovative enterprises to enter established markets by introducing types of service the core of which no longer consist of the software itself but of a bundle of services packaged such that the final result interests the consumer. And part of that bundle may be in-house-developed software, thrown into the open-source ring so that others may not only use but also improve on it. A main incentive for placing stuff in the ring is the conviction that open-source software has greater integrity and security than closed, proprietary products because many more eyes are scrutinising the code. Large corporations have much to lose from the shift in interest from proprietary to open-source software. What is the secret of the success of open-source? Paul Ramsey explains the success as follows (see Lemmens, 2008c):

> The success of open-source has been built on a combination of enthusiasm and economics. The enthusiasm comes from the technology geeks who are asked to build systems; they like to use software components that are transparent, so that they can talk easily to its authors, get defect fixed very quickly, rather than working around it. And open-source software provides all these things. The economics comes from the zero capital cost aspect of open-source. Of course, open-source isn't zero cost, because it takes staff time to integrate and manage it, but it *is* zero capital cost. It also incurs zero for things like managing licensing and royalties. Anyone who has built a system on top of a proprietary library knows the pain of getting to the rollout stage and trying to figure out how many licenses they are going to need. Four CPUs, or eight? When a CPU licence costs $20,000, and the penalty for contravening licence terms is a lawsuit, these are not trivial questions. For new organisations in particular, or for those building a new system, the low start-up cost and lack of licensing restrictions are particularly tempting.

Together with his team Paul Ramsey developed PostGIS, an open-source spatial extender for the PostgreSQL relational database, which is also open source. PostGIS has gained much interest and credit from researchers and companies, big and small, all over the world and is being used by organisations requiring a consistent, intelligent and reliable database for spatial objects.

10.3 Exchanging Geo-data over the Web

Vast amounts of digital geo-data for use in a variety of application domains, such as urban and rural planning and disaster fighting, have been created in the past 30 years. The formats in which this data is stored are plenty and are most proprietary to the various data producers and GIS vendors. As a result GIS users found it very difficult to share data unless they used the same software. In the early 1990s people began to worry about the waste of resources arising from so much incompatibility. Interoperability became the recipe, resulting in 1994 in the founding of the Open GIS Consortium (OGC), a decade later renamed the Open Geospatial Consortium. This platform brings together geospatial software providers, integrators, government agencies and academic organisations to facilitate a consensus process for defining, testing and approving standards that promote the integration and use of geospatial data and services. The US Department of Agriculture and the US Army Corps of Engineers were founding members. The US Geological Survey joined early on, and national and regional mapping agencies around the world have joined the 'major user' stakeholder community. Standards are an integral part of most of today's enterprise information systems and mapping agencies contribute to the standards effort because they wish to influence the process by which standards are made so as to meet their own requirements. In October 2007 Microsoft joined the organisation as a Principal Member. Also other major companies, including Google and Oracle, as well as major national communities, including the European Union, India and China participate in OGC. Since the internet and the World Wide Web have rapidly become increasingly important in the field of geo-information technology most OGC standards have been developed for the web environment.

OGC primary concern is thus the worldwide implementation of standards. What are the benefits of adopting standards for cadastres and National Mapping Agencies? David Shell, former Chairman and CEO, Open Geospatial Consortium communicated to us (Lemmens, 2008a, b):

> National Mapping Agencies ask for standards-compliant products in their procurements because they want to get the most out of the investments they have made in their legacy systems and out of new investments they are making in geospatial technologies. It is a question of "future-proofing" their systems and not wanting to be bound to a single vendor. They see the value of modular systems in which different capabilities can be provided by a variety of vendors possessing diverse and frequently special expertise. For these reasons standards are an integral part of most of today's enterprise information systems.

Implementing standards in products is expensive. Why should companies manufacturing hardware, software and services adopt standards if implementing them is expensive and it is not clear how to evaluate resulting benefits relative to immediate business requirements? The reply of David Shell to this question reads:

> The answer can be found by looking at vendors who have very recently implemented OGC standards in their products. They are already benefiting and have found that the costs associated with developing standards and then implementing them in their products are very similar to their usual product development costs. And they are by definition accustomed

to investing in this way. Our vendor members may have had to wait a year or two until demand for standards-compliant products became a significant market driver, but our process has been fast enough to ensure that in reality the impact on product life-cycles has not been too significant for them. Now it seems apparent that we are past the tipping point in this regard, and users generally find their vendors can deliver the standardised products they require.

The remaining part of this section treats several languages for encoding and exchanging geo-information, more specifically Geography Markup Language (GML), Keyhole Markup Language (KML), Geographic Data Files (GDF) and AJAX.

10.3.1 GML

Geography Markup Language (GML) is an open software interface for geo-information specified by the OGC. Today GML, which has been co-operatively adopted by the main GIS vendors, plays a key role in both geo-information encoding and dissemination and in the description of geographic objects for geo-information web services. GML is a geo-information exchange language based on XML (eXtensible Markup Language), the standard language of the internet. GML is not tied to any proprietary GIS or database software but is specifically designed for feature-based geo-information. It is an open standard, which anyone can use and which has been designed to support interoperability. OGC design goals include more specifically (www.iso.org; Quak and Lemmens, 2004)

- providing a means of encoding spatial information for both data dissemination and data storage, especially in a wide-area internet context
- being sufficiently extensible to support a wide variety of spatial tasks, from portrayal to analysis
- establishing the foundation for internet GIS in an incremental and modular fashion
- allowing for efficient encoding, such as data compression of geospatial geometry
- providing easy-to-understand encoding of spatial information and spatial relationships, including those defined by the OGC Simple Features model
- enabling separation of spatial and non-spatial content from data presentation (graphics or otherwise)
- permitting easy integration of spatial and non-spatial data, especially for cases in which the non-spatial data are XML-encoded
- enabling ready linkage of spatial geometric elements to other spatial or non-spatial elements
- providing a set of common geographic modelling objects to enable interoperability of independently developed applications.

To make a map from the actual geo-data, this needs to be converted into symbols, colours line and area styles: for short, a Digital Cartographic Model (DCM). Therefore, in addition to the actual data a GML consists of a second type of document that provides information how this data are structured: schema definition. This tells the computer how to read and interpret the data, basically referring to the DELM. In 2007 GML became an ISO standard: ISO 19136: 2007.

10.3.2 KML

Keyhole Markup Language (KML) is a file format based on XML and used to display geo-data. Originally developed by Google Inc and thus, unsurprisingly, employed by Google Earth and Google Maps, KML 2.2 has since 14 April 2008 been a worldwide industry standard for geobrowsers; further developments are being co-ordinated by OGC. The reference system used by KML is based on geographic coordinates (latitude and longitude) in WGS84 (World Geodetic System, 1984), which is the only reference system supported. KML makes it possible to put features, images, 3D models, text information, and so on in geo-browsers, and view the scene by virtual camera. The view given by the camera can be steered by moving it along x, y and z (height) axes, and the viewing direction by rotating along these axes: heading, tilt and roll.

10.3.3 GDF

Geographic Data Files (GDF) is a file format for exchanging geographic files. It originally came into existence as a European standard developed by the European Committee for Standardisation (CEN) in co-operation with digital map providers, automotive and electronic-equipment manufacturers. The file format makes it possible to describe and transfer information about roads and related data. The CEN GDF 3.0 has been the main input for the world standard ISO GDF 4.0. GDF is a flat plain-text file, and this impedes its use for large-scale geo application, inducing the need for conversion into a more efficient format. GDF is predominantly used for (car) navigation systems. Map vendors such as Navteq and Tele Atlas, who were in on its dawning, are now providing maps in GDF. The format is also applied in fleet management, dispatch management, road-traffic analysis, traffic management and automatic vehicle location.

10.3.4 AJAX

AJAX stands for Asynchronous Javascript And XML, a group of techniques for web development. Introduced in 2005, its primary aim is to increase functionality, speed and usability of interactive web applications, and thus improve user friendliness with an eye to boosting web use. There is no transmission of entire pages from server to client in response to user request, only parts are exchanged. 'Asynchronous' means that additional data are loaded at the client side and kept in reserve in anticipation of coming requests. This feature tremendously improves responsive efficiency and the user does only notice the presence of this feature by fast responses. This is because the data are loaded in background, without interfering with the data the user sees on display. The language in which function calls are made is Javascript. XML (Extensible Markup Language), the last function abbreviated in AJAX is used to encode documents and is a language which facilitates sharing of structured data across the internet.

As so often, developing and implementing standards such as GML is a necessary but not sufficient condition for exchanging map products over the internet. Indeed, standards and technological capabilities form the fundament. But it will be the awareness of decision makers and their commitment to incorporating new technology in existing work processes that will make it all work.

References

Aydinoglu AC, Yomralioglu T, Quak W, Dilo A (2009) Modeling emergency management data by UML as an extension of geographic data sharing model: ASAT approach. Proceedings of TIEMS 2009 annual conference, Istanbul, pp 87–98

Booch G, Rumbaugh J, Jacobson I (2005) The Unified Modelling Language user guide, 2nd edn. Addison Wesley, Reading, MA, USA

Fowler M, Scott K (2000) UML Distilled: a brief guide to the standard object modeling language, 2nd edn. Addison-Wesley, Reading, MA, USA

Hespanha J, van Bennekom-Minnema J, van Oosterom P, Lemmen C (2008) The model driven architecture approach applied to the Land Administration Domain Model version 1.1 with focus on constraints specified in the Object Constraint Language. FIG Working Week, Stockholm, Sweden, 19 p, 14–19 Jun 2008

Kresse W, Fadaie K (2004) ISO Standards for geographic information. Springer, Berlin, Heidelberg, New York. ISBN 3-540-20130-0

Lemmen C, van Oosterom P (2002) Towards a standard for the cadastral domain: proposal to establish a Core Cadastral Data Model, 3rd International Workshop 'Towards a Cadastral Core Domain Model' of COST action G9 'Modelling Real Property Transactions', Delft, October, 18 p

Lemmen C, van Oosterom P (2003) Further progress in the development of a core cadastral domain model. FIG Working Week, Paris, April 2003

Lemmen CHJ, van Oosterom PJM (2006) Distributed cadastral systems: FIG core cadastral domain model version 1.0. GIM Int 20(11):43–47

Lemmen C, van Oosterom P, Uitermark H, Thompson R, Hespanha JP (2009) Transforming the Land Administration Domain Model (LADM) into an ISO Standard (ISO19152). Proceedings of the FIG working week: surveyors key role in accelerated development, Eilat, 24 p

Lemmen C, van Oosterom P, Thompson R, Hespanha JP, Uitermark H (2010a) The modelling of spatial units (parcels) in the Land Administration Domain Model (LADM). Proceedings of the XXIV FIG international congress 2010, Sydney, 28 p, April 2010

Lemmen C, van Oosterom P, Eisenhut C, Uitermark H (2010b) The modelling of rights, restrictions and responsibilities (RRR) in the Land Administration Domain Model (LADM). Proceedings of the XXIV FIG International Congress 2010, Sydney, 40 p, April 2010

Lemmens M (2005) Unified modelling language. GIM Int 19(10):11

Lemmens M (2007) Open source software. GIM Int 21(5):13

Lemmens M (2008a) A new age (I). GIM Int 22(1):7–9

Lemmens M (2008b) A new age (II). GIM Int 22(2):7–9

Lemmens M (2008c) Open source success. GIM Int 22(10):8–9

Lerner J, Pathak PA, Tirole J (2006) The dynamics of open source contributors. Am Econ Rev 96(2):114–118

Letellier F (2008) Open source software: the role of nonprofits in federating business and innovation ecosystems. http://flet.netcipia.net/xwiki/bin/download/Main/publications-fr/GEM2008-FLetellier-SubmittedPaper.pdf

Osterloh M, Rota S (2007) Open source software development: just another case of collective invention? Res Policy 36(2):157–171

Pradeepkumar AP, Radhakrishnan T (2008) FOSS GIS: the future of GIS, Free Software, Free Knowledge. Free humanity: national conference on free software, CUSAT, Cochin, India, 15–16 Nov 2008

Quak W, Lemmens M (2004) Exchanging geo-data on the web: geography Markup Language: a 'de facto' standard. GIM Int 18(12):35–37

Rossi MA (2006) Decoding the free/open source software puzzle: a survey of theoretical and empirical contributions. In: Bitzer J, Schröder P (eds) The economics of open source software development, Elsevier, Amsterdam, The Netherlands, pp 15–55. ISBN 978-0-444-52769-1

Schiff A (2002) The Economics of open source software: a survey of the early literature. Rev Netw Econ 1(1):66–74

Schwarz M, Takhteyev Y (2010) Half a century of public software institutions: open source as a solution to the hold-up problem. J Public Econ Theory 12(4):609–639

Šumrada R (2005) UML in use case modelling: property transactions in Slovenia. GIM Int 19(10)

Uitermark H (2010) The ISO/TC 211 Land Administration Domain Model, Conference on Cadastre, Public Policies and Economic Activity, Madrid, 1–2 June 2010

Chapter 11
Quality of Geo-information

Ease of use of many geo-data collection devices is high and anyone who can handle a mobile phone can push the right buttons on a mobile GIS device within a few hours of training. This is a consequence of the shift from labour-intensive analogue methods to digital geo-data collection and processing technology. After being collected much geo-information can today be processed in a cheap way by virtually anybody who buys a computer and a GIS package and who is sufficiently skilled to read manuals. However, the processing steps can not be performed blindly. The collection and processing of geo-information involves a chain of activities and not only the data collection itself or information extraction is central but also quality assurance. Measuring and information extraction can be done by anybody, but to arrange the measuring and the processing of the data in such procedures that errors and inaccuracies are avoided with the minimum effort in terms of labour, time and money is a key activity. Pushing the buttons of geo-information technology tools alone cannot produce high-standard geo-information. It is professionalism that brings in quality. Quality assurance has always been high on the agenda of surveyors and geomatics specialists. This chapter treats the basic quality issues in collecting and processing geo-data.

11.1 Basics

The data used in any geo-related discipline stem – directly or indirectly – from measurements captured at some moment in time. All measurements contain errors; this tenet is an empirical reality. In other words, no measurement is exact, every measurement contains errors, the true value of a measurement is never known and the exact sizes of the errors present are always unknown (Wolf and Ghilani, 1997). Accordingly, quality of geo-information is an error issue, concerned with minimising the size and effects of errors, or, even better, to avoid their presence. The presence of measurement errors can be described in terms of accuracy and precision. The general public uses the words accuracy and precision interchangeably, but within the field of geomatics these terms have specific, distinct meanings. Accuracy describes how good the measurement resembles a quantity which is called the 'true'

M. Lemmens, *Geo-information*, Geotechnologies and the Environment 5,
DOI 10.1007/978-94-007-1667-4_11, © Springer Science+Business Media B.V. 2011

Fig. 11.1 Accuracy versus precision

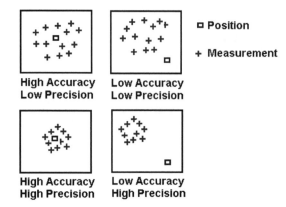

value. Although the true value will never be known, the term is used as an abstraction to identify in a conceptual way what the actual value of the measurement should be. How nearer a measurement is to the true value the more accurate the measurement is said to be. Precision is a measure of the variability in the measurements of a certain quantify, the nearer the values of a sequence of repeated measurements are to each other the more precise the measurement. Figure 11.1 schematically demonstrates the difference between accuracy and precision.

Quality of geo-information is a subject which has been approached from both the surveying domain and from the GIS domain. From the surveyor's perspective the gravity of quality is on the measurements, while for the GIS user this will be the quality of the datasets he or she is using for solving an Earth-related problem or for supporting decision making.

11.2 Surveyor's Perspective

Surveyors will emphasise subjects like the following: did I make sure that all blunders and systematic errors have been eliminated? What is the accuracy of the coordinates or other derived quantities in terms of standard deviation (σ; sigma) or Root Mean Square Error (RMSE)? Surveyors will confirm the statement, 'It is almost as important to know the accuracy of a result as to know the result itself' (Allan, 2007). The mathematics a surveyor uses to calculate coordinates from measurements are taught at length in secondary school and most people will understand the mathematical basics used by the surveyor. However, surveying is not a paper exercise. It is not just a matter of multiplying a given length with a given width to calculate a rectangular-shaped area. The length and the width are not givens in the surveying exercise; they have to be measured and the measurements are performed in a physical environment, either on-site where conditions can be harsh sometimes or in the office using, for example, Digital Photogrammetric Workstations. A surveyor's mind is predetermined to prevent occurrence of measurement errors and once they occur to cope with them. And they will occur; making errors is unavoidable. The surveyor knows that his measurements will be prone to error, not because he would be careless but because, as stated earlier, making measurement errors is an experimental reality.

Allan (2007) provides a simple but insightful classroom example on measurement errors. Using a 100-mm ruler each student in the classroom has to draw a straight line 300 mm in length. What could go wrong? A lot! The length of lines drawn by the students varied and this could be verified by eye-judgement alone. Measuring the lines in centimetres provided the same value for all students all the time, but using a ruler with millimetre divisions resulted in varying outcomes. Some rulers provided systematically lower outcomes than others. One student had miscounted and gave the length as 400 mm. The exact or 'true' value should be 300 mm, but only a few students really obtained this value from their measurements. This example shows all the particularities a surveyor is confronted with in a nutshell.

All students had received the assignment to draw a 300-mm straight line resulting in a lot of redundancy. Would drawing just one line not be enough? Redundancy enables independent checking and thus improving reliability. The various outcomes when using a ruler with a millimetre division enables high precision but shows that taking high-precision measurements never gives identical numerical values, the measurements will be noisy. Repeated measurements will slightly deviate because of random measurement errors. Some rulers systematically provided lower values than others. This demonstrates that measuring equipment needs *calibration* at regular times, for example, annually. Calibration means comparison with a known length which acts as a standard. The standard length should have been measured with a device of which the precision is at least ten times better than the device to be calibrated. Finally, one of the students measured a length that deviated a lot from the specified length; he made a blunder yielding a gross error or outlier.

11.3 Random, Systematic and Blunder

It is common knowledge to all surveyors that every measurement, whether collected using conventional surveying equipment, GNSS receivers or any other geo-data collection technique, contains error. The above discussion revealed that measurement errors can be categorised into three types: random errors, systematic errors and blunders. Errors can be introduced by the measuring instrument, the observing procedures or the environment in which the measurement equipment and the surveyor operate. The measuring instrument is a physical device that reacts on changes in temperature, humidity and air pressure, resulting in unpredictable variations in measurement values, called random errors. Also unpredictable changes in atmospheric conditions and the observing procedures employed may cause random errors.

Systematic errors are typically introduced by imperfections in the instruments, such as lens distortions. They can be quantified and compensated for by careful calibration procedures. Forgetting to change the default height above the ground of a

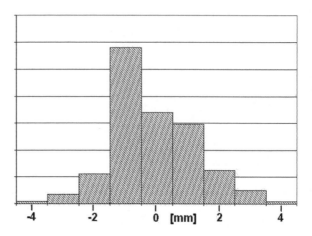

Fig. 11.2 Deviation histogram indicating that random errors have approximately a Gaussian distribution; there is a systematic error present too

GNNS receiver mounted on a pole also causes a systematic error in the measurements. This type of systematic error can be easily rectified during post-processing because it is just one value. Adding this value to all measurements will yield the correct values. When carrying out many height measurements on the same point using a precise levelling device one can compute the mean height and the deviation of each measurement from the mean height. Next one may plot these deviations, which are as many as the number of height measurement, in a histogram. Figure 11.2 shows that the random errors are approximately Gaussian distributed that is the histogram can be mathematically represented by the well-known bell-shaped curve, called Gaussian function. However, the peak of the curve should be located at zero. This is not the case, indicating a systematic error is present in the height measurements.

Blunders, also called gross errors or outliers, are typically caused by carelessness. It is extremely important that blunders are detected and removed from measured data; else the results will be unreliable. Introducing observation workflows, in which taking redundant measurements is standard procedure and using proper processing methods based on sound statistics, blunders can be detected and sometimes eliminated (see Section 11.8). But they cannot always be eliminated because the presence of blunders identified during post processing not necessarily results in pinpointing the erroneous measurements themselves. The only remedy will be re-measurement, and often this means going back into the field and this can be an expensive activity. A standard processing procedure used to identify blunders and possibly get rid of them is based on Least Squares Adjustment (see Section 11.5).

11.4 Error Propagation

It will seldom occur that data are used in their pure form as measured or delivered by a provider. Surveyors measure a combination of angles and distances to compute coordinates or image coordinates using Digital Photogrammetric Workstations to compute object coordinates in a national reference system. GIS users combine Digital Elevation Data, soil characteristics and rainfall data to compute landslide risks. If errors are present in the input data – and they are – they will also be present in the output in some form; errors propagate from the input to the output. What is the effect of random errors in the input data on the calculated parameters? This question can be answered when the random errors of the input data are known in statistical terms in advance. Given the statistical description of errors present in the input data and the computations carried out on the input to generate the output, the statistical properties of the output data can be computed. Usually the random errors are described as variances and co-variances. The mathematics involved in error propagation are complicated and described at length in a number of textbooks, see, e.g. Wolf and Ghilani (1997), Wolf and Ghilani (2006), Burrough and McDonnell (1998), Heuvelink (1998), and Zhang and Goodchild (2002). The first two books mentioned are directed towards the land surveying community and the latter three books aim to serve the community of GIS users.

Basically, the methods involved are based on Taylor series expansion of the mathematical functions applied to the input data. But in the GIS domain also *Monte Carlo simulation* is often used especially if the mathematical functions are not (well) differentiable. A simple example of error propagation is the addition of two parameters ($R = P + Q$) of which the standard deviations (SDP and SDQ) are known. Than the standard deviation of the outcome of the addition, R, which we call (SDR) reads square root of ($SDP^2 + SDQ^2$). When SDP and SDQ are similar in size (i.e. SDQ = SDP) then SDR reduces to the square root of 2 SDP^2, which equals 1.4 SDP.

Figure 11.3 visually demonstrates the error effects propagated through vice versa operations of vector-to-raster transformation as adopted from Rawashdeh (2006). First a polygon is converted to raster and next, in a subsequent step, the raster polygon is converted back to vector format. The degradation depends on the size of raster cells. A point in the vector database, for example, becomes a square of the size of one pixel or raster cell. A very powerful apparatus to compute the statistical properties of the output generated from redundant noisy input data using differentiable mathematical functions is Least Squares Adjustment.

11.5 Least Squares Adjustment

Having many applications in science and engineering Least Squares Adjustment (LSA) is a mathematical procedure that uses redundant measurements to determine a 'Most Probable Value' for some unknown parameter(s) by 'minimizing the sum of the squares of the residuals of the measurements'. Redundant means that in the mathematical model used to link observations with unknown parameters employs more observations than strictly necessary to compute the unknown parameters. Such an approach results in an over-determined system of equations. LSA is frequently used in photogrammetry and surveying. Most users of GNSS receivers rely on LSA for reducing random error and blunders in their measurements. The user, however,

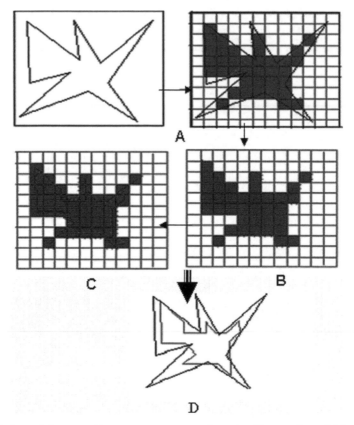

Fig. 11.3 Degradation caused by vector-to-raster-transformation. Vector polygon (**a**) is converted to raster (**b**) and back again to vector mode. The resulting polygon overlaid on the original (**d**) shows the effects on shape and size

will not be aware that LSA is employed because the calculations are done instantaneously by software in a black box environment. LSA can be adapted to any mathematical model, whether it relates measurements to unknown parameters in a linear, quasi-linear or non-linear sense.

LSA also takes account for the quality of the input data, which may be generated by a broad spectrum of data-collection techniques. Some measurements may have a better quality (usually indicated by a smaller standard deviation, which is the square root of the variance) than others. Weighting the input data according to their quality, their contribution to the determination of unknown parameters can be controlled. The unknown parameters receive some portion of the error that relates to the quality of the input data. The lower the quality of the input data, the lower weight will be assigned to the data in the adjustment process. Quantification of the weight is often done by using the inverse of the squared standard deviation (variance). Another advantage of LSA is the ability of analysing the input data on the presence of errors and to detect and possibly remove blunders in the input data. The method has been proven to be superior in detecting blunders in survey measurements. When a geodetic reference network has to be densified by lower order reference points, LSA helps to determine the best distribution of reference points over the area, i.e. it is an aid in pre-survey planning.

11.6 GIS User's Perspective

In contrast to surveyors, GIS users themselves will not very often collect data in the field; they mainly make use of existing data sets. Sometimes it will be necessary to digitise well-identifiable points from large-scale topographic maps for obtaining ground control points (GCP) to geo-reference a satellite image or other dataset (When the procedure is designed well, the operator is requested to measure more GCPs than strictly necessary to compute the unknown transformation parameters; the redundant data collected is used to compute the transformation parameters based on Least Squares Adjustment, enabling quality assessment and blunder detection.). Sometimes the GIS user will have to conduct a field visit to sample 'ground truth' enabling him to carry out a supervised classification of multispectral satellite images. In general, however, the work of a GIS user aims at analysing geo-data for solving a problem or for managing a part of the urban or rural environment and taking measurements is just required as far it supports his major activities.

For example, a forest manager may use topographic maps, satellite imagery and mobile GIS for monitoring wood production. A census specialist may use maps to find out why the population in a certain area is decreasing at a disproportional rate. The data may stem from purchase in whole or in part. Consequently, a GIS user is interested whether the dataset he buys is fit for his particular use; he wants to know how useful a given dataset will be for his domain of application and therefore fitness for use will be his major quality criterion. Examining fitness for use can by done by considering the following primary quality parameters, positional accuracy, temporal accuracy, thematic accuracy, completeness, consistency and resolution (Veregin, 1998).

11.6.1 Positional Accuracy

Positional accuracy is also called spatial accuracy. The determination of this essential quality parameter is the specialised working area of the surveyor. The positional accuracy metric is usually expressed in standard deviation (SD) or two times standard deviation. The single SD indicates that around 67% of all data lie within the given bounds. When the coordinates of a point have SD = 5 cm, there is a 67% change that the actual values will lie in the range –5 to +5 cm around the given value, assuming that the data are Gaussian (normal) distributed. When the accuracy measure is given as 2 SD there is 95% probability that the actual values are in the range –2 to + 2 SD around the given value.

11.6.2 Temporal Accuracy

Temporal accuracy aims at describing the discrepancy between the actual date of capturing of the data and the date as recorded in the metadata of the dataset. When a provider of satellite imagery delivers a dataset, the client may be inclined to consider the delivery date as data of collection while the actual date can be years back. Google Earth permanently shows copyright claims of the image(s) shown on screen; however, dates associated with copyright are not the dates of collection.

11.6.3 Thematic Accuracy

Thematic accuracy concerns the accuracy of attribute values. Examples of attribute values are land use classes resulting from multispectral classification of satellite images and soil pollution values. The metrics used to describe thematic accuracy depend on the measurement scale of the attributes, whether they are measured in nominal scale, ordinal scale, interval scale or ratio scale.

11.6.4 Completeness

Suppose a dataset should contain all the buildings present in a municipality which have an area larger than nine square meters. However, a number of schools and one historical building are missing. These are errors of omissions; something is not documented while it should be. This is an example of spatial incompleteness. Incompleteness can also concern time or theme. Space, time, theme and scale are the four basic dimensions of geo-data. When the dataset does not contain buildings erected after the year 2005, the dataset is said to be temporal incomplete, if the producer claims the dataset is complete up to September 2006. When no schools at all are captured, the dataset suffers from thematic incompleteness. The reverse of incompleteness is over-completeness, for example, also buildings smaller than nine square metres have been stored in the dataset.

11.6.5 Consistency

A dataset is consistent when contradictions are absent. Most important is topological consistency. If a dataset only may contain polygons, for example, a cadastral register of properties, the presence of lines of which the endpoints are not connected to the endpoints of other lines indicates topological inconsistency. Inconsistency may also occur when the class 'family house' is assigned to a polygon, while this class does not exist as an attribute value in the dataset specifications; the dataset is said to be thematic inconsistent.

11.6.6 Resolution

Resolution, also called Level of Detail (LoD), is related to how much detail is present in the data and may refer to space, time and contents. The level of detail visible in satellite images is directly related to spatial resolution or ground sample distance (see Figs. 9.1 and 9.4). The temporal resolution of satellite images defines how often the same area is captured by the sensor and determines how quickly scene changes can be detected. Quick scene changes occur in the event of disasters such as earthquake, landslide or hurricane. Monitoring of such high dynamic phenomena requires images of both high spatial and high temporal resolution. Often the damaged areas have to be captured on a daily basis (high temporal resolution). However, high temporal resolution goes at the cost of high spatial resolution. There is an invincible tension between the two; high spatial resolution means low temporal resolution and vice versa. As a result individual Earth observation systems with high spatial and temporal resolution are non-existent. Hence, disaster managers have to make use of several Earth observation systems.

Different Earth observation systems have also different spectral resolutions. Spectral resolution refers to the number of electromagnetic bands, which can be recorded by the sensor; the more spectral bands the more specific classes can be assigned. For example, when using six bands instead of three, the class vegetation may be further subdivided into the classes: forest, meadows, cereals and barren.

Sometimes too much detail may be present in a dataset; one cannot see the forest for its trees. This happened to the 1:20,000 map created by the National Geographic Institute, Belgium, aimed at serving walkers and bikers. This map is glued together from four reduced-size 1:10,000 maps. Both size and semantic content are identical to the base-map, but the users complained that the maps are overcrowded with symbols and difficult to read. The information content of a map is always in competition with its worth as a communication tool, but now the information had been overemphasised; the number of symbols on the 1:20,000 were nearly twice as high as on the 1:25,000 map, while the scales were similar. Although scale is a measure of reduction, scale also determines the level of detail, which may be presented in a map. Reducing the Level of Detail of a map is called generalisation. Generalisation degrades shape, number of features and area (Fig. 11.4).

In the creation of 3D landscape models five levels of detail (LoD) are distinguished. The simplest level is LoD0: a DEM with superimposed ortho-rectified aerial or satellite imagery or a map. Such models may be sufficient for rural areas but do not provide enough above ground details for being useful in cities. The starting level of creating 3D city models is LoD1, which refers to uniform block models. LoD 2 augments LoD1 by adding detailed shapes of roofs, including roof elements (Fig. 11.5). LoD 3 adds to LoD 2 facades, pillars, house fronts and other details and projects texture on all the outside walls. LoD 4 is the most detailed level; one can not only 'walk around' the lifelike virtual building from the outside but one may also visit the inside of the building; one can virtually look at the paintings at the walls and furniture in the rooms.

Fig. 11.4 Three types of
generalisation: elimination of
the small islands (*top*),
regrouping the small islands
into one island (*middle*) and
regrouping the small islands
to become part of the large
island (*bottom*)

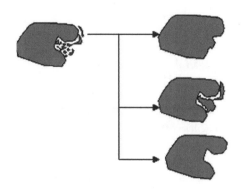

Fig. 11.5 3D model of 'Brandenburg Tor' and vicinity, Berlin, Germany. The model is created by combining large scale city plans (building outlines), airborne Lidar (heights) and aerial images and Terrestrial Laser Scanning

11.7 Quality Assessment

Suppose a municipality wants to buy a land use map, generated from a high-resolution multispectral satellite image using a maximum likelihood classification method. The date of acquisition of the original image appears to be fine and also the selection of land use classes is according to the specifications set out by the municipality. However, the officers are unsure whether the positional accuracy and the thematic accuracy are sufficient for the task at hand. How should one proceed? Basically, assessing positional accuracy or thematic accuracy requires field samples, in remote sensing usually called 'ground truth', in other disciplines check points or reference values. After collection the field samples are confronted with the dataset. The apparatus depends on the measuring scale (nominal, ordinal, interval or ratio) on which the features have mapped. Land use classes are typically mapped on a nominal measurement scale; the values differ but no ordering or other relationships exist between them. Coordinates are mapped on an interval scale; the difference between two sequential values (e.g. expressed as centimetres) is equal but the origin of the reference system can be put anywhere; its location can be chosen arbitrarily.

11.7.1 Nominal Scale Data

Classes are attributed to objects – in the case of multispectral classification of satellite images the objects will be raster cells – but the only relationship between the classes is that they differ from each other; forest land use differs from residential land use. No computations can be carried out on class data; one may not add the class of the one object to the class of another object. Such an operation would result in a non-sense result. However, GIS systems do not prevent the user from performing such types of mathematical operation. Land use classes are often stored as numbers for convenience purposes. Residential area, for example, is represented as number 1, forest areas as number 2 and so on. A GIS will allow addition or multiplication of these numbers, but the outcomes have no meaning. The numbers 1 and 2 are class attribute values and do not indicate that a raster cell with value 2 would be better or bigger than raster cells with value 1. This would only be the case if the numbers would have been mapped on rank order scale, interval scale or ratio scale. Values on a nominal scale may only be examined on difference; one may count the number of occurrences of each class, e.g. the number of raster cells attached with the tag residential area relative to the total number of raster cells. If the number of classes is only two (e.g. male/female) the nominal scale is called binary scale.

The conventional way to obtain an accuracy value of land use maps generated from satellite images is by constructing an error matrix, also called confusing matrix or contingency table (Lillesand et al., 2004). An error matrix is actually a table; the number of rows equals the number of columns. The class of each pixel with a ground truth value is compared with the corresponding pixel in the classified image. Not all pixels in the image will have the same class as given by the ground truth and these

Table 11.1 Example of a confusion matrix

	Resid	Forest	Water	Agri	Waste	Total
Residential	77	0	3	3	2	85
Forest	2	83	4	11	6	106
Water bodies	5	5	65	6	1	82
Agriculture	4	11	4	89	9	117
Waste land	7	3	2	12	87	111
Total	**95**	**102**	**78**	**121**	**105**	**501**
Overall accuracy	80%					

pixels are counted as errors. Table 11.1 gives an example of a confusion matrix and shows that in total 85 pixels are classified as residential, but only 77 pixels are residential according to the ground truth; water bodies (3 pixels), agriculture land (3 pixels) and waste land (2 pixels) have also been classified as residential. These are errors of commission, in total $85 - 77 = 8$; pixels which have been classified as residential, while they have another land use.

The ground truth data contains 102 forest pixels, but 106 pixels have been classified as forest. Five pixels, which are forest on the ground, have been classified as water bodies, 11 pixels as agriculture land and three pixels as waste land. These are errors of omission, in total $102 - 83 = 19$; these are the number of pixels which are not identified as forest, while they are covered by forest. The error of commission for forest is $106 - 83 = 23$. The overall accuracy is the number of pixels which have received the right land use class divided by the total number of pixels, in this example 80%, which is not bad at all.

The use of overall accuracy as a quality measure for classified multispectral satellite imagery has been criticised because assignment of the right class to a pixel may just happen by chance. The probability that one out of five classes is assigned to a pixel by chance is 20%. Therefore, one has gone in search for other measures, which can cope with assignment by chance. Such a measure is the kappa coefficient, which is also expressed as a percentage like the overall accuracy and has the advantage that the chance component has been removed. Its computation involves not only the main diagonal elements of the error matrix but the values in all cells. When the classification would be entirely based on class assignments by chance, the value of the kappa coefficient would be zero. Since the chance component has been removed the value will be lower than the overall accuracy.

11.7.2 Interval Scale Data

How to determine the positional accuracy of the land use map? The values of coordinates are not mapped on a nominal scale, as the above land use classes, but on an interval scale. This is a higher type of scale on which calculations can be conducted, such as addition, squaring and averaging. To determine the positional accuracy of the planar (xy) coordinates of the satellite image, the municipality decides to select

Fig. 11.6 Discrepancies
between planar coordinates
measured on ground with
GNSS receiver and in satellite
image; horizontal axis:
x-coordinate, vertical axis:
y-coordinate; units on both
the horizontal and vertical
axis are metre

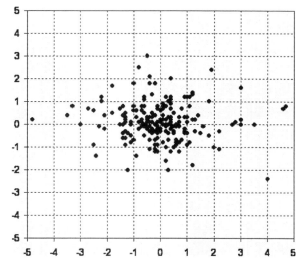

80 well-identifiable points well distributed over the entire image. Their coordinates
are measured in the terrain with GNSS receivers, with accuracy at centimetre level.
As a measure of accuracy the *root mean square error* (RMSE) of the x and y coordinates is applied. From a practical point of view RMSE and the statistical measure
of standard deviation are very similar and RMSE is therefore used as a substitute
for the standard deviation of a dataset. First the individual error of each of the 80
measurements are determined by subtracting the value measured in the field from
the value measured in the image, for all points in the test set. Figure 11.6 graphically
presents the distribution of the discrepancies between the planar coordinates measured with GNSS receivers and the corresponding planar coordinates in the satellite
image. Dealing with a graphical representation for documenting the accuracy of a
dataset is not quite feasible; it is much more convenient to use *one* single value,
indicating the overall accuracy and such a measure is RMSE.

Therefore the individual errors are further processed. First each individual error
is squared. Next the sum of the squared values is computed and the outcome divided
by the number of points resulting in the mean. Finally the square root of the mean
is calculated, resulting in the root mean square error. RMSE of the x coordinate and
the y coordinate are determined separately. Next the planar RMSE can be computed
by applying the well-known formula of Pythagoras: add the squared RMSE of the x
coordinate to the squared of RMSE of the y coordinate and next take the square root
of the outcome.

Table 11.2 gives an example of computing RMSE for a limited set of planar
points. With the computation of RMSE all errors are squeezed to fit into one single
value and this overall measure does not take into account possible spatial variations in the error distribution. The use of RMSE as a measure of error presumes
that the errors are randomly distributed all over space (spatial invariant). Usually

Table 11.2 Example of computing RMSE of x-coordinates and y-coordinates as well as planar RMSE

Point	x image	y image	X GNSS	Y GNSS	$x - X$	$y - Y$	$(x - X)^2$	$(y - Y)^2$
1	194,184.81	469,054.43	194,186.12	469,053.24	−1.31	1.19	1.73	1.41
2	194,286.47	469,550.98	194,283.64	469,549.89	2.83	1.09	8.00	1.20
3	194,289.02	469,945.65	194,288.34	469,943.46	0.68	2.19	0.46	4.79
4	195,292.21	470,539.12	195,294.12	470,542.21	−1.91	−3.09	3.65	9.53
5	196,502.49	470,941.38	196,501.84	470,942.73	0.65	−1.35	0.42	1.83
						Sum	14.26	18.76
						Mean	2.85	3.75
						RMSE	1.69	1.94
					Planar	RMSE	2.57	

Note: Measurement unit is metre

Fig. 11.7 Example of an error vector plot; errors are not randomly distributed, they tend to point into the same direction, while they are getting bigger from *left* to *right*

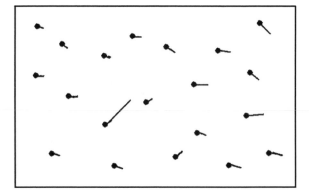

this assumption is justified. The error distribution in Figure 11.6 is approximately random; there are some x-coordinates and y-coordinates, which are both too big resulting in dots in the right-upper quadrant but others are too small resulting in negative values. However, the assumption of randomness may sometimes be violated and therefore generation of an error vector plot is highly recommended.

Figure 11.7 shows an example of such an error plot. The dots represent the location of the check points in the reference dataset and the vectors connect the reference locations with the locations as measured.

11.8 Quality Control

In the above we treated which types of error may occur when acquiring a dataset and how to assess the level of errors present in the data. Figure 11.7 shows that one of the error vectors, which is pointing to the north-east, is larger than the other error vectors. The size of the error indicates that the probability is high that it is not just a random error, as the other ones, but due to a blunder, also called outlier or gross error. The gross error may be present in the GNSS reference data or in one or more observations. So, a next step in the quality assurance process is to get rid of gross errors or, at least, to reduce their influence. This is done by means of statistical testing.

To illustrate how surveyors utilise redundancy (more observables are measured than strictly necessary to compute the unknown parameter) for performing statistical testing let us take as an example the computation of the area of a table. When being asked how to arrive at its area, most people, remembering their high-school mathematics, will be inclined to state: let's measure length and width of the table, multiply the two values and there you are. However, this is a typical scholar solution. How do you know you did not make an error while measuring? And similarly important is the question: how do you know that the table can be really modelled as a rectangular figure? To increase the certainty that an error, if present, will be detected at least one additional measurement, along length and width, has to be taken. An obvious candidate is the diagonal of the table. By using the well-known Pythagorean theorem (in a right-angled triangle the square of the side opposite the right angle equals the sum of the squares of the other two sides), the diagonal can be calculated from length and width. The calculated diagonal can be compared with the measured one and if the difference is too large, a gross error has been introduced during the measurement process or the assumption that the table can be modeled as a rectangle has been violated. Because there is not enough redundancy to decide whether the length, width or diagonal has been wrongly measured, you have to measure the table once again. In reality you will capture geo-data in the field using, for example, a mobile GIS system and the calculations are done on the office PC, which is often a long distance away from the site. So, discovering errors present in the dataset which you cannot resolve in the office means a trip back to the site. Therefore, one should use software which is smart enough to detect errors during data collection while you follow a predescribed data-collection procedure, which enables the software to carry out the calculations to detect and identify errors.

How large should an error at least be to classify it as a gross error? This depends on the accuracy of your measurements. The threshold for measurements taken with a millimeter precision is smaller than for measurements with a centimeter precision. The threshold can be computed assuming the following:

– the precision of the measurement device can be indicated as one measure: the standard deviation, or its squared value called variance
– the random errors can be modeled by means of the famous bell-shaped curve of Gauss
– one has defined a significance level indicating the percentage of measurements of which the error should be below the threshold.

Next, by using error propagation procedures the error present in derived quantities can be determined and compared to the threshold. A thorough theoretical and mathematical underpinning of testing procedures suited for attribute data, such as soil pollution, can be found in Heuvelink (1998). For geometric geo-data, such as coordinates, refer to Baarda (1968) and Koch (1999). Gross error detection is still a vivid research area in the land surveying community; this observation is convincingly proven by the work of Gökalp et al. (2008); Erenoglu and Hekimoglu (2010) and Gullu and Yilmaz (2010). Also in the field of remote sensing accuracy issues are still imperative; see, e.g. Almutairi and Warner (2010).

11.9 Concluding Remarks

The above explored the basic issues of quality assessment and quality control and introduced associated concepts including error propagation and least squares adjustment. The wide variety of error issues studied in the environmental sciences – whether hydrology, soil science, physical geography, geomorphology, geology, oceanography, forestry, meteorology, climatology or geo-ecology – becomes clear when looking at the massive volume of papers collected in Heuvelink and Lemmens (2000). The subjects and research questions treated in this volume include amongst others:

– Which off-the-shelf DEMs are best suited for catchment analysis and other environmental applications?
– Error propagated to slope, aspect and other parameters derived from DEMS
– How to develop an optimal sampling methods for a particular application
– Quality assessment of land use maps, soil acidification models and remote sensing change detection
– Unconventional quality assessment methods using Bayesian networks, Monte Carlo simulation or fuzzy set theory
– Validation of land slides models and other models under development describing environmental processes

References

Allan AL (2007) Principles of geospatial surveying. Whittles Publishing, Dunbeath, Caithness, Scotland, UK. ISBN 978-1904445-21-0

Almutairi A, Warner TA (2010) Change detection accuracy and image properties: a study using simulated data. Remote Sens 2(6):1508–1529

Baarda W (1968) A testing procedure for use in geodetic networks. New Series vol 2, no 5. Netherlands Geodetic Commission, Publications on Geodesy, New Series, vol 2, no 5, Delft, The Netherlands

Burrough PA, McDonnell RA (1998) Principles of geographical information systems. Oxford University Press, Oxford, UK

Erenoglu RC, Hekimoglu S (2010) Efficiency of robust methods and tests for outliers for geodetic adjustment models. Acta Geod Geoph Hung 45(4):426–439

Gökalp E, Güngör O, Boz Y (2008) Evaluation of different outlier detection methods for GPS networks. Sensors 8(11):7344–7358

Gullu M, Yilmaz I (2010) Outlier detection for geodetic nets using ADALINE learning algorithm. Scientific Res Essays 5(5):440–447

Heuvelink G (1998) Error propagation in environmental modelling with GIS. Taylor & Francis, London, UK

Heuvelink GBM, Lemmens MJPM (eds) (2000) Accuracy 2000. Proceedings of the 4th international symposium on spatial accuracy assessment in natural resources and environmental sciences, Delft University Press, The Netherlands

Koch K-R (1999) Parameter estimation and hypothesis testing in linear models, 2nd edn. Springer, Berlin, Heidelberg, New York. ISBN 3-540-65257-4

Lillesand ThM, Kiefer RW, Chipman JW (2004) Remote Sensing and Image Interpretation, 5th edn. Wiley, New York, NY

Rawashdeh, SA (2006) Spatial data quality degradation: real-life examples from Jordan. GIM Int 20(11):13–15.

Veregin H (1998) Data quality measurement and assessment, NCGIA Core Curriculum in GIScience, ncgia.ucsb.edu/giscc/units/

Wolf PR, Ghilani CD (1997) Adjustment computations: statistics and least squares in surveying and GIS. Wiley, Chichester, West Sussex, UK

Wolf PR, Ghilani ChD (2006) Elementary surveying: an introduction to Geomatics, 11th edn. Pearson Prentice Hall, Upper Saddle River, NJ. ISBN 0-13-148189-4

Zhang J, Goodchild M (2002) Uncertainty in geographical information. Taylor & Francis, New York, NY. ISBN 0-415-24334-3

Chapter 12
Applying Geo-information Technology

The previous chapters have clearly demonstrated that numerous types of massive amounts of geo-data have become and continue to become available for researchers and practitioners. Geo-information technology is an excellent aid in solving a great variety of Earth-related problems. However, all users of geo-data will agree that no single geo-data type can provide the optimal solution for a research-related, development-related or some other problem. This chapter will demonstrate that most applications require fusion of data stemming from multiple geo-data sources. Specifically, this chapter focuses on the following fields of application: urban planning, reconstruction of heritage sites, vulnerability assessment of urban areas, biodiversity, forest biomass mapping and optimal placing of solar panels on roofs. The next three chapters will treat in greater detail use of geo-information in the fields of census taking, disaster management and land administration. We start in the next section with a general view on managing the environment in the context of planning and decision making.

12.1 Management of the Environment

A century ago large portions of the Earth surface showed up as white areas on paper sheets. Today every hook and cranny of the Earth has been explored and is being exploited in order to feed and nurse a rapidly growing human population. Human beings are getting increasingly aware how their capacity to manipulate planet Earth and everything what is living on it, in short the environment, may affect the quality of life of humans and other species in the long run. As a result, the last decades showed the emergence of a brand new paradigm never been seen in history before; the Earth is not anymore considered as a hostile environment but as a vulnerable system, which deserves kind stewardship and vigilant care. For example, the concerns about climate change are omnipresent, and many measures are taken to reduce carbon emission in an attempt to relieve or even reverse the trend.

M. Lemmens, *Geo-information*, Geotechnologies and the Environment 5,
DOI 10.1007/978-94-007-1667-4_12, © Springer Science+Business Media B.V. 2011

12.1.1 Alternative Planning Scenarios

During the last 50 years geo-information technology has proven to be crucial for understanding environmental processes and tackling Earth-related issues. The use of geo-information technology enabled geographers and other geoscientists to develop mathematical models of natural processes taking place in the air, in the water and in the solid part of the Earth. The efforts resulted in great improvements in understanding the effects of human activities on the environment, and allowed determination of the types of measures needed to prevent aggravation or possible reversing the downward trend. All around the world geo-data, GIS and other software packages are used for monitoring the environment, for safe guarding the natural environment and avoiding further deterioration, for reconstructing natural environments and for maintaining or improving quality of life in urban areas. Most types of analysis carried out with GIS systems could have never been performed by manual means alone using paper maps, drawing tables and pencils. Using manual means only would simply take too much time and too many skilled labour forces.

One of the main advantages of using GIS and geo-data in digital format is the ability to calculate alternative scenarios from which the best can be chosen according to some optimisation criterion. For example, when one aims to build a high-speed railway between two capitals, a multitude of issues and interests have to be taken into account. The selection of the trajectory has to be optimised in terms of costs, damage to ecosystems and biotopes, and quality of life of the people living and working in the vicinity of the railway once constructed and exploited. To determine the optimal trajectory land use, land value, soil and rock properties, geo-tectonic characteristics and many more parameters have to be quantified along a broad strip between the two major cities. The outcome will not only depend on the set of criteria but also on the quality of the data. Therefore, the selection of the data should be primarily steered by considerations of the quality, i.e. fitness for use. To warrant procurement of good quality data, the data should be assessed prior to use as is common in most production processes. Also manufacturers of bikes, computers, cars and washing machines check the quality of the separate parts prior to assembling the final product and the same should be true for the use of geo-data as building components of the analysis framework of any planning activity.

12.1.2 Monitoring

The above concerns a planning example in which the effects of a future human activity on the environment are taking into account. Once the construction of the railroad is finished and trains are running according schedule, the effects on the environment should be properly monitored. Monitoring involves regularly taking measurements and comparing these measurements with each other to identify trends and to understand the causes of change. Once the real world process, e.g. the acoustic pressure level in urbanised areas through which the high speed trains run, is not aligned anymore with the desired or anticipated standards because the connection is so popular that the intensity had doubled in a few years, decision makers, non-governmental organisations (NGO) or pressure groups will likely come into action. The pressure groups may force government to erect sound walls along the sensitive parts of the trajectory. In many countries the finding that quality of life of people is affected by a work of infrastructure may result in heavy disputes in parliament and society as a whole. As an alternative to sound walls, government may decide to provide financial compensation to dwellers, especially in less densely populated areas, enabling them taking provisions to make their houses noise proof.

Often it will appear unavoidable to build a small part of a long distance railroad close to a vulnerable ecosystem. To track the possible effects a monitoring program may be established, using, for example, as hypothesis that the environment of some rare birds is negatively affected by the presence of the railroad. A quantifiable measure has to be defined and such a measure can be found in the number of nests in the breeding season. For example, a decision rule could be that when the number of nest halves in a 5-year period action has to be taken. Such action may be the artificial creation of a similar biotope further away from the railroad. The location of the nests can be measured with GNSS devices and next stored in a GIS. With the help of time series analysis trends can be identified and quantified.

12.1.3 Changing Habits

The above example shows that the use of geo-data and GIS are closely connected to decision making. Since the world is becoming increasing complex, due to an ever growing world population, urbanisation and improved living conditions in many parts of the world, the role of GIS and geo-data in decision making and policy development will accordingly increase. Geo-information technology helps to establish good quality of life in urban conglomerates where tens of millions of people have to share the facilities established in a relatively small area with high concentrations of buildings, cars, plants and so on. Maintaining and improving quality of life in urban areas is particularly challenging because the needs and habits of human beings change over time; one can not assume that they are givens. For example, up to the turn of the millennium, cars were thoughtlessly used without taken environmental effects into account. The car was even used to bridge short distances by the majority of people in developed countries. However, this habit is gradually bending mainly as a result of two developments. The first is the acknowledgement that quality of life and health are marching hand in hand, while the insight that health is strongly intertwined with physical strain has become common knowledge. Travelling by car is only mentally stressful but requires no muscular strain. The second development is that fuel consumption of cars is becoming increasingly expensive. As a result, people are moving from using cars to using bikes to bridge short distances to working places or shopping centres and also for leisure. Some countries stimulate the use of bikes by tax facilities, not only to improve public health but also to reduce air pollution. Decision makers have to take account of such changing habits by ensuring that bikers can travel safely without their lives being endangered by cars and trucks driven recklessly and without being forced breathing particulate matter emitted by heavy traffic and being exposed to intolerable noise. This can be best achieved by construction of separate cycle paths. Here again geo-data and GIS come in for determining the best locations of cycle paths and once constructed to maintain them. Recognising trends may be considered as modelling the future and geo-information technology can be of great help to find out whether an environment is future proof, how the environment should be adapted to these needs and the costs of these measures.

12.1.4 Crisp Boundaries

Pollution is any form of emission which causes harm to human beings, animals and the environment at large (Middleton, 2008). The emission can be smoke blown into the air by plants and cars, chemicals drained off through streams and channels, and noise coming from machines. Particularly important is traffic noise. Standards for acoustic pressure levels accepted for residential areas range from 50 to 70 decibels. The decibel scale is logarithmic. An increase of about three decibels means a doubling of sound level. As a result of lessons learned in the past new residential areas will not be built in the vicinity of roads with intensive traffic, such as highways. To determine the areas that are suited for constructing dwellings sound models have been developed. The acoustic pressure at a certain location in the vicinity of a road will depend on a multitude of parameters, including the number of cars and trucks passing per unit of time, maximum speed allowed, type of asphalt, muffling of the sound by the presence of constructions between the road and the location, such as plants, sound walls and office buildings and, foremost, distance to the road. Using a geo-data set containing outlines of the road, outlines of buildings and their heights, types of trees and their heights, the model allows computation of acoustic pressure level for any location. Usually, the sound levels will be represented by introducing a raster as data model. For each cell the acoustic pressure level will be computed. Next one can derive from the raster the contour lines at both sides of the roads indicating the transition boundaries, e.g. 55 decibels, depending on national or local regulations. The contour lines define a buffer zone where building of dwellings will not be allowed. Such contour lines form crisp boundaries, at one side of the boundary building permits will be granted at the other side of the boundary one will not be permitted to build dwellings. It is obvious that a slight change of the value of the parameters in the sound model applied will result in other locations of the boundaries, while also data quality may significantly affect the outcome.

However, decisions makers often have to taken decisions on a binary scale: yes or no, allowed or not allowed and there is no room for introducing gradual transitions. No wonder that the outcome of such models often is subject to heavy disputes between government, citizens and pressure groups. Although, decisions bring clarity, which is beneficial from an economical perspective (better a bad decision than no decision at all) it may happen that a government decides to compensate those living close to highways or airports and consequently suffer from extraordinary sound levels. Using acoustic models to determine the buffer zones around roads and airports may result in the peculiar situation that people living inside the buffer zones receive compensation while their neighbours who suffer from the same inconvenience most of the time, but accidently live at the other side of the crisp boundary, are given the cold shoulder. This is rightfully experienced by many as injustice and enhances the impression of the public at large that government is unreliable and is acting arbitrarily. Applying such models for taking yes/no or go/no-go decisions should be done with utmost caution.

The above provides a brief introduction how geo-information technology constitutes a sound foundation for planning and decision making and how it forms an indispensable aid to model and understand the environment. Gatrell and Jensen (2009) treat in greater detail the use of geo-information technology as an aid in spatial modelling, planning and decision making associated with the spatial expansion of urban areas and economic development in rural environments.

12.2 Urban Planning

In many parts of the Earth, cities are losing their attractiveness for living due to rapid and uncontrolled – also called organic – growth resulting in air and water pollution, traffic noise, loss of green areas and congestion of roads. For example, the new capital of Nigeria, Abuja, has been designed on the drawing tables as an attractive garden city. The master plan shows a plethora of green areas well distributed over the city area. However, the planned green areas have never come into existence because the identified areas have been occupied by shopping centres and low quality dwellings, which are rented to privateers, which want to make quick money in a city housing the fortunate. Second-hand cars imported from all parts of the world rush over the broad and well-designed roads converting traffic participation in an insecure adventure.

The focus of this section is on planning issues of a country where cities are still growing organically and the movement of rural dwellers to urbanised areas result in rapid erection and expansion of informal settlements. The country concerned is Turkey. Urban centres in developing Turkey are challenged by a population leap, which began in the 1950s as a result of industrialisation. From 23.6 million in 1985 urban population expanded to 44.1 million in 2000, which is an increase from 45 to 65% over a 15-year period. The rapid urban population growth caused changes in size and scale of cities, type of housing and street lay-out. Also land-use changed and the need for open and green areas – the city lungs – increased. Most worrying is the uncontrolled settlement of low-income families in squatter areas, which may harm the ecological balance and cause deterioration and demolition of historical and natural heritage sites. Essential in urban planning is determination of the consequences of alternative scenarios – including the zero-scenario, i.e. no action – from ongoing and anticipated demographic processes.

The scope of the section is limited to demonstrating how geo-information technologies can be applied in urban planning by discussing practical cases. An introduction to the different theoretical and methodological approaches for measuring and analysing the complex processes and impacts of suburbanisation can be found in Besussi et al. (2010). The authors introduce methods for measuring the spatial pattern of urban sprawl by integrating remotely sensed imagery with spatial socioeconomic data and how to interpret the resulting information to understand structure and form of urban settlements.

12.2.1 Eskisehir

To demonstrate how urban planning may benefit from geo-information technology we take a tour through the city of Eskisehir, Turkey, located in the northwest of Anatolia. Figure 12.1 shows a street view in the very centre and Fig. 12.2 depicts the city as captured from space. When travelling by train from the primary metropolis Istanbul to the capital Ankara, a trip of 6.5 h bridging 562 km, one will arrive within 4 h in Eskisehir, at least when you take the Baskent Express, which is the fastest train connecting Turkey's two major cities. Our guides are Aksoylu and Uyguçgil (2005), Ulu et al. (2005) and Uz et al. (2006). Around half a million people are living in the city of Eskisehir, 30% in squatter settlements. After the 1970s, rapid

Fig. 12.1 View on the modern downtown Eskisehir, Turkey

Fig. 12.2 Panchromatic Ikonos image of Eskisehir, Turkey

industrialisation triggered property developers to construct informal houses around industrial sites and along the main roads. Not concerned with living standards or environmental concerns – their main interest was and continues to be profit – they did not put much effort into constructing basic services.

In the year 2000, nearly 169,000 people were living in 16 squatter settlements and to get insight in the living conditions a study was conducted on the in-house conditions and physical access to public services, in particular health care and schools. As descriptive parameters for the in-house conditions floor area coefficient and connection to utilities (electricity, running water, sewer system and natural gas) were used. The parameters to quantify the access to public services were determined as ratio of services to the number of people and walking distances from the place they lived to the services. Such an analysis requires a lot of (geo-)data and computations using GIS to arrive from the data at numbers for the descriptive parameters. The resulting number may trigger local authorities to take measures and to invest money to improve quality of life. The data required include the following:

– Size of the population in the 16 squatter settlements
– Number of people living in each house
– Location, footprint area and number of storeys of each house
– Utility services present in or in the neighbourhood of dwellings
– Location and capacity of kinder gardens, primary schools and high schools
– Location and capacity of health services

All data should be available in digital format in order to be processed in a GIS environment and to arrive at that it was necessary to scan all analogue maps and plans present in the municipality, to digitise on the scanned maps the outlines of blocks, houses and service buildings and to store attribute data in tables. Furthermore, extensive surveys of land use and population statistics had to be conducted. To ease and accelerate analysis by computer, all data should be properly arranged in a database. From the outlines of houses, the footprint areas can be computed and combining the resulting numbers with population statistics and number of storeys the floor area coefficient (number of people living on a unit of area) can be computed. From the location of houses and the services walking distances can be computed, and so on. The results of the analysis can be visualised by maps. What do the numbers tell about quality of life in Eskisehir? Public services are inadequate according to standards, there are not enough services to cover the needs of all the people and the walking distances to their locations are too high. While 98% of the houses in 1989 had a septic tank, in 2002 60% of the houses were connected to the sewer system, 39% had a septic tank and 1% did not have any disposal system. Nearly all buildings have electricity and indoor running water: 99.2 and 91.5%, respectively, which is more than Turkey's average. None of the districts are connected to a source of natural gas.

At the end of this part of the tour our guides Aksoylu and Uyguçgil (2005) conclude: 'The use of GIS techniques should be stimulated to support settlement development by using planning models and scenarios and proper data in digital format'.

12.2.2 Heritage Site Under Threat

Rapid industrialisation and urbanisation confronted Eskisehir also with the problem of destruction of the Odunpazari heritage site. The area has a size of around 50 hectares and its mediaeval ground plan has been largely conserved. The street pattern is shaped organically, following the boundaries of property plots, which are randomly shaped. The streets – paved with large stones – lead to squares with a fountain or a tree and their width and slope are determined such that a man driving a loaded animal can pass. Houses are surrounded by a garden or built around a courtyard and the majority has two floors. The most important buildings are Kursunlu Mosque, constructed in 1516, and the tomb of Sheikh Edebali. Until the late 1960s the governmental focus was on conservation of these types of spot monuments not on groups of houses or quarters, as was already common in many European countries. In the 1970s regulations were issued to protect groups of historical buildings but they were not effective because of inadequate planning, unqualified staff and insufficient financial resources.

Planning is the apt manner to sustain historic city characteristics; however, when applied inappropriately planning can also irreversibly destroy valuable remains of the memorable past. What are the consequences of implementing a development plan on a heritage site? Geo-information technology provides the tools to evaluate such a question. To do so, a building register is required in which basic characteristics are stored such as quality and type of construction, comfort status and economic lifetime. The best way to collect such data is by visual inspection during field survey. The data can be drawn on maps, enlarged aerial photographs or satellite images or, alternatively, directly stored in the computer by using Mobile GIS. Also the outlines of buildings, streets and other features have to be collected. That can be done from existing (base) maps or from aerial images using photogrammetric techniques. In Odunpazari 2,075 buildings were digitised from existing maps and the roads were digitized from scanned 1:5,000 scale base maps. Also cadastre data can be of use. All these data describe the present situation and to arrive at insight how the present situation might be affected by implementation of the development plan the last has also to be stored in a database. The paper maps presenting the proposed situation were scanned with an A0 scanner and transferred to the reference system of the 1:1,000 base map using affine transformation.

The required data from raster maps was manually digitised and stored as separate vector layers. In total 15 thematic maps were created and stored in a GIS environment. By overlaying the present situation with the proposed development plan buildings that would be destroyed could be determined.

12.2.3 Lungs of the City

The quality of urban areas has a direct influence on socio/economic development. Quality of living in an urban environment depends on a multitude of parameters, including level of air pollution, acoustic pressure and the presence of parks. The importance of having green areas within the borders of heavily populated areas not only as places of recovery, recreation and leisure but also as sources of clean air is well recognised. Already in the nineteenth century, Hyde Park and Central Park were baptised 'Lungs of London' and 'Lungs of New York', respectively. To arrive at insight of the adequacy of open and green areas for sports and play in Eskisehir, researchers at the Anadolu University developed a method that relies on using satellite images, digital map data, GIS and image processing software. The satellite images were Land TM images recorded in September 1999, ground sample distance (GSD) 30 m and Ikonos images from 2002, GSD 1 m. Figure 12.2 shows the panchromatic band of the Ikonos image. The Landsat images were rectified and corrected for radiometric and atmospheric distortions. Map data used included Esksehir City Plans and maps of the city neighbourhoods. The analogue maps were scanned and manually digitised by heads-up digitisation. Other data included location of playground and sport fields, 2004 population data and classes of green areas. Since the population data were arranged according to city neighbourhoods, all other data were rearranged and reclassified on the same basis. The spatial distribution of green areas was determined by three methods: normalised difference vegetation index (NDVI), supervised classification and visual classification (see for an introduction to classification of remotely sensed imagery Jensen (2005)). The NDVI was computed from Landsat TM images. The same images were used for supervised classification using the minimum distance classifier. Since the resolution of Landsat

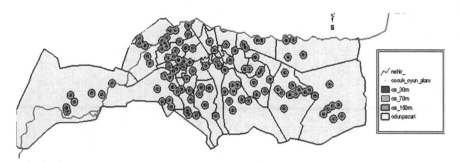

Fig. 12.3 Accessibility of playgrounds for 3- to 6-year-old children determined by using a GIS buffer zone operation; radii of buffer zones: 30, 70 and 150 m

was not high enough to determine all green areas Ikonos images were manually digitised to complete the mapping of green areas. Ikonos images were also used to create 3D models for some parts of the study area. The analysis results were visualised in nearly a hundred maps. For example, determining the accessibility of sports fields and playground areas was done by representing them as point features to buffer accessibility zones around these points according to age groups, one such being 3- to 6-year-old children (Fig. 12.3).

12.3 Recreating the Past

Architectural heritage is threatened by many manmade and natural influences, including war, population pressure, poorly managed tourism, neglect or inappropriate use of sites, air pollution and climate change (Rüther, 2007; Rüther et al., 2009). Fortunately, the relevance of our cultural heritage and the need to protect and preserve it appears to be widely accepted.

Geo-information technology helps to create realistic three-dimensional pictorial scenes to generate accurate digital records of historical and archaeological objects while reducing overall costs. These scenes are presented in such a way that the user experiences what it is like to be present in another world. This technology bears the tag Virtual Reality (VR). Meanwhile, liberated from the media-fostered mystification through which phase some technologies inevitably seem to have to pass, VR is rapidly developing into a practical and powerful 3D-visualisation tool for a wide variety of applications. Historians and archaeologists are increasingly interested in using VR for creating artificial environments of past realities: it enables researchers, students and tourists to obtain a much better understanding of the past than they might get from text and pictures alone. VR also enables historians to create realisations of scenes that never existed in the reality of the historical past but which were however depicted long ago in paintings, plans and sketches or described in documents. For example, by using computer visualisation of Hans Holbein's The Ambassadors (1533), Hart et al. (2005) revealed that the painting was possibly originally hung on a staircase and encoded a cross formed by the figure of one of the ambassadors: crucial information for art historians. VR also enables a glimpse of spectacles such as Napoleon's triumphal route through Paris and his coronation at Notre Dame.

When creating a fantasy world such as is used in games and films, it may suffice to use fantasy data. However, when creating 3D virtual scenes of a past reality true data have to be acquired. How to arrive at this? Broadly speaking, true data can be collected along two lines. The first is by using old maps, documents and other historical sources. The second uses the present as reference through retrospectively interpreting current geo-information such as maps, aerial images and Lidar scans. When a historical site still partially exists in the form of ruins or as a group of conserved buildings reconstruction of the site as it was is very important and requires 3D documentation. In the past this has been mainly done using land surveying equipment such as levelling instruments and total stations, GNSS technology and photogrammetric. But these are laborious; terrestrial 3D laser scanning (TLS) combined with digital imagery is more effective and is of great help in creating models of the present situation upon which models of the past situation can be superimposed (Fig. 12.4). However, yielding cost-effective, highly detailed and accurate photo-realism of complex architectural structures is not possible by a single technique; modelling complex and large-scale architecture requires the integration of multiple techniques, including terrestrial and aerial photogrammetry, laser scanning, survey and GPS (El-Hakim et al., 2008).

On top of these models the past may then be reconstructed using old maps, pictures, other documents and (why not?) the fantasy and sophisticated guesswork of the historian. Indeed, for modelling the past along the second line, historians and researchers involved in archaeology are increasingly valuing the potentials of TLS. This is because the costs of fieldwork have become relatively modest as a result of scanner developments that enable carrying out full vertical and horizontal coverage in one set-up. Office processing is not as expensive as it was a few years ago either, thanks to advancements in software allowing the time-efficient extraction of 2D drawings and 3D models from point clouds. Apart from declining costs contributing to the popularity of laser scanning for capturing buildings, the dense point cloud

Fig. 12.4 Djingereyber Mosque, Timbuktu, Mali, heritage site reconstructed by the Aluka Heritage Documentation Group at the University of Cape Town using Terrestrial Laser Scanning

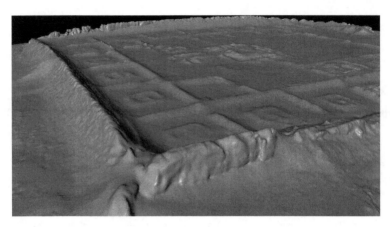

Fig. 12.5 TIN model of Por-Bajin Fortress site, testimony to the former presence of ancient civilisation in the wildness of Siberia: perspective view from south-west

generated by the technology is very well suited to modelling complex geometry – especially at detail level – of heritage buildings and ruins. In one of the largest-scale projects in modern archaeology carried out at the Por-Bajin fortress site, testimony to the former presence of ancient civilisation in the wildness of Siberia, one work-day of laser scanning with a Trimble GS200 gave 254,000 points over 5.6 hectare, resulting in about 4.5 points/m^2 (Anikushkin and Kotelnikov, 2008). Figure 12.5 shows a Triangular Irregular Network (TIN) representation of the data collected. The more the data collected, the better the archaeological features that can be localised, and precise localisation helps to save time and money in the expensive and time-consuming excavation stage.

Archaeological survey data can be obtained by surface, aerial or geophysical methods; the latter are non-destructive, cheap and fast (Anikushkin and Kotelnikov, 2008). Geophysical methods are many and widespread, electrical-resistance and magnetic-anomaly surveys being the most popular. Combining geophysical with elevation data allows marking of anomalies, which indicate underground archaeological objects. An illustrative example of how geomatics technology can be used to reconstruct the past is the work done by Al-Ruzouq and Al-Zoubi (2007), researchers from Jordan. They used photogrammetry, DEMs, GIS, 3D modelling and ground-penetrating radar to document the Baptism area, a site of early Christianity, where John the Baptist lived and performed baptisms during the time of Christ. Requirements for creating the 3D models included high-level geometric accuracy, detailed reconstruction and efficiency in model size and photo-realism.

Photogrammetry was used to produce DEMs, orthophotos and 3D modelling of existing archaeological structures. For DEM generation, conjugate points in stereo-images were automatically extracted using adaptive stereo matching techniques that take into account the relief of the terrain, so that more points are collected in hilly terrain than in flat. Validation was done with checkpoints of which the coordinates were measured with GNSS receivers. Orthoimages were created from the aerial

Fig. 12.6   Church, rendered with texture (*left*), reconstructed from photogrammetric images

images using the DEM to correct for relief displacement. The 3D coordinates resulting from photogrammetric adjustment procedures were used to create 3D models of structural elements and of a church (Fig. 12.6). Textures added to the surfaces of 3D models give a 'real world' appearance of ruined heritage sites, which is much appreciated by architects and renovation experts because they require realistic views for further inspiration. The 3D model can be digitally rotated so that the scene can be studied from many perspectives.

The assets of a ruined heritage site are often partially or entirely buried beneath the ground. Ground-penetrating radar (GPR) enables to capture the subsurface and detect structures. GPR emits into the ground electromagnetic (EM) waves of frequency band 10–1,000 MHz and is used in applications ranging from construction and archaeology to forensic science. GPR was used to map buried walls, graves, buried ruins, cavities or chambers, crypts, marble plates and other buried antiquities, with the purpose of directing onward excavations. Figure 12.7 shows a radargram taken along a 23-m profile using the 40-MHz frequency band; the anomaly between 15.5 and 17.5 m indicates the presence of a buried wall. Boundary maps, point location of archaeological sites, transportation layer, hydrological maps, DEMs, orthoimages, satellite images, land classification image, aquifer profiles and geology were brought together in a GIS.

El-Hakim et al. (2007), see also El-Hakim et al. (2008), developed a method to create detailed 3D models of castles by integrating terrestrial and aerial photogrammetry, TLS, surveying and GNSS (Fig. 12.8). Using floor plans that exist for most castles a low-accuracy model with approximate heights is created, which acts as departure for more accurate modelling and determining the most appropriate techniques in the next stages, which will depend on location, size, surface, shape, texture and required level of detail. Control points have to be strategically placed and measured with a total-station to scale the models, assess final accuracy and register

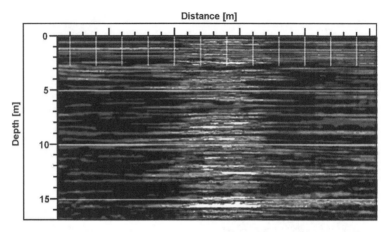

Fig. 12.7 Radargram showing ground structure up to 15 m deep. Vertical pattern of erratic colours in centre of radargram indicates presence of buried wall

Fig. 12.8 3D model of Valer in Trentino-Alto Adige, northern Italy, reconstructed by using a combination of terrestrial and aerial photogrammetry, TLS, surveying and GNSS

all models in the same reference system when overlap is not sufficient. With a few GNSS measurements the model can be placed within a global coordinate system, necessary for integration with a DEM.

From a low-flying helicopter image are recorded capturing exteriors, walls, courtyards, roofs and main grounds and the derived models substitute most of the floor-plan model. Occluded parts or areas for which geometrical detail is required, digital cameras or laser scanners should be used. When surface complexity is high TLS is preferred, and this is also the case when rooms are narrow. When texture requirements are essential a camera can be better chosen. Only for rooms with complex structures TLS was used; for all others digital cameras were used. A good network configuration (good breadth to depth or B/D ratio and image distribution) is often difficult to achieve in narrow spaces or courtyards. When occlusions are the major impediment 3D information can be obtained from single images using geometric constraints. Dense, multi-image matching is employed to capture the small details of low relief or interior decoration. Finally all models should be brought into the same reference system by using control points, common points and redundant data in overlaps. Remondino and Zhang (2006) provide a suite of surface reconstruction algorithms aimed at modelling large scale objects captured by ground-based devices. El-Hakim et al. (2007) conclude that the automation level of all modelling steps is rather low and smart and efficient techniques for seamlessly combining models, removing overlaps and filling gaps are lacking.

GIS and Virtual Reality together with advanced geo-data acquisition techniques such as TLS enable archaeologists and historians to create a vivid reconstruction of the past. That is the nice thing about new technology: when it is in its inception and initial stages nobody can foresee which applications may benefit from it in the long run. This is because professionals far beyond the field of view of the developer will become potential clients once they are aware of the technology and the costs become affordable.

12.4 3D City Models

Planning, designing and managing the urban environment require appropriate decision-making. For example, the realisation of a general zoning plan, approved by authorities, requires detailed architectural designs of all the future building, roads and other objects and their location and orientation in space. Architects and planners should be enabled to quickly assess feasibility, errors or conflicts of alternative designs. As a vital basis for developing and assessing designs, architects are in need for accurate topographic and land use maps of the area. If possible and affordable they would not only want to utilise planimetric maps but also 3D models of the area in which they can place alternative 3D designs of buildings, bridges, roads and other objects. For many applications the availability and use of three-dimensional (3D) geo-information in the form of 3D city models is crucial.

12.4.1 Level of Detail

Traditionally, the term 'mapping' refers to geo-information technologies in which real-world objects are two-dimensionally represented on a physical medium, usually on a sheet of paper. The map is both medium of storage and visualisation, and its scale determines the features of the latter; the larger the scale the more detail the map producer is able to show. Anyone asking what the scale is actually wants to know how elaborate objects are represented.

Today technology and geo-datasets are in place, which enable the creation of 3D landscape models: satellite imagery with resolution as high as 50 cm (and even higher, if US government regulation on civilian imaging would allow), digital 2D topographic/cadastral maps, airborne and terrestrial imagery, airborne Lidar and terrestrial laser scanning. 3D city models are the most widely created and used landscape models. They consist of Digital Elevation Models (DEM) of ground surface overlaid with structure and texture of buildings and possibly other objects. Such models may be created at five levels of detail (LoD). The simplest level is LoD0: a DEM with superimposed ortho-rectified aerial or satellite imagery or a map. At LoD1, basic block-shaped depictions of buildings are placed over LoD0. LoD2 adds to LoD1 detailed roof shapes. LoD3 represents further expansion by adding to LoD2 structural elements of greater detail, such as facades and pillars, and draping all objects with photo-texture. The highest level, LoD4, is achieved when buildings can be virtually 'visited' and viewed from the inside. Real-scale models can be created from digital models using high-speed professional 3D printers, such as the Z510 Spectrum 3D printer from Z cooperation (Fig. 12.9). This technology creates a 3D object by collating successive layers of material and allows for full colour printing and is fast. Each layer is printed by deposition of a fine powder, which is subsequently fixed by the binder deployed in standard inkjet technology. A detail of the real scale model of the Croatian capital, Zagreb, created by the company Geofoto

Fig. 12.9 Z510 Spectrum 3D printer from Z cooperation, dimensions: 107 × 79 × 127 cm; 204 kg

Fig. 12.10 Detail of the real scale model of Zagreb's city centre created using 3D printing technology

using 3D printing is shown in Figure 12.10. Probably by the year 2020 3D colour printers will have made their appearance on the home desktops of the general public, joining the colour inkjet printers already there.

12.4.2 Manual Editing

Apart from imagery (satellite, aerial and terrestrial), other data sources are used to create 3D city models: airborne and terrestrial Lidar, and topographic and cadastral 2D maps holding outlines of buildings. Using a mix of data sources allows a 3D city model to be created highly automatically. At smaller scales an automatically generated 3D model may look nice and realistic. But when zooming in, one will see that parts of some roofs may be projected on the street, buildings seem partly to rest on trees or railways, garages have green roofs and appear to hang like boxes in the air above a playground; trees are glued onto facades.

   One cause of such incongruities is the limited accuracy of the data. Inaccuracy may be introduced as a shift between airborne imagery and Lidar data used to create the skeleton of the 3D model; it comes into view as roofs projected on streets, and buildings resting on other objects. Then there is the influence of time. A recently constructed building may be captured in the Lidar data but absent from earlier recorded aerial images and digital maps, causing it in the 3D model to hang in the air, while the roof is decked with the land cover captured in the aerial image. Laymen, who usually experience artificial 3D worlds through animations and games, often believe that the creation of 3D landscape models must be a piece of cake. The 3D world of games can be modelled perfectly, but creating realistic 3D landscape models requires much more craftsmanship and time, lots of time. As usual, the devil is in the detail, and it takes huge manual editing efforts to eradicate the faults.

12.4.3 Examples

This subsection presents the process of the creation of 3D city models of Zagreb (Novakovic, 2011) and Teheran (Parmehr et al. 2011). The 3D city model of Zagreb has been entirely created from digital aerial images using photogrammetric techniques. In September 2008 Geofoto, a Zagreb based photogrammetric firm, recorded 4,000 aerial images (GSD = 8 cm) using Vexcel UltraCam X digital camera. The high level of overlap, 80% along track and 60% across track, allowed the creation of true-orthoimages. Testing various methods of data extraction revealed that automatic creation of digital surface models (DSM) resulted in disappointing accuracy and produced surfaces that did not correspond with those wanted.

Moreover, alternative methods to operator-based photogrammetry, such as automatic DSM creation or airborne Lidar, provide only geometry of the Earth's surface and no information whether an object is a house, high-rise building or road. Such semantic attributes have to be attached in an additional data enrichment stage, which is time consuming. Therefore, the DSM and the outlines of objects were manually extracted from stereo-images using digital photogrammetric workstations. Buildings were mapped by delineating the rooflines, from which, after orthogonal projection on the ground using software, building facades and footprints can be generated. Roofs which were smaller than 25 m^2 were not measured. The resulting 3D vector models of buildings were imported in software together with the true orthoimages and DSMs. Aerial images were imported too as source of texture to be draped over the building facades. The virtual 3D city model was materialised using a 3D printer. The printing of each of the 527 blocks of maximum size 25.4 × 35.6 × 20.3 cm, together constituting the scale model of Zagreb, took approximately 3 h. The real scale model was exhibited in the town hall of Zagreb. The public appreciated its appeal while professionals were impressed by its great level of detail. City planners understood the value and the model was used to present strategic infrastructure projects to the city council.

In 2005 Iranian authorities approved the general zoning plan to develop Abasabad, northern Tehran, into an area where the millions of people living in the metropolis can undertake cultural, recreational and leisure activities. The realisation of a general zoning plan, approved by authorities, requires detailed architectural designs of all the future building, roads and other objects and their location and orientation in space. Architects and planners should be enabled to quickly assess feasibility, errors or conflicts of alternative designs. As a vital basis for developing and assessing designs, architects are in need for accurate topographic and land use maps of the area. If possible and affordable they would not only want to utilise planimetric maps but also 3D models of the area in which they can place alternative

3D designs of buildings, bridges, roads and other objects. This was the aim of creating LoD1 and LoD2 3D models of an area of 623 hectare in Abasabad using high resolution digital aerial imagery and scale1:500 maps, created in 2004 from accurate land surveying data. In 2006 the Iranian national mapping agency recorded aerial images (GSD = 10 cm) using a Vexcel UltracamD digital camera. A geodetic reference network consisting of 30 easily identifiable and well distributed ground points was established over the area. The points were partly used as ground control points in the areotriangulation procedure and partly as check points to detect the possible presence of outliers and to estimate accuracy.

The X, Y and Z coordinates of the reference points were measured with dual frequency GNSS receivers operating in static mode with an accuracy of 5 cm root mean square (RMS) error in all three coordinates. After bundle adjustment the coordinates of the checkpoints showed a RMS error of less than 10 cm. Next to the above preparatory work, a digital elevation model (DEM) was automatically generated by applying the image matching techniques available in the digital photogrammetric workstation used. Since points resulting from automatic matching do not necessarily represent the bare ground editing, carried out by a human operator, was required. Next an orthoimage mosaic was created, which was enhanced by outlines of buildings, transportation lines, trees and other objects as vector data extracted from the scale 1:500 map. To create the 3D city model, first the orthoimage mosaic was draped over the DEM resulting in a photo realistic surface of the area resulting in a LoD0 representation. Next 3D building models and 2D urban plans were superimposed onto the model resulting in a LoD1 representation. In addition to vector information 3D objects of important buildings and landmarks were integrated into the 3D model to increase its visual attractiveness and to support the design process better (LoD2 representation). Designs developed by city planners where interactively merged with the 3D city models employing software tools (Fig. 12.11). The models created at different scales during the planning and design process supported the architects to identify design errors or conflicts of interest effectively.

12.4.4   Use of 3D City Models

Based on the work of Li and Wu (2006) we give some examples of how 3D city models can be used for applications including determination of intervisibility between objects, Location Based Services (LBS), sunlight/shadow analysis in densely populated areas, and air and noise population. These applications usually require, in addition to 3D city models, appropriate mathematical methods, including geo-statistical analysis, time-series analysis and analysis of moving platforms (dynamic analysis).

Intervisibility means that one object is visible from the other and vice versa. Intervisibility is used in urban planning, for example, for analysis of sunlight and shadow of buildings, determining the locations of telecommunication antennas and creating LBS in cities. Sunlight effects human wellbeing a lot; availability of

Fig. 12.11 3D city model showing a design how Abasabad, northern Tehran, Iran, may look like in the future

sunlight depends not only on the length of day as function of position on Earth and season but also on the presence of shadow-generating objects, such as terrain relief and (high-rise) buildings. Location, size and shape of residential buildings, office buildings and recreational facilities have therefore to be planned with great care. Computation of the area covered by a shadow is done by projecting the 3D building surfaces onto a 2D plane on the ground along the direction of a ray of sunlight. The extent of the shadows of surrounding buildings can be superimposed on the designed building to determine sunlight hours and sunlight interval.

To compute whether an antenna can receive signals from a receiver in a certain position use is made of the following information: height of receiver, antenna and objects obstructing the line of sight between antenna and receiver, and elevations of their base at ground level (Fig. 12.12, top). All this information can be extracted from 3D city models. An example of LBS is base-station broadcasting information picked up by a vehicle. This application requires the base-station should reach every street of the city. Buildings and other obstacles will obstruct the propagation of EM signal by reflection, absorption, refraction or penetration. The creation of LBS requires analysis of how EM signals interact with city objects. Also, the strength of an EM signal weakens with increasing distance from the base-station, so factors to be used in the analysis include

– distance between base-station and vehicle
– antenna height of base-station and height of moving antenna
– heights of buildings and other obstructing objects
– distances between buildings in the street and the moving platform.

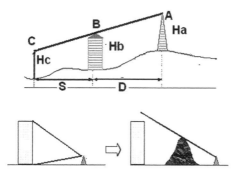

Fig. 12.12 Intervisibility between receiver at location C and wireless communication antenna at location A obstructed by building B (*top*). Ground elevations of A, B and C and their heights above ground can be obtained from a 3D city model. Obstacles change strength and direction of noise spread; position and height above ground of obstacles can be extracted from 3D city models (*bottom*)

Human sensory organs are very sensitive: too few sunlight affects health but also too much noise. In today's cities main noise producers are traffic/transportation, industry and public activity. Noise spreads in certain directions and is emitted at a certain strength that weakens with increasing distance from the source. The presence of obstacles also influences direction and strength of noise at a certain location (Fig. 12.12, bottom). The noise pollution index is determined by many factors, including terrain relief, presence of sound walls, vegetation and buildings and their surface structure, and the height of bridges, streets and railways.

12.5 Weed Control from Space

Weed infestation in young forest stands can severely affect tree growth, reducing volume production and decreasing seedling survival. Assessing the conditions across whole stands from the ground is often problematic, with assessments usually based on sample observations. Airborne and satellite imagery have frequently been used and some key forestry operations can be monitored with high resolution satellites such as QuickBird and Ikonos. Norris-Rogers and Ahmed (2007) developed a weed control monitoring method based on combining textural analysis and classification of multispectral bands for use on high-resolution satellite imagery. Assessing the amount of weed requires differentiating the tree crop from weed. Multispectral classification alone is not reliable because weed and tree crop have the same spectral characteristics. However, plantation trees are generally planted in rows and linear features can be identified by using textural analysis, particularly on the panchromatic band because of its high spatial resolution, which is usually four times higher than of the multi-spectral bands of the same satellite. The presence of vegetation, whether wood or weed, is identified by examining the multispectral bands using unsupervised classification, which ranks the degree of vegetation into four classes: from bare soil to very strong vegetation growth. By assuming that the vegetation present in the rows as identified by textural analysis represents the tree crop and vegetation outside the rows would represent weed it is possible to identify and quantify weed infestation.

This method was tested in South African on a site used for the production of wood chips for the pulp and paper industry. The study demonstrated that tree crop could be separated from weed, but the result is highly dependent on stand age. One to three months after planting crop, soil and weed can be insufficiently differentiated while between 12 and 16 months after planting the best differentiation occurs. After 16 months, canopy has closed and weed monitoring is no longer essential. As with most remotely sensed information, ground verification would still be necessary, but accuracy assessments exceeded 80%.

12.6 Biodiversity Monitoring

Biodiversity monitoring aims at ensuring conservation and sustainable use of resources and has received increased attention in the wake of the growing awareness on climate change. The major steps adopted globally for the conservation of biodiversity are

– establishment of protected areas (national parks, wildlife reserves and conservation areas)
– enforcement of conservation laws
– awareness creation.

Monitoring involves taking repeated measurements and comparing them to understand the causes of change. Thus, biodiversity monitoring involves

– carrying out repeated surveys to find out the size and extent of the population of a species and the quality of their habitat
– analysing the results to find out the trend and rate of change

Monitoring needs to be done against a predefined objective, for example, to increase the population of Blue Sheep (*Pseudois nayaur*) to a pre-specified level in a certain area in a given time. The population of the current year is required to define the target for the future.

12.6.1 Nature Conservation in Nepal

The following is based on the work done by Tucker et al. (2005), environmental researchers from King Mahendra Trust for Nature Conservation, Nepal, as published in Chapagain (2005). In developing countries local people heavily depend on biodiversity resources for living. In Nepal large tracts of forests have been cleared for human settlement in hills and mountains since the 1950s. Today few biodiversity monitoring systems exist. Suppose one wants to monitor Snow Leopard and information about their habitat enables to identify their living areas. By map overlays of distribution criteria within a GIS, areas, which match all the criteria, can be identified saving much fieldwork especially in the mountainous areas where most of the wildlife areas are only accessible on foot. GIS also enables to define locations for taking samples, randomly or systematically, for validating data sets. Next the sample locations are staked out in the field with GNSS. After sampling the data can be stored, displayed and analysed in a GIS to reveal spatial patterns. Visualisation of field data on maps allows determining occurrence of features and how they are spatially grouped. This type of investigation can only be done in a practical way with GIS since manual methods would be time consuming and produce less convincing results. The uniform addressing format in the form of latitude, longitude and sometimes also altitude, in a GIS eases the location of sample plots and transects for repeated measurements. Satellite remote sensing can further add value to the use of GIS in biodiversity monitoring. Archived satellite images analysed with GIS can provide habitat information for the years prior to the start of a biodiversity monitoring program.

12.6.2 Mangrove Monitoring in Bengal

The threats to mangrove ecosystems are many and result from human exploitation of biological resources and natural hazards such as land erosion and accretion. Satellite images and GIS techniques are suited for biodiversity monitoring of mangroves, as Singh et al. (2006) demonstrated by studying an area of 9,630 km^2, located in West Bengal, along the coast of India. Threats to this area include habitat alteration, overexploitation, water pollution, climate change and an abysmal increase in human and cattle. Essential parameters in monitoring biodiversity of mangroves are disturbance index and fragmentation level. The disturbance index provides a measure for the change over time from mangrove to other types of land cover. Fragmentation level provides a measure for the division of mangrove areas: high levels of human activity and natural processes such as erosion cause mangrove areas to become divided into small patches embedded within areas of other types of land cover. Conservation priority can be determined from disturbance indices and fragmentation level, together with other landscape parameters including soil type, drainage density, land use/land cover and geomorphology. Landsat-TM and IRS-1D/1C, LISS-III data from 1987, 1998, 2000 and 2002 were used enabling time-series analysis over three periods: 1987–1998, 1998–2000 and 2000–2002. They were corrected for geometric and radiometric distortions. Survey of India 1:50,000 topographic maps were used for geo-referencing and extracting topographic details such as village location, roads, and ferry network and drainage information. Field survey provided ground truth about vegetation types, land use and land cover and these were manually delineated by heads-up visual interpretation of the satellite images.

12.6.3 Monitoring Fish Habitat in Washington State, USA

How do changes in river flow effect downstream fish habitat. This is a question the United States Bureau of Reclamation is confronted with and to answer the question accurate hydraulic models of the river through time have to be created. This is conventionally done using vessel-based acoustic methods and/or land survey. Whilst both methods are reliable and accurate, they show, in addition to irregular coverage due to surveying along a series of cross-sections and profiles, so requiring data interpolation, other major limitations:

– time-consuming and labour intensive
– dangerous, especially during periods of high flow
– susceptible to significant problems of access
– offering no full bottom coverage.

Could Airborne Lidar Bathymetry (ALB) address all these limitations? While ALB technology has been used with great success in 'clear water' coastal environments for more than 20 years, it has always been thought to have limitations in shallow-water river areas. Major limitations include lack of water clarity caused by transport of sediments, and difficulty in determining depths when these are less than 50 cm. Millar et al. (2005) conducted a case study in the Yakima River in Washington State and demonstrated that ALB offers accuracy and precision at least as good, if not better than that achieved in typical surveys by boat equipped with sonar and GNSS receivers. They captured two river reaches with 2×2 m point density from a flying height of 300 m and swath width of 60 m (Fig. 12.13). Kinematic GNSS was used to position the aircraft using one GNSS base station per site. A spacing of one point per 4 m^2 is much higher than can be achieved with sonar. Lower accuracy of ALB may be compensated by a more detailed coverage of the riverbed. The costs of ALB survey are similar to those of conventional sonar surveys when the reach is longer than 25 km; the larger the project, the more feasible ALB becomes.

12.6.4 Wetland Conservation

To conserve wetlands in accordance with regulations and policy, they have to be monitored (Jensen, 2007). An essential part of the monitoring process concerns the accurate and detailed mapping of wetlands on a regular basis. Aerial colour and false-colour photographs as well as multispectral images have been the conventional means of capturing environmental areas. However, in wetland areas covered

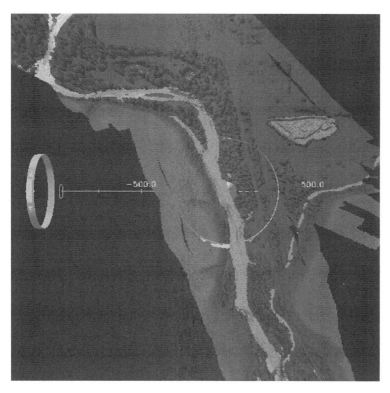

Fig. 12.13 Map of the Kittitas reach, Yakima River, Washington State, USA, created with airborne Lidar bathymetry

by forest, which are the most common type of wetland in the US and most likely to be lost in the future, image-based remote sensing technologies have severe limitations. As a new and alternative approach Lang and McCarty (2009) investigated the suitability of airborne Lidar to map forested wetlands. A 33-km^2 study area on the eastern shore of Maryland was captured with the Optech ALTM 3100 airborne Lidar sensor (wavelength 1,064 nm) at an altitude of 610 m and 20° scan angle during a period of inundation (March 2007). The Lidar sensor used allows the emission of 100,000 pulses per second. The backscattered signals were recorded with up to four samples per return. Simultaneously, multispectral images (green, red and near infrared) were recorded with a digital camera, 12 cm GSD. Outlines of inundated areas were mapped from both the Lidar intensity data and the multispectral imagery collected simultaneously and compared with in situ measurements. The latter were sampled at locations with spacing of 5–10 m along transects, which were not pre-specified, using a Trimble mobile GIS device. To improve accuracy at least 15 GNSS measurements per point were taken and these were enhanced by real-time WAAS corrections. Each point was classified as being inundated or not. Comparison between outlines of inundated areas obtained from the GNSS ground truth and the outlines extracted from the digital images revealed an overall accuracy of 70%. The comparison with the outlines extracted from the airborne Lidar data showed an overall accuracy of 96%; a considerably better accomplishment.

12.7 Access to Geo-data for Citizens and Tourists

For managing the Bay of Halic (Golden Horn) area, Istanbul, Turkey, a geo-database has been created, containing maps and data on cultural and historical places, important buildings, schools, hospitals, police stations, bridges, wharfs and shipyards (Alkan et al., 2011). The database can be accessed from desktop computers connected to the internet as well as from handheld mobile GIS devices, which can wirelessly communicate with the database. Initially the users were limited to city managers and urban planners but after a successful rehabilitation project, carried out in the first decade of this century, which reshaped the Golden Horn area in a centre of trade and culture, the database has been opened for use by citizens and tourists. The map data, created from aerial photographs, acquired in 1996, consists of 1:5,000 digital orthophoto mosaic overlaid with vector data extracted from the same aerial photos. Cultural and historical places along with other important objects have been indicated on the base map and attributed with name, use, type, address and the like. Precise bathymetry was conducted over the Halic bay; positioning was performed with dual frequency GNSS receivers and depth measurements with an echo sounder. After processing with hydrographic and mapping software a DEM of the sea bed was produced and integrated into the database using GIS software. The web-based GIS created in this way gives people all over the world access to orthoimage maps, bathymetric data and historical data of the Golden Horn area. As a further development, the Golden Horn database has also been made accessible to handhelds equipped with GNSS receiver, digital camera, pocket PC and mobile GIS software. The communication and data transport are carried out wirelessly. The software enables users to create built-in and custom-made toolbars as well as edit and identify forms. The edit form enables rapid collection of geo-data and the identify form enables to retrieve information about objects stored in the database in pre-designed templates. The handhelds, equipped with customised software tools, are used by government field workers as low-cost devices for easy and efficient geo-data acquisition as well as citizens and tourist for retrieving orthomaps and other geo-data of the Golden Horn area and its surroundings in real time for navigation purposes (Fig. 12.14).

Some restrictions and weakness of mobile GIS devices, which emerged during design and implementation of the project, include

– screens of handhelds are small which affects the display quality
– low processor speed
– battery may be rapidly exhausted
– cost of wireless connection and fees of software licenses may be prohibitive.

Fig. 12.14 Display of Map
on a mobile GIS device while
in the field (B/W
representation)

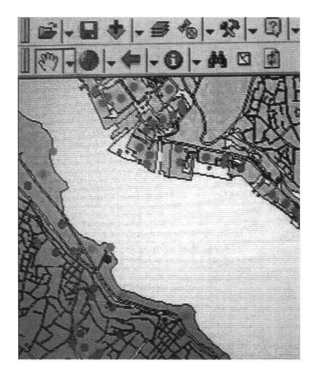

12.8 Forest Biomass Mapping

Forest biomass relates to the mass of the above ground parts of living trees and is
a fundamental parameter for estimating forest productivity and carbon storage at
national level. Mapping of forest biomass improves our knowledge of the carbon
cycle and facilitates prediction of the future climate on Earth. However, national
governments also estimate the biomass of forests growing on their territory to meet
reporting requirements internationally ratified in the Kyoto protocol. Assessing of
forest biomass is also required for reducing the emission of greenhouse gasses from
deforestation and degradation. This so-called REDD process uses financial incen-
tives to reach the aim of reducing greenhouse gasses. Consequently, many countries
collect geo-data from they can estimate forest biomass and compute from this the
amount of carbon storage in forests. It goes without saying that these methods
should be efficient and cost-effective.

A diversity of geo-data acquisition technologies, either alone or in combined
use, has proven suitable for forest biomass mapping. Airborne Lidar is able to
capture the structure of forest canopies in their full three dimensions and is therefore an efficient technique for determining forest stand volume and forest biomass
(Naesset et al., 2006). The earliest research on using airborne Lidar data for forest
biomass and volume estimation dates back to the mid 1980s. Nelson et al. (1988)

used forest canopy height data, acquired with an airborne-pulsed Lidar, to predict ground-measured forest biomass and timber volume. Popescu (2007) developed a method for assessing above ground biomass and component biomass for individual trees using airborne Lidar data in forest settings typical for loblolly pine stands. Koch (2010) provides a state-of-the-art review of geo-data acquisition technologies focussing on biomass estimation including (full-waveform) airborne Lidar, polarimetric InSAR, very-high-resolution multispectral images, hyperspectral images and digital photogrammetry.

12.9 Renewable Energy

A stream of evidence collected over the past five decades proves that the temperature of the Earth is rising (Esty and Winston, 2006). For example, the eastern side of the Antarctic continent is nearly 3°C warmer today than it was half a century ago. There is confirmation enough that the presence of greenhouse gasses in the atmosphere, including carbon dioxide, methane and HFC-23, has an effect on temperature: the higher the concentration of greenhouse gasses, the higher the global temperature (Middleton, 2008). There is also plenty of scientific proof that greenhouse gasses have risen from 280 to 385 parts per million (ppm), an increase of 37.5% since the beginning of the Industrial Revolution around 250 years ago.

The effects of global warming are also well known. The most hazardous are a rise in sea level, frequent, heavy rainfall, and an increase in the incidence of hurricanes. Sea-level rise would transform Notting Hill in the UK and Amersfoort in the Netherlands into seaside resorts and submerge coastal megacities, including New York and London. The change in weather pattern also causes more heavy storms to sweep inland from the oceans over coastal zones. Such typhoons most affect lowland countries with poor infrastructure. In early 2008 Cyclone Nargis took the lives of over 130,000 human beings in Burma and left the same number homeless. The World Health Organisation estimates that since the turn of the millennium climate change has taken an annual toll of 150,000 casualties. The majority of citizens and governments now accept that climate change is real and that something must be done. 'We have the resources, the knowledge and the technology to fight climate change, and we will do it!' declared European Commission Vice-President Margot Wallström in August 2007, presenting a novel solar-powered car in Brussels. The technology to fight climate change is based on generating electricity without emitting carob dioxide: solar power, wind energy and nuclear power.

German researchers have carried out calculations suggesting that two thirds of the power consumed by households in Berlin and other major cities in the country can be yielded from photovoltaic and thermal solar panels placed on building roofs (Ludwig et al., 2009; Ludwig and McKinley, 2010). To harvest such high-energy quantities the solar panels have to be placed such that they catch sun radiation in an optimal manner. Sun panels should therefore not be placed in shadow areas and should have such an orientation with respect to the sun that the angle of incident radiance is on average as close to 90° as possible, all day long, and year round. The calculation of optimal placement of solar panels requires algorithms to determine the essential roof parameters including effective roof area, shaded areas, roof inclination and roof orientation. Dormers, chimneys and other small roof structures reduce the effective area on which solar cells can be mounted and have thus to be determined too. Furthermore, shadows from vertical objects diminish the amount of incident energy, even significantly more than wrong orientation of the panels. Therefore the size, shape and location of all trees, high-rise buildings and other shadow-creating objects have to be measured.

In a project called Sun-Area the city of Berlin has convincingly demonstrated that the above essential roof parameters can be derived from digital surface models created from accurate, detailed airborne Lidar data (Ludwig and McKinley, 2010). The accuracy of the Lidar data was 15 cm planar precision and 15 cm height precision. The point density varied from 1 to 15 points per m^2. The project resulted in a digital solar atlas of Berlin freely accessible via the internet through web-based GIS technology as an incentive for home and business owners to place photovoltaic and thermal solar panels on their roofs (www.3d-stadtmodell-berlin.de).

References

Aksoylu S, Uyguçgil H (2005) Benefits of GIS in urban planning: investigating the squatter settlements in Eskisehir, Turkey. GIM Int 19(2):51–53

Alkan RM, Kalkan Y, Erdoğan C (2011) Golden Horn GIS: access to geo-data for the millions. GIM Int 25(3):21–25

Al-Ruzouq R, Al-Zoubi A (2007) Archaeology and GIS in Jordan: Geo-IT in reconstruction of baptism area. GIM Int 21(1):23–25

Anikushkin M, Kotelnikov S (2008) 3D-Model of Siberian fortress. GIM Int 22(7):13–15

Besussi E, Chin N, Batty M, Longley PA (2010) The structure and form of urban settlements. In: Rashed T, Jürgens C (eds) Remote sensing of urban and suburban areas. Remote Sens Digit Image Processing 10(Part 1):13–31. Springer

Chapagain NR (2005) GIS and GPS for biodiversity surveying and monitoring: experiences gained in the Annapurna Conservation Area, Nepal. GIM Int 19(7):53–55

El-Hakim S, Gonzo L, Voltolini F, Girardi S, Rizzi A, Remondino F, Whiting E (2007) Detailed 3D modelling of castles. Int J Archit Comput 5(2):199–220

El-Hakim S, Remondino F, Voltolini F (2008) 3D modelling of castles: integrating techniques for detail and photo-realism. GIM Int 22(2):21–25

Esty DC, Winston AS (2006) Green to gold: how smart companies use environmental strategy to innovate, create value, and build competitive advantage. Yale University Press, New Haven and London

Gatrell JD, Jensen RR (eds) (2009) Planning and socioeconomic applications. Springer, 223 p. ISBN 978-1-4020-9641-9

Hart V, Day A, Robson J (2005) How historians are now using computer technology: investigating spatial effects in architecture and painting. GIM Int 19(9):13–15

Jensen JR (2005) Introductory digital image processing, 3rd edn. Prentice Hall, Upper Saddle River, NJ

Jensen JR (2007) Remote sensing of the environment: an earth resource perspective, 2nd edn. Prentice Hall, Upper Saddle River, NJ

Koch B (2010) Status and future of laser scanning, synthetic aperture radar and hyperspectral remote sensing data for forest biomass assessment. ISPRS J Photogramm Remote Sens 65(6):581–590

Lang MW, McCarty GW (2009) Lidar intensity for improved detection of inundation below the forest canopy. Wetlands 29(4):1166–1178

Li C, Wu J (2006) Applying 3D city models: intervisibility, LBS, sunlight/shadow analysis, air and noise population. GIM Int 20(2):45–47

Ludwig D, McKinley L (2010) Solar atlas of Berlin: airborne Lidar in renewable energy applications. GIM Int 24(3):17–22

Ludwig D, Lanig S, Klärle M (2009) Towards location-based analysis for solar panels by high resolution remote sensors (Laser Scanner), ICC2009. 24th international carthography conference, Santiago, Chile

Middleton N (2008) The global casino: an introduction to environmental issues, 4th edn. Hodder Education, London

Millar D, Gerhard J, Hilldale R (2005) Using airborne Lidar bathymetry to map shallow river environments. Proceedings of the 14th Biennial coastal zone conference, New Orleans, LA, 17–21 Jul 2005

Naesset E, Gobakken T, Nelson R (2006) Sampling and mapping forest volume and biomass using airborne Lidars. Proceedings of the eight annual forest inventory and analysis symposium, pp 297–301

Nelson R, Krabill W, Tonelli J (1988) Estimating forest biomass and volume using airborne laser data. Remote Sens Environ 24(2):247–267

Norris-Rogers M, Ahmed F (2007) Separating the trees from the weeds: satellite imagery for forest management in South Africa. GIM Int 21(1)

Novakovic I (2011) 3D model of Zagreb: real and virtual. GIM Int 25(1):25–29

Parmehr EG, Afary AR, Basiri B (2011) 3D city models: supporting tools for urban planning and design. GIM Int 25(2):29–31

Popescu SC (2007) Estimating biomass of individual pine trees using airborne Lidar. Biomass Bioenergy 31(9):646–655

Remondino F, Zhang L (2006) Surface reconstruction algorithms for detailed close-range object modelling. IAPRS&SIS 36(3):117–121. Bonn, Germany

Rüther H (2007) Geo-info and cultural heritage. GIM Int 21(2):65

Rüther H, Chazan M, Schroeder R, Neeser R, Held C, Walker SJ, Matmon A, Horwitz LK (2009) Laser scanning for conservation and research of African cultural heritage sites: the case study of Wonderwerk Cave, South Africa. J Archaeol Sci 36(9):1847–1856

Singh SK, Kushwaha SPS, Joshi PK (2006) Mangrove mapping and monitoring: RS and GIS in conservation and management panning. GIM Int 20(3):61–63

Tucker G, Bubb P, de Heer M, Miles L, Lawrence A, Bajracharya SB, Nepal RC, Sherchan R, Chapagain NR (2005) Guidelines for biodiversity assessment and monitoring for protected areas. KMTNC, Kathmandu, Nepal

Ulu A, Aksoylu S, Çabuk A, Anadolu (2005) GIS for urban conservation: change detection and priority determination in historical Turkey. GIM Int 19(4):48–51

Uz Ö, Çabuk A, Çabuk SN (2006) GIS and RS in city planning in Turkey: open and green areas in cities. GIM Int 20(5):46–47

Chapter 13
Census Taking

Accurate, reliable, detailed and up-to-date data on where and how a country's population lives, its age structure, sex ratio, internal migration, condition of health, level of employment, and level of literacy is essential for planning, policy intervention and monitoring of development goals. Such demographic data are fundamental and key for determining location, type and size of schools, hospitals and factories, and future numbers of teachers and medical doctors. It is also vital for the planning and construction of roads, railways, bridges, ports and so on. The most appropriate procedure to acquire data about the population of a country and how they is by performing a population and housing census is because of its great relevance to the economic, political and socio-cultural planning of a country as well as its essential role in public administration, understanding and planning socio-economic developments, and supporting public and private activities census taking has been promoted internationally since the end of the nineteenth century, when the International Statistical Congress recommended that all countries in the world conduct them (United Nations, 2008).

Census taking on a regular basis (usually every 10 years) is also the main source for the determination of the changes in socio-economic structure of the population over decades and centuries. Demographic data on how many people are living in the administrative units of a country are also a basis for allocating central government budgets to these areas. The housing condition of the population provides a reliable indicator of citizens' well being and information on housing is indispensable in estimating housing needs and formulating housing policies. In short, census data are invaluable for planning and carrying out any economic, political or socio-cultural activity in any country.

A census is a huge, comprehensive and complex undertaking as its preparation, conduction and processing of the massive amount of data collected require large amounts of resources in terms of humans, training, technology and money (Hardin et al., 2007). It is said to be the biggest of peacetime operations in terms of planning, funding, logistics and execution, affecting as it does the entire population living within the territory.

Producing detailed statistics for small areas and small population groups is the foundation of any population and housing census and the necessary data to be

M. Lemmens, *Geo-information*, Geotechnologies and the Environment 5,
DOI 10.1007/978-94-007-1667-4_13, © Springer Science+Business Media B.V. 2011

collected to arrive at such statistics concern composition, characteristics, spatial distribution and organisation (United Nations, 2008). This chapter first treats the four different methods used in census taking and next considers how land survey-ing, maps and satellite imagery can help to define the boundaries of enumeration areas, taking the 2006 census of Nigeria, in the preparation and execution of which the author has been actively involved in the period February 2005–July 2006, as an narrative example. The last section of the chapter considers how airborne and satel-lite imagery are useful resources to arrive at population estimates in urbanised areas. The approaches discussed are of particular interest for developing countries where cities continue to grow organically often without authorities having the means to intervene.

13.1 Methods

The methods used for carrying out a Population and Housing Census can be divided into four main categories (United Nations, 2008):

– Group assembly method
– Self response method
– Direct interview/canvasser method
– Virtual census

In the *group assembly* method, heads of households are requested to come together at a certain venue, where they are asked to submit authorised persons pre-specified information on members of their household. In the *self-response* method question-naires are posted to persons or households who are expected to fill them in and submit them to a nearby census office. Both methods are time-efficient and require no expensive resources. But each has major defects (Daugherty and Kammeyer, 1995). The group assembly method results in only limited information. The self-response method requires the filling in of written questions, which requires a basic level of literacy among the whole of the population. What is more, both methods face the problem of how to verify the information provided.

The direct *interview/canvasser* method is a de facto method whereby every indi-vidual physically present in the country is interviewed face to face. This is the most reliable but also the most expensive method because it requires for a period of at least 2 weeks enumerators must be appointed at the ratio of 1 to every 500 of the population. In a *virtual census* the main portion of the data necessary to compile census statistics is taken from existing registers, the backbone being the population register. Furthermore, samples are taken using the self response method. Conducting a virtual census is feasible in countries with a well-established tradition in collecting demographic figures and other statistics. In such countries a virtual census can be a cheap but reliable alternative to the expensive de facto census.

The remaining part of this section elaborates upon characteristics of virtual census taking, using the Netherlands as an example. Further, the de facto census is detailed, taking Nigeria as an example.

13.1.1 Virtual Census

In the Kingdom of the Netherlands, traditional head counting started in 1829 and have been taken place every 10 years ever since. The exception is 1941 due to the break out of WWII. A population census had already been conducted in autumn 1795 in the then recently established Batavian Republic, formerly the Republic of the Seven United Netherlands. The 1971 census, the fourteenth, was the first that made use of automated processing of responses, using optical mark recognition. The data of each person, collected during face-to-face interviews, were recorded on individual enumeration charts. The 'Big Brother is Watching You' syndrome was well developed and mainstream all over Europe and many people feared privacy violation and also the abuse of population data during WOII was still fresh in memory in the early 1970s. As a result census taking became a controversial activity although the non-response in 1971 was just 0.2% (van Maarseveen, 2005). However, during the trial census conducted in preparation of the 1981 census, public revulsion had risen to such a level that the non-response appeared to be 26%, leading to the decision to waive the census, which was followed by the legal discontinuance of traditional census taking in 1991.

As a substitute for the 1981 census, 1991 census and 2001 census data were collected using population registration and samples. When preparing for the 2001 census the term virtual census was assigned to this type of census taking. In the 2001 virtual census 40 tables were used, a lot more than used in the 1981 and 1991 census and leading to more numerically consistent results. Thirty of the tables were filled with demographic data on occupation, level of education and economic activity, eight tables concerned housing and three tables concerned commuting (Schulte Nordholt, 2005). The reliability of the results is comparable to that of a conventional enumeration in which the entire population is interviewed face-to-face. However, the cost savings are huge. Would a conventional census cost 300 million Euros, a virtual census costs three million Euro, a difference of a factor 100! A virtual census would be impossible in Nigeria because of the lack of sufficient and reliable register data to compile all census variables from.

13.1.2 De Facto Method

A narrative example of a de facto census is the countrywide Population and Housing Census held in Nigeria in early spring 2006. Nigeria is often called the giant of Africa because of the extent of its land mass, abundance of natural resources and also because it is the biggest country on the continent in terms of population. The census was for the greater part financed from donor money to which the European Community contributed a very substantial part: over 100 million Euros (Lemmens, 2006). The generosity of the European Community was especially motivated by the need to provide a sound foundation for acquiring demographic statistics for making reality good governance in general and more particularly to gather reliable population figures for supporting a fair course of the general elections scheduled for taking place in 2007. (The part of the general elections devoted to voting a new president got threatened in the 2005–2006 period due to a proposed amendment to the constitution to allow residing President Olusegun Obasanjo to pursue a third term. The proposal was blocked by the Nigerian Senate on 16 May 2006.)

Although in the mind of the general public a census is no more than counting people – and what's so difficult about that? – anyone ever involved in census-taking knows what a huge and comprehensive operation it is. Indeed, it is said to be the biggest of peacetime operations in terms of planning, funding, logistics and execution, affecting as it does the entire population living within the territory. The census period in Nigeria formally covered 5 days in the early spring of 2006, but the number of days necessary to count the population had to be extended to seven. During that period over 600,000 functionaries were on the move from one building to another, from one household to another and from one individual to another. They interviewed every individual present in the country, whether of Nigerian origin or foreigner. Nigerians who stayed outside the country during the census period were not enumerated. Census questionnaire forms, photocopy of the original enumeration area map and B2 pencil in hands, the enumerators collected data on age, sex, occupation, educational level, employment status, marital status, school status and so on in the EA assigned to them.

To enforce confidentiality enumerators had to take an oath of secrecy, prescribed by law. The oath reads as follows: 'I [name] do swear that I will faithfully and honestly fulfil my duties as [function] in conformity with the requirements of the National Population Commission Decree of 1989 as amended and that save as provided in that decree, I will not disclose or make known any matter or thing which comes to my knowledge by reason of my employment by Commission to an unauthorized person'. All these functionaries had to be trained, accommodated, transported and paid by the end of the census period. The questionnaire booklet, specially designed for recording the responses, consisted of four A4 pages. Each page was machine readable so that the collected data could be automatically captured by scanning in one of the seven Digital Processing Centres distributed over the country using OMR/OCR/ICR technology.

OMR (Optical Mark Recognition) captures data by detecting patches marked grey with a pencil, using the principle that marked patches will reflect less light than parts of the paper left blank. To avoiding questions and other information on the questionnaires would be recognised as marked patches, they are printed in a backdrop colour (red ink). OCR (Optical Character Recognition) transfers handwritten, typewritten fonts or printed text into digital text and is an achievement of the many research carried out in fields of pattern recognition and computer vision the past decades. ICR (Intelligent Character Recognition) is OCR plus; the plus being that the computer software has a learning facility to improve accuracy and recognition levels during processing. Scanning paper forms and next to extract data from the resulting digital images by computer is a delicate technology and requires careful handling of the forms. The interview conditions were often not optimal because they had to be conducted in the open air or other hard situations. Furthermore, enumerators used cello tapes, pins and other adhesives to tie the forms, which caused mutilation and damage. Many such forms could not be automatically processed.

13.2   Enumeration Area Demarcation

Many pre-census activities are required to divide the 923,768-km^2 landmass of Nigeria into small units: enumeration areas (EA). Enumeration areas form the nucleus of any census, the unit for collection of all demographic data. An EA is an area carved out of a bigger locality or a group of localities with well-defined and identifiable boundaries. It is an area that a team of enumerators is expected to cover during the census period. An EA that consist of more than one locality is referred to as multi-locality EA. The essence of carving out these units is to avoid any omission or duplication of count during enumeration (Lemmens, 2006). To obtain accurate and reliable data it is essential that no individual is left out of the enumeration process and that none is enumerated more than once. To achieve this goal EA maps need to cover every corner of the country so that neither gaps nor overlap emerge between areas.

At the crux of establishing such a rigorous data-collection framework lays the use of detailed geo-information. The number of people living within an EA should be fairly small – no more than 500 – enabling a team of enumerators to interview every individual present within the EA during the days assigned to census taking. The carving up of the country into EAs has been done for earlier censuses, but during the 1991 census and prior ones, EA maps were primarily drawn as sketches; they were neither scaled nor geo-referenced. In sketches gaps and overlaps in enumeration areas can easily occur and this was the main reason for adopting the strategy to move from sketches to maps at scale. The EA mapping in preparation of census 2006 was performed with the help of traditional surveying methods. The boundaries were measured with compasses and measuring wheels according to the traverse method: turning points on closed polygons are connected by angle and distance measurements and next the size and shape of the boundary of each EAs are mapped to scale by hand drawing on grid sheets. The grid sheets are not the end product; they are used as back drop for drawing a refined representation on transparencies; one EA per transparency. The selection of the scale depended on the size of the EA; the mapping criterion was that EA map should fit to A3 paper format, which is 30 × 40 cm.

The layout of transparencies is predefined in preprint (Fig. 13.1). The map section of the transparencies contains not only the outlines of the EA but also identification of the location of the starting point, scale indicated by a bar and/or a ratio, for

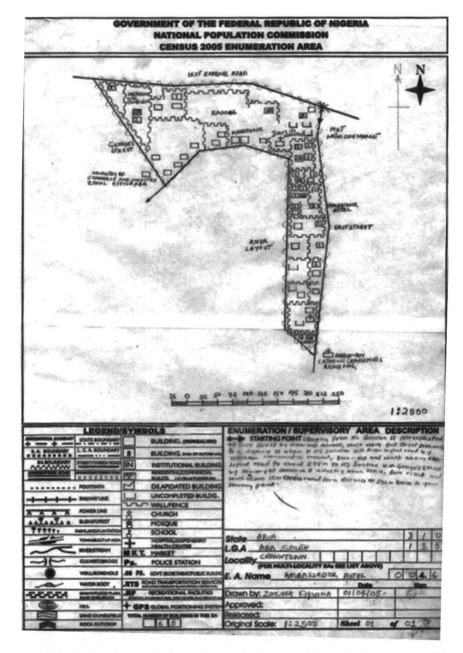

Fig. 13.1 Enumeration area in Local Government Area Aba South (identification code 135), in Abia State (identification code 310), redrawn manually from a grid sheet after stake out and measurement using measuring wheels and compasses

example, 1:750, and North Arrow. In addition to the map section, the sheet contains other sections:

– Administrative section, containing names and identification codes of state, local government area (LGA), locality and EA
– Description of the boundary of the EA
– Legend explaining the meaning of graphic symbols used to represent features

The boundary description in Fig. 13.1 reads as follows:

> Starting from the junction of Ikot-Ekipene road and East street by Water Soe market, walk along East street southwards for a distance of 404 m to its junction with River layout water road by Watchman Catholic Charismatic Renewal turn right and walk along River Layout road to about 575 m to its junction with George's Street by Ministry of Commerce & Industry Zonal office, turn right and walk along Ikot-Ekipene road for a distance of 393 m back to your starting point.

With the help of the North Arrow, names of streets, streams, buildings and so on (annotation) and the symbols representing features on the ground the enumerator can orientate himself in the EA. Broadly, features can be distinguished in man-made and natural features. Furthermore, features are either linear or cover an area. Examples of linear man-made features are roads, railways and channels. Examples of man-made area features are buildings, agricultural fields and orchards. Both man-made and natural features may bear names. Man-made features have regular shape: a building is often rectangular or at least shows straight boundaries, which will meet each other perpendicularly, this is also often the case for agricultural fields and orchards. Linear man-made features are also characterised by regular shape like straight lines and smooth curves. Natural features are often characterised by irregular shape. A natural linear feature, such as a stream, may appear as a jagged line, which will be also the case for the boundaries of area features.

The hand-drawn EA map on A2 transparency and its descriptive features shown in Fig. 13.1 is rather sketchy and it was anticipated that they would be fair drawn with the help of computers and software. The hand-drawn EA maps were scanned and served as backdrop for drawing vector plots of each EA. However, due to lack of equipment and manpower only 7% of hand drawn maps were converted into better readable fair drawn representations. Figure 13.2 shows an example of a fair drawn EA map. Lack of sufficient skilled people also caused that up to 6% of the maps used during census taking never passed the stage of grid sheet.

13.3 Satellite Images

Also very-high-resolution satellite images were used, in particular Quickbird and Ikonos images processed and delivered by Infoterra (Fig. 13.3). They were a gift of the Department for International Development (DFID) of the UK Government. DFID manages UK's aid to poor countries and works to get rid of extreme poverty.

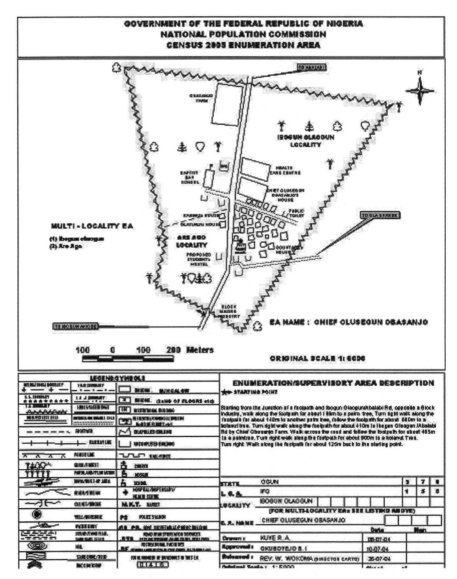

Fig. 13.2 Fair drawn EA map drawn on computer using scanned hand drawn EA maps as backdrop. Chief Olusegun Obasanjo, former President of Nigeria, resides in this EA

The very-high-resolution images covered nearly all major cities, towns and urban conglomerates, that is around 5% of the total land mass of Nigeria but the major part in terms of population. For some areas, particularly located in the tropical rain forest part, no useful very-high-resolution satellites images could be acquired due to nearly permanent cloud cover.

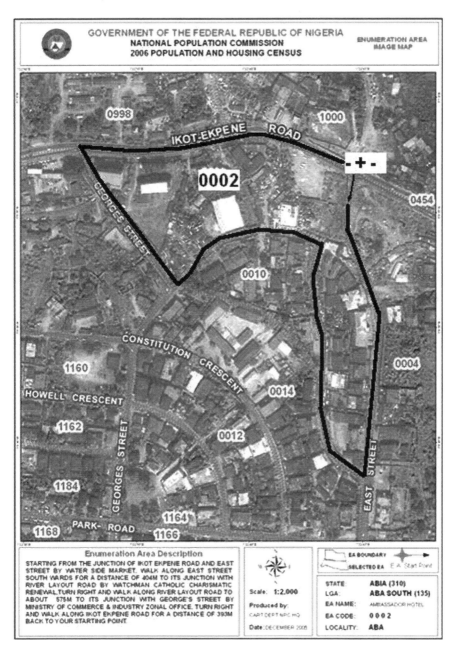

Fig. 13.3   Enumeration area carved out on a high-resolution satellite image

The images were pansharpened and resampled to a ground sample distance of 1 m. SPOT 5 images, GSD 5 m, were also delivered as a DFID gift. Aimed at covering the whole country but due to cloud cover problems, again in the south part of the country, cloud-free SPOT 5 images of only around 90% of the total area could be acquired. Also the SPOT5 imagery were pansharpened, combining the panchromatic band (GSD 5 m) with multispectral bands (GSD 10 m) resulting in colour images. All satellite images were ortho-rectified using the worldwide Digital Elevation Model (DEM) from the Shuttle RADAR Topography Mission (SRTM) made from the C-band radar data. Here worldwide means all the land masses of the globe between 60° North and 56° South latitude. Each grid cell, size 3 arc seconds (90 × 90 m), of this DEM provides an elevation in meters above mean sea level. Most of the images were recorded between 2003 and 2005 and none were acquired before the year 2000. They were thus reasonably up-to-date with respect to new building developments.

The first objective of the very-high-resolution Quickbird and Ikonos images was identifying and outlining the boundaries of EAs in urban areas during the pre-census stage. For that purpose templates for creating image maps were specially designed on which the boundaries of EAs could be demarcated, described and annotated by operators intensively trained by Infoterra (Fig. 13.3).

Ultimately, the majority of EAs should be captured by satellite image maps, but due to lack of capacity the final number did never exceed 40,000 or 7% of the total number of EAs. These images were printed on large format colour plotters and during the census period used by enumerators during fieldwork. A second aim of the satellite images was to act as a means for checking suspicious outcomes in the post-census stage. Population overcounts (imaginary population) are persistently created by local political leaders as a result of allocating national funds to the Local Government Areas and States in proportion to their population size. Politically, Nigeria is a sensitive country and the multicultural population keeps a sharp eye on whether the scarce wealth is properly and fairly distributed and whether the own tribe is not betrayed. So, the fear was that rulers of hamlets and wards, either located in remote areas or encapsulated by urban conglomerates, would create phantom citizens by confiscating blank questionnaire forms and hire some accomplices with a well-developed imagination and a quick hand of writing. Counting the number of houses and buildings in an EA enables to verify the fairness of population figures.

13.4 How Many Are We?

In a period of 2 years, 662,000 EAs were created and these were assigned to 332,800 enumeration teams; that means one team had to cover two EAs. After a preparation period of more than 3 years, including a pre-test carried out between 11 and 13 April 2005 in 20 randomly selected EAs per state to test the accuracy and reliability of EA maps and the census instruments, and a trial census as a full dress rehearsal conducted between 29 August and 3 September 2005, the census was held from 21 to 27 of March 2006. The countrywide census covered all 774 Local Government Areas of the 36 states of Nigeria and the Federal Capital Territory (FCT Abuja).

The provisional results were released on the 29 December 2006 by the National Population Commission and the answer to the question 'How many are we?' could finally be given. During the census moment Nigeria was populated by 140,003,542 people, 71,709,859 males and 68,293,683 females, giving a gender ratio of 105 males to 100 females. The previous census, conducted in 1991, which was only a population census not a housing census, revealed a population of 88.9 million, which means that the average annual population growth rate is 3.2%. The 2006 census was in so far unique that it included a housing census for the first time in the history of Nigeria. On Tuesday 9 January 2007 the President announced that the census results had been accepted and made official the population figures.

A successful census requires that the data collected and statistics derived from the data are trustworthy (Lemmens, 2006). The term 'trustworthy' implies that on the one hand data and statistics are reliable and accurate, and on the other hand that they are acceptable to national and international stakeholders. For example, before committing themselves to any country, overseas investors are likely first to ensure that there is no defect in the socio-economic and demographic figures provided. The same will be true for international donors. A first requirement of trustworthiness is that all processes involved are transparent and all methodologies clearly defined and scientifically sound. The recruitment of enumerators and purchase of equipment and materials should also go according to transparent procedures. Any lack of clarity in methodology, recruitment or procurement processes will affect the ultimate acceptability and credibility of the census.

13.5 Lessons Learned

When looking at the 2006 census from a distance the following issues pop up. Enumeration Area Demarcation was impeded because the boundaries between LGAs and states were not well defined in many parts of the country, resulting in many disputes, which had to be resolved by the National Boundary Commission. Also international boundaries were disputed. Some EAs were poorly drawn or not correctly described leaving uncertainty at the side of the enumerators about the territory to examine. Sometimes the enumerators could finish quickly because only a few hundred people were living in the EA assigned to them, while others were overloaded with work because they had to interview up to one thousand people, twice as many as was considered the optimal number. Awareness rising through publicity campaigns and public enlightenment and advocacy programs was high on the agenda. Publications in electronic and print media, posters, handbills, pamphlets and stickers in four prominent languages aimed at making clear the relevance of census data to the general public. However, disputes on social issues and socio-demographic discrepancies overshadowed the main message that the census is a prerequisite for sustainable development and can be an important catalyst for good governments. This message came not through to large parts of the population and ignorance was widespread.

13.6 Population Estimation

According to the US Census Bureau the world counted 6.9 billion inhabitants by the end of the year 2010. Since 2008 more than 50% of the world population is living in urban areas. Population increase is often associated with urban sprawl, resulting in a decrease of agricultural land and forests and producing problems such as loss of security of food production, loss of biodiversity and deterioration of the environment (Lu et al., 2006). The role of the census, discussed above, is to collect data on people and how they live at the scale of small areas. The most important question a census should answer is 'How many are we?' But in urban fringes with rapid population growth, the population counts recorded every 10 years during census taking become progressively less representative as the decade progresses and therefore interim estimates of small areas are required for planning and business purposes; a difficult demographic task (Hardin et al., 2007). Airborne and satellite images provide a feasible and cost-effective means to arrive at interim estimates of the urban population, which is especially valuable in developing countries as almost all growth of the world's population will accumulate in their cities; after 2017 over 50% of the population in developing countries will be urban dwellers (Jensen et al., 2007). The methods used in the GIS and remote sensing literature for estimating population can be subdivided into statistical methods and interpretation of aerial or satellite images (Wu et al., 2005). Here we focus on the latter.

Population estimation from aerial and satellite imagery can be performed at various scales: from individual dwelling units to land use zones. Using an appropriate mathematical model the number of people living in a dwelling unit can be derived from its size and shape and land use provides a clue for population density. To arrive from the observable parameters – size and shape of the house or house density in a built-up area – to the number of people, a transformation has to be performed in the form of, for example, a linear function: $P = aS + b$, with P the number of people as unknown parameter and S the size of the house or another measurable parameter used as substitute of people. Determination of multiplier a and gain factor b requires the availability of reference values (ground truth), which may be acquired with the help of fieldwork. Ground truth is also necessary to check the accuracy of the results.

The approaches of population estimation applied by researchers using aerial or satellite imagery can be subdivided into four categories (Lo, 2006; Hardin et al., 2007):

– Counting individual dwelling units in high-resolution aerial (stereo) images (e.g. Green, 1957; Collins and El-Beik, 1971; Lo, 1995)
– Measuring the area of urbanised land (e.g. Sutton et al., 1997; Lo, 2001)

– Dividing an area in different land use zones (e.g. Kraus et al., 1974; Adeniyi, 1983)
– Computer supported classification of multispectral satellite images (e.g. Harvey, 2002)

The remaining part of this section will focus on counting houses, land use zones and multispectral classification and is based on the review of population estimation given by Hardin et al. (2007) and the references cited therein.

13.6.1 Counting Houses

Identification and collecting measurable quantities of houses in an administrative unit or other area of investigation requires large scale aerial photos, a skilled operator, a catalogue indicating what should be measured and measurement tools such as a digital photogrammetric workstation. When volume is used as measurable parameter, stereo images should be preferably used. An approximation of volume can be derived from orthophotos using the length of the shadow of the building to arrive at an estimate of its height, which provides, together with the area of delineated footprint, the volume. This approach requires information about date and daytime of image capture – together they determine sun angle – and is not very accurate. The interpretation of large scale aerial photos is labour-intensive. Furthermore, the scale of the photographs should be sufficiently large. For example, scale 1:20,000 aerial photos of Lagos, until 1991 the Federal Capital of Nigeria, did not show enough detail to indentify and measure individual dwelling units (Adeniyi, 1983). These shortcomings triggered the development of less accurate approaches based on identification of land use zones.

13.6.2 Land Use Zoning

The basic assumption underlying the land use zoning approach is that land use provides a clue for house density and this in turn for population density. First one selects the essential land use types, for example, three residential categories (single family, multi-family and trailer park) and one commercial/industrial category. The generation of a land use schema, which is fit for purpose, is critical for success. For each of the categories the population density is determined using some form of ground truth, which can be obtained by counting the actual population in randomly selected sample areas and dividing the outcome by the size of the area. Next the categories are identified and delineated on the images. The decisive parameter to be determined from the resulting zones is area which gives, after multiplication with the population density of the concerning category, the population of the zone. By summing up the population per zone the total population of the administrative unit is calculated. An experiment conducted by Kraus et al. (1974) revealed that the land use zoning approach could easily produce 7% overestimation due to an exaggerated ground truth value, and severe underestimation in the multifamily residential category, due to problems identifying isolated apartments and residences in older business districts, resulting in an overall 7% underestimate. Using the land use zoning approach in a study covering Ilorin, capital of Kwara State, Nigeria, Olorunfemi (1984) concluded that the method is suited for administrative units where the area of residential land is known a priori or population data are unavailable due to the remote location of the administrative unit, political problems or census taking has not been performed because of lack of resources. The applicability of the land use zoning approach is not limited to aerial photos. Also satellite images, which are less expensive, can be used.

13.6.3 Multispectral Classification

Since the launch of Landsat-1 in July 1972, the availability of image content in the form of pixels has triggered the development of population estimation approaches based on multispectral classification. The spectral signature of each pixel is the integrated result of the reflections of the diverse land uses covered by the pixel. So, the spectral signature or derived quantities, such as texture measures, provide a clue for land use. Part of the land use will be residential, i.e. covered by houses. Since the spectral signature of houses differs from the spectral signature of other types of objects, such as vegetation, the density of houses can be determined from multispectral pixels and this gives an indication for population density and hence number of people living in the area. Per-pixel population estimation seems to produce appropriate results for larger areas as errors of over- and underestimation tend to average to zero. Nevertheless, as Li and Weng (2005) noted: 'using remote sensing techniques to estimate population density is still a challenging task both in terms of theory and methodology, due to remotely sensed data, the complexity of urban landscapes, and the complexity of population distribution'.

Many researchers working in the field of airborne Lidar are developing and testing algorithms to extract size and shape of buildings from dense airborne Lidar data (see, e.g. Lemmens et al., 1997; Maas and Vosselman, 1999; Zhang et al., 2006; Meng et al., 2008). Demir and Baltsavias (2010) tested four methods combining information from aerial images and airborne Lidar data. The disadvantage of Lidar data is that their acquisition is rather expensive and hardly affordable for many countries.

References

Adeniyi PO (1983) An aerial photographic method for estimating urban population. Photogramm Eng Remote Sens 49:545–560

Collins WG, El-Beik AHA (1971) Population census with the aid of aerial photographs: an experiment in the city of Leeds. Photogramm Rec 7:16–26

Daugherty HG, Kammeyer KCW (1995) An introduction to population, 2nd edn. Guildford Press, New York, London, 244 p. ISBN 0 89862 616

Demir N, Baltsavias E (2010) Combination of image and Lidar data for building and tree extraction. In: Paparoditis N, Pierrot-Deseilligny M, Mallet C, Tournaire O (eds) Int Arch Photogramm Remote Sens Spat Inf Sci XXXVIII(Part 3B):131–136

Green NA (1957) Aerial photogrammetric interpretation and the social structure of the city. Photogramm Eng 23:89–99

Hardin PJ, Jackson MW, Shumway JM (2007) Intraurban population estimation using remotely sensed imagery. In: Jensen RR, Gatrell JD, McLean DD (eds) Geo-spatial technologies in urban environments: policy, practice and pixels. Springer, Berlin, Heidelberg, New York, pp 47–92. ISBN 978-3-540-22263-4

Harvey JT (2002) Population estimation models based on individual TM pixels. Photogramm Eng Remote Sens 68:1181–1192

Jensen RR, Gatrell JD, McLean DD (2007) Applying geospational technologies in urban environments. In: Jensen RR, Gatrell JD, McLean DD (eds) Geo-spatial technologies in urban environments: policy, practice and pixels. Springer, Berlin, Heidelberg, New York. ISBN 978-3-540-22263-4

Kraus SP, Senger LW, Ryerson JM (1974) Estimating population from photographically determined residential land use types. Remote Sens Environ 3(1):35–42

Lemmens M (2006) Carving up a country. GIM Int 20(1):11

Lemmens M, Deijkers H, Looman P (1997) Building detection by fusing airborne laser-alimter DEMS and 2D digital maps. Int Arch Photogramm Remote Sens 32(Part 3–4/W2):42–49

Li G, Weng Q (2005) Using Landsat ETM+ imagery to measure population density in Indianapolis, Indiana, USA. Photogramm Eng Remote Sensing 71:947–958

Lo CP (1995) Automated population and dwelling unit estimation from high resolution satellite images: a GIS approach. Int J Remote Sens 16:17–34

Lo CP (2001) Modelling the population of China using DMSP operational linescan system nighttime data. Photogramm Eng Remote Sens 67:1037–1047

Lo CP (2006) Estimating population and census data. In: Ridd MK, Hipple JD (eds) Remote sensing of human settlements: manual of remote sensing, vol 5, 3rd edn. American Society for Photogrammetry and Remote Sensing, Falls Church, VA, pp 337–377

Lu D, Weng Q, Li G (2006) Residential population estimation using a remote sensing derived impervious surface approach. Int J Remote Sens 27(16):3553–3570

Maas H-G, Vosselman G (1999) Two algorithms for extracting building models from raw laser altimetry data. ISPRS J Photogramm Remote Sens 54:153–163

Meng X, Wang L, Currit N (2008) Morphology-based building detection from airborne Lidar data. Photogramm Eng Remote Sens 75(4):437–442

Schulte Nordholt (2005) The Dutch virtual Census 2001: a new approach by combining statistical sources. Stat J UN Econ Comm Eur 22(1):25–37

Sutton P, Roberts D, Elvidge CD, Meij H (1997) A comparison of nighttime satellite imagery and population density for the continental United States. Photogramm Eng Remote Sens 22:3061–3076

United Nations (2008) Principles and recommendations for population and housing censuses; revision 2, Statistical Papers, United Nations, New York, series M, no. 67 (Rev. 2)

van Maarseveen J (2005) Twee eeuwen volkstellingen: de virtuele volkstelling 2001 vergeleken met haar voorgangers, Sociaal Economische Trends, 1ᵉ kwartaal 2005, pp 33–38

Wu SS, Qiu X, Wang L (2005) Population estimation methods in GIS and remote sensing: a review. GIS Remote Sens 42(1):58–74

Zhang K, Yan J, Chen S (2006) Automatic construction of building footprints from airborne Lidara data. IEEE Trans Geosci Remote Sens 44(9):2523–2533

Chapter 14
Risk and Disaster Management

> *The case for GIS at all stages of disaster management is overwhelming, from preparedness through to response, recovery and mitigation. Yet at the same time almost nobody outside the realm of geospatial professionals recognises this.*
>
> Michael F. Goodchild, Professor of Geography at the University of California, Santa Barbara, USA

Natural disasters – floods, drought, hurricanes, landslides, earthquakes, tsunamis and volcano eruptions – increasingly cause casualties and damages to infrastructure and property. In addition to huge suffering, the economies of many countries, especially the developing ones, where 95% of all fatal casualties from natural disaster occur, get severely affected as a disaster strikes the region. The role of geo-information technology to support risk and disaster management has been convincingly demonstrated and there is no doubt about its importance (Zlatanova and Fabbri, 2009). The number of geo-information technologies that can support risk and disaster is steadily growing, particularly in developing countries. For example, the Asian Disaster Preparedness Center, Bangkok, Thailand, operational for over 20 years, established a regional network – involving 26 countries – of stations for capturing data for signalising multiple types of hazards.

In absolute terms, population growth and urbanisation has never before reached such high rates as today, which has resulted in the peculiar situation that the number of people being alive today is larger than the number of people ever died during the entire history of human mankind. From the data provided by the US Census Bureau it can be estimated that the world population is seven billion people by the end of 2012. Mankind is putting increasing pressure on the one and only Earth his race has to share, not only by sheer numbers but also, unfortunately, wasteful lifestyle. Those numbers and that lifestyle are causing many areas to become rapidly more vulnerable to a wide range of man-made and natural hazards. Disasters caused by earthquakes, volcano outbreaks, hurricanes, explosions, drought, flood, major fires and major oil spills are daily news items. Indeed, we are living in a world full of risk, one that has to cope with the hectic, multitudinous manifestations of our being in it; a world which behaves in a sometimes hostile and cruel way, as taking revenge for our own careless behaviour. This chapter elaborates upon the different types of disasters, the five stages of disaster management and gives examples of the use of

M. Lemmens, *Geo-information*, Geotechnologies and the Environment 5, DOI 10.1007/978-94-007-1667-4_14, © Springer Science+Business Media B.V. 2011

geo-information in the different stages of natural and man-made disasters. Since we explore practical experiences gained during real life situations most of the literature used stems from professional journals.

14.1 Natural and Man-Made Disasters

Natural disaster risks stem from sudden energy release in one of the three basic environmental compartments: air, water and land. Natural disaster can be defined as any occurrence causing decay of usability and productivity of land and structures built upon it, and injury or loss of human life. Disasters also have longer term consequences for economic growth, development and poverty reduction – disaster, poverty and development are intertwined. Disasters can considerably set back development efforts by making financiers reluctant to invest, further limiting development of the area. Annual economic loss associated with natural disasters doubled every decade of the last half century. Economic loss resulting from natural hazards even tripled in the course of the nineties and now amounts to an annual one million millions Euros. Nonetheless, many people living in developed regions remain quite confident that they are living in the most secure part of the world. To some extent this is true. The United Nations estimated that 85% of those exposed to earthquakes, tropical cyclones, floods and droughts live in countries classed as medium or low in development terms.

Over half of the Earth's population – about 3.5 billion people – live in urban conglomerates. The motivation that moves people to cities is the expectation of a better life. And as Dr. Anna Tibaijuka (2008), Under Secretary General and Executive Director UN-HABITA states, 'This process of urbanisation cannot be reversed; sending people back to their villages simply does not work; it never did and it never will'. As a result, what was a city three decades ago has now turned into a metropolis, and metropolises have turned into megalopolises. Many rich towns are jam-packed and suffer from traffic congestion. Rescue teams on their way to a road accident, a building on fire or any other time-critical situation are often confronted with deadly delays. In many overcrowded metropolitan areas selection of the fastest route to an emergency is left to the ambulance driver or fire engine navigator. Emergency information systems, in which GNSS technology and digital map data are key components, may help not only to guide an emergency team along the fastest path but also to allocate available teams to different stations to maximise area cover; co-ordinated use of geo-information technology saves human lives. Geo-information technology can be successfully applied because all natural and man-made disasters have a spatial component to deal with.

14.2 Phases of Disaster Management

The Yangtze River summer flood of 1998 and the 2001 and 2003 earthquakes hitting Gujarat, India, and Bam, Iran, respectively, the Christmas 2004 tsunami disaster, the Sichuan earthquake 2008, the Haiti Earthquake January 2010 and many others have shown that management of disasters is urgent and therefore it has been given high priority on the international agenda. Disaster management involves many diverse activities, which may be categorised into two subsequent states: pre-disaster and post-disaster activities. Pre-disaster activities include the following (Lemmens, 2005b, 2007):

1. *Recording* the sensitivity of the region to certain types of catastrophes.
2. *Risk-reduction:* making provisions to reduce the region's vulnerability to catastrophes.
3. *Readiness or preparedness:* planning emergency aid and development of scenarios and monitoring systems in case a disaster happens.

Post-disaster activities include the following:
4. *Response* to save lives: injured people are provided with medical and nursing care, food support and provisional housing.
5. *Recovery:* revitalising and reconstructing the area: rebuilding housing, reconstructing roads and railways.

14.2.1 Recording

The first stage – *recording* – concerns assessing the sensitivity of the area to certain types of catastrophe. In this stage the risks and danger for human life and environment are determined. Sur and Sokhi (2006), for example, demonstrate how geo-information technology can be used to assess the vulnerability of cities for exploding petrol filling stations using as study area a suburb in New Delhi with much recent building development and activity. The risk appeared to be particularly big due to high traffic volume. Many buildings are vulnerable to explosions at two or more petrol stations (see for more details Section 14.4).

14.2.2 Risk Reduction

The second stage concerns *risk reduction*. When the risks and danger are known, one may start to take provisions to make the region less vulnerable to the occurrence of the catastrophes to which the area is sensitive. Proper land-use planning and management, and taking strengthening provisions are the actions to be carried out here. An example is the method developed by Klingseisen and Leopold (2006) to determine landslide vulnerability of hilly and mountainous areas in Austria (Fig. 14.1). Lack of risk awareness combined with the pressure of population growth causes governments to allow the building of houses in potential landslide zones. The factors determining the possible occurrence of landslides include geology, land use, slope, and aspect and terrain roughness. Geo-data used in the study included 1:200,000 geological maps and airborne imagery at scale 1:15,000 captured in 2001. From the latter Digital Colour Orthoimages with a Ground Sample Distance (GSD) of 0.25 m were created and by using stereo matching techniques a Digital Elevation Model

Fig. 14.1 Typical landslide morphology with subsidence, very steep slopes and bent trees. The permanent movement of the ground demolished the building at the foot of the slope

(DEM) with 10-m raster spacing. Mapping was supported by national topographic base-map scale 1:50,000 and topographical vector data containing, amongst other, rivers and roads. Land-use data stemmed from the European Corine Landcover dataset. Information on geology and landslide occurrence was obtained through interviews with local authorities, land-use managers and geologists, and from the literature. During fieldwork landslide extent was recorded with a Mobile GIS consisting of a GNSS receiver equipped with ESRI ArcPad 6 software (accuracy of 10 m enabled generation of 1:50,000 hazard map) and also data on morphological features, tree bending – a landslide indicator – and damage to manmade objects was collected. With the help of a Geographical Information System (GIS), the probability of the occurrence of landslides over the whole area was computed on 10 m raster cells. About 50% of the region is predicted to be not endangered, in about 33% of the region landslides may occur and about 17% is definitely endangered.

14.2.3 Readiness

In the third stage – *readiness* – planning of emergency aid and development of scenarios and monitoring systems are central together with the establishment of early-warning systems. Especially in the creation of early-warning systems geo-information plays a key role. To prevent human beings from death or injury early-warning systems have been developed based on using advanced geo-information technology, such as earthquake prediction using GNSS technology. Indeed, GNSS is increasingly becoming a key technology for earthquake monitoring. For example, the subcontinent of India has developed a GNSS-based early-warning strategy for Earthquake Hazard Assessment in 2005. India regularly witnesses devastating natural disasters that cause great human suffering and loss of goods. In the Himalayan area, more than 650 earthquakes of a magnitude exceeding 5 occurred over the past 100 years while the same area is also vulnerable to landslides. The Rivers Ganges and Brahmaputra regularly flood the plains of north-east India, through which 60% of the country's river water flows. Not only is the sudden presence of abundant water a threat, but the lack of water is also a threat; 68% of sown area is prone to drought. There are also recurrent forest fires that harm the vegetation dynamics of ecosystems, affecting tropical structure and contributing to an increase of greenhouse gasses in the air. The east and west coasts are affected by severe cyclones.

The National GPS Programme for Earthquake Hazard Assessment in 2005 initially launched in 1998 by the Government of India Department of Science & Technology, includes the establishment of a network consisting of 43 permanent GNSS stations, about 700 semi-permanent (campaign-mode) GNSS stations in the Himalayan regions and the peninsular shield of India, and many more field-GNSS stations being established and monitored for local campaigns. The Indian Earthquake Hazard Assessment programme will contribute to the understanding of plate motion and crustal deformation in the region and to the development of models for earthquake-hazard assessment (Kulkarni, 2006).

14.2.4 Response

The urgency of the activities in the above three stages is often hard to understand because they have to be carried out when the sky is still seemingly cloudless. Nothing has happened yet; there is no urgency to force authorities to put effort into these activities. Given the necessary financial resources and all the other priorities and issues many countries are facing, authorities will thus often feel no drive to come into action up to the moment that a disaster severely hits the country. Then one gets triggered to take measures. This is apparently different from the fourth and fifth stage. Now, the catastrophe has actually struck. Thus authorities cannot afford to keep their hands crossed; the gravity of the catastrophe creates an obligation to respond rapidly. The fourth stage – response – is the most dramatic stage. The catastrophe caused unthinkable human suffering and environmental damage. Rescue teams will attempt to save lives, injured people will be cured and nursed, and relief will be offered to sufferers by food support and provisional housing. Damage-inventory systems are important in helping rescue teams to find their way through devastated built-up areas.

What can we do to mitigate disaster? Most important is providing sufficiently detailed and timely geo-information to disaster managers, emergency teams and foremost to the victims and other people affected (Tran-Vinh, 2010). They need rapid answers to the *What*, *Where* and *When* questions to design and execute warning and evacuation plans and to disseminate information on the actions that they want to undertake to all involved. A decision support system built on GIS platforms and nourished with appropriate data is a prerequisite to help answering the above 'W' questions. GIS platforms contain the proper tools to collect, manage, analyse, model and display spatio-temporal data and are able to disseminate information to the public through the internet. The development of such systems is a challenge because near real-time data from larger regions have to be collected and interpreted. A natural disaster usually rapidly diverges causing major casualties and damages over larger regions. Consequently, continuous monitoring of ampli-. tude and divergence direction is important. Disasters do not stop at boundaries of provinces or states; data collection should thus surpass administrative boundaries. Therefore, (inter)national linking of monitoring networks is necessity. Furthermore, on-site collected data have to be integrated with other, existing geo-data, such as base-maps or orthomaps, to be useful. Since the data are usually distributed over several institutions and organisations within a country a well-functioning spatial data infrastructure (SDI) at both national and regional level should be in place.

The near real-time data fostering the above GIS-based decision support system may stem from post-event aerial images, Lidar data or optical and radar satellite imagery. Airborne remote sensing, GIS technology and ICT, for example, played an important role in post-disaster damage assessment after Hurricane Katrina – the costliest and one of the deadliest hurricanes in the history of the United States of America. The response stage is world news for a few weeks and given high priority by all news stations all over the world. The dramatic images displayed during these days are sometimes all that the general public will remember of the disaster for many years. When the general public has gone back to the order of the day, the fifth stage arrives: recovery.

14.2.5 Recovery

In the recovery stage, actions are undertaken so that survivors can, in the foreseeable future, pick up their daily lives again. This stage thus mainly consists of revitalisation and reconstruction. Houses are rebuilt; roads and railroads as well as other works of infrastructure are repaired. An important although often neglected part of this stage is strategic development; this means tackling the question of how to prevent the area from future disasters in order to secure a safe and sustainable future. An essential part of this tail stage is thus that it acts as a driving force and fosterer for starting and keeping vivid the initial stage: assessment. Prevention is better than cure and developing early-warning systems aims at preventing human suffering. The Indian government has understood this. After the Christmas 2004 tragedy, a tsunami early-warning system was put in place with surprising speed.

14.3 Tsunami Early-Warning System

The mighty submarine earthquake lifted the floor of the Indian Ocean, triggering a vast rush of water what surged towards the coasts of India, Indonesia, the Maldives, Sri Lanka and Thailand (Lemmens, 2008a). Nearly a quarter of a million people lost their lives and the damage was unprecedented. The devastating oceanic waves constituting a tsunami originate from submarine earthquakes mainly generated by the forces released by subduction of oceanic tectonic plates. In South Asia two subduction zones periodically release huge forces: one along the Andaman-Nicobar-Sumatra Island Arc, east of India, and the other in the Makran zone, north of the Arabian Sea, west of India. Rupture along the zone east of India caused the gigantic submarine earthquake leading to the tragedy of 26 December 2004. The trail of destruction resulting from a huge wall of water slamming up over her coasts demonstrated to the Government of India the imperative of setting up a National Early Warning System (NEWS) designed to guard against tsunamis induced by the two subduction zones described above.

The project was completed in just 2 years and involved about 150 scientists and engineers from fourteen organisations. NEWS costs Rs.125 Crore (US $30 million) to put in place, executing authorities being the Indian Ministry of Earth Sciences as nodal ministry, in collaboration with the Department of Science and Technology, the Department for Space and the Council for Scientific and Industrial Research. NEWS was formally inaugurated on 15 October 2007. The three main observation components of the system enable estimation of hazard risk posed by a submarine earthquake; one component is for detecting the cause of a tsunami, earthquake; two are for detecting consequence, sea-level rise.

Real-time seismic data from (inter)national seismic networks enables detection of earthquakes greater than magnitude 6 on the Richter Scale, as they occur in the Indian Ocean and within 20 min of the birth of such forces. Pressure sensors forming the core of the system are positioned on the bottom of the ocean and act as sentinels to detect sea rise; four are installed in the southern Bay of Bengal, and two in the northern Arabian Sea. Time-series analyses allow computation of changes in water pressure; increase indicates a passing tsunami. The measurements are passed on to a buoy, which then sends the information in real-time via satellite to the Indian National Centre of Ocean Information Services (INCOIS) in Hyderabad, where a state-of-the-art National Tsunami Early Warning Centre (NTEWC) has been set up. At this hub, real-time data from all sensors are collected and analysed, and alerts issued to the control room at the Ministry of Home Affairs for further dissemination to state and local administration, media and the public. Communication with the Ministry of Home Affairs is via a satellite-based, virtual, private net; authorised officials are also contacted by phone, fax, SMS and e-mail.

The second observation component involved in detecting sea-level rise consists of a series of 30 tide-gauges that monitor the progress of tsunami waves. All data generated by the three sensor systems in the network are continuously transmitted via satellite to the Hyderabad centre and here monitored. The travel time and magnitude of a tsunami can be estimated from a large database of model scenarios, using location and magnitude of earthquake derived from the sensor data. Historical earthquake data provide a basis for ascertaining areas under threat of inundation and, by combining this information with high-resolution DEMs and cadastral maps at scale 1:5,000, it becomes possible to identify communities and infrastructure at risk. A high level of redundancy is built into the communication system to prevent it failing in the case of technological breakdown. The large submarine earthquake of magnitude 8.4 that occurred on 12 September 2007 in the Indian Ocean proved the effectiveness of the system. But technology alone provides just part of the solution. The weakest component in any early-warning system is not technology but the human factor. Particularly crucial are preparedness of government agencies and the capability of people in vulnerable regions to respond swiftly to alert. Periodic workshops guarantee the user community becomes familiar with the use of geo-information. Further, the Indian government tries to increase public awareness by distributing easily understandable publicity material. Early-warning systems cannot prevent disasters but may reduce harm, damage and human suffering.

14.4 Petrol Station Vicinity Vulnerability Assessment

Petrol stations contain high concentrations of highly flammable substances. Fire and explosions may cause huge damage, injury and loss of human life. The main causes of petrol filling stations catching fire include sparks from a vehicle or electrical short-circuit, cigarette lighting and reservoir leakage after earthquake and subsequent fire. For example, on 28 September 2005 in Dehradun, north of New Delhi, a petrol tanker caught fire during emptying. A spark produced by a refilled scooter setting off caused the accident. A motorcycle near the tanker also caught fire and a car fuelled by LPG exploded. Two people were badly injured. Sur and Sokhi (2006) show how GIS can help to assess the vulnerability of the vicinity of filling stations by a case-study conducted in Dehradun.

The city of Dehradun is an area of much recent building development and economic activity and accommodates a total of 17 petrol stations, 14 of them in and around the central business area, situated along four main roads carrying heavy traffic. Six petrol pumps are present on just one of these roads, all situated within 200 m distance from the next. So if an explosion occurs many people may be injured or lose their lives. A rough estimate of the population living in the vicinity of a filling station is obtained by taking as rule of thumb 20% of the population residing within the 200-m buffer zone around each petrol pump. During field survey building use within a radius of 200 m around the petrol pumps was determined and stored in a database and the positions of the pumps were determined by GNSS fieldwork and marked on Ikonos satellite imagery.

Five categories of building use are distinguished: residential, commercial, commercial/residential, institutional and medical (Fig. 14.2). Since commercial buildings are mostly congested and more inflammable materials are present, their vulnerability is higher than that of other buildings. In the 200-m buffer zones a total of 2,418 buildings were counted. Next, estimation was made of the population of each of the buffer zones; the total number is around 14,000 persons. Both building density and population density are divided into three classes (Table 14.1). On the basis of building density, land use and population present within the 200-m buffer zone, the zones around the petrol pumps are grouped into three categories:

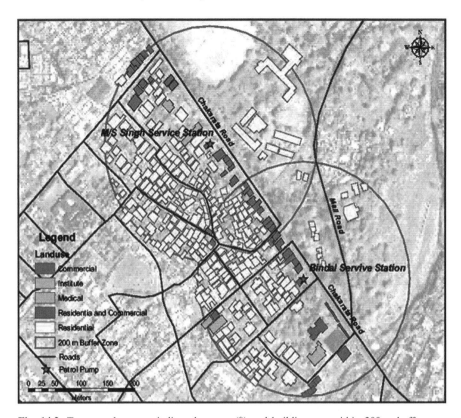

Fig. 14.2 Two petrol pumps, indicated as star (*) and building use within 200-m buffer zone (*circles*) determined by using a GIS

Table 14.1 Building density classes and population density classes

Class	Building density [buildings/hectare]	Population density [persons/buffer zone]
High	>15	>1,500
Moderate	10–14	1,000–1,500
Low	<10	<1,000

less hazardous, moderately hazardous and very hazardous. Four of the 14 buffer zones appeared to be very hazardous, eight are moderately hazardous and two are less hazardous. In the four very hazardous zones the buildings are mainly used for commercial and commercial/residential purposes and highly vulnerable to fire risk, whilst very high traffic volume increases the risk. Many buildings are vulnerable to explosions at two or more petrol filling stations.

The method can be refined by including other parameters, such as wind direction, daytime and night-time population, number of vehicles on the road and level of risk of people residing in varying buffer zones (50 m, 100 m, etc.). Combined with map data such as streets, pipelines, buildings, residential areas and storage facilities, emergency managers can use the method for purposes of mitigation, preparedness, response and possible recovery.

14.5 Fire Fighting

Is our municipality effectively covered by fire stations and, if not, how should fire stations be distributed? How should optimal distribution of stations be determined? What traffic problems may be faced at various moments of the day? These are all questions confronting fire brigades. The Centre for Sustainable Tourism and Transport, NHTV Breda University of Applied Sciences, the Netherlands carried out a case study in the Dutch city of Zutphen and demonstrated the capabilities of GIS in optimising location of fire stations (Landré, 2008). Time is the decisive factor in any fire fighting. The time between convey and actual fire fighting may be differentiated into five subsequent stages:

– Dispatch time: time between call and notifying fire station.
– Turnout time: time between notification and leaving fire station; may depend on type of vehicle and moment of the day.
– Drive time: time between leaving station and parking fire engine in vicinity of fire; turnout time plus drive time gives response time.
– Excess time: time necessary to move from fire engine to fire.
– Set-up time: time necessary to set up equipment.

Fire risk and drive time are the two key parameters for optimal location of fire stations. Residential areas generally have lower fire risk than city centres, allowing for longer response times. The vulnerability of areas can be mapped and stored as a layer in a GIS. Drive time is computed as distance divided by average speed; the latter depends on traffic density (largely a function of moment of the day) and type of road: highway, collector or local road. Optimal location of fire stations thus also requires use of a detailed road map. Next possible locations of fire stations can be identified and along a trial-and-error procedure their optimal distribution determined. One can test, for example, whether addition of a fire station to the east of an existing one would substantially improve response time. By including a cost-benefit analysis feasibility can be checked.

14.6 Flooding

Many civilisations all over the world flourish as a result of being located on river deltas. To keep all feet dry in the lowlands, two enemies have to be combated: flood threat from the sea and flood threat from the river(s). On the other hand, water is a vital source of life. To stay on friendly terms with your biggest enemy requires an almost divine level of management. No wonder that in a country like the Netherlands the heir to the throne has chosen water management as major professional speciality through which to safeguard the heritage of his co-citizens. In many other countries too the development of flood forecasting systems stands high on the government agenda. Insurance companies show a high interest in flood-risk assessment systems.

In the aftermath of Hurricane Katrina Aerial photogrammetry has proven of great help in assessing damage, prioritising relief efforts and planning the recovery of flooded areas. Hurricane Katrina formed over the Bahamas on 23 August 2005 and made landfall on the Louisiana and Mississippi coastlines on 29 August. Katrina flooded 80% of New Orleans and many other areas in the Mississippi Delta. Many levees failed and the water could freely stream over the land (Fig. 14.3). With Katrina still days from landfall authorities decided to arrange a photogrammetric survey to map the damage that was sure to come (Corbley, 2006). Starting on 3 September over 30,000 square kilometres of the most heavily hit parts were captured with four ADS40 digital cameras, a development of Leica Geosystems, at GSD 30 cm during 16 days. Since all ground control points were underwater and none could be put in place, georeferencing of the images was solely done using the onboard GNSS receiver and inertial navigation system. On the images homes, which had been rendered uninhabitable, could be identified and linked with zip codes and street addresses databases. An essential part of each post disaster activity is providing financial aid to property owners whose homes have become uninhabitable.

Fig. 14.3 Levee failure on 17th Street Canal, New Orleans, USA, caused by Hurricane Katrina, August 2005, captured by ADS40 digital camera, Leica Geosystem

With hundreds of thousands of homes feared lost the usual damage assessment by field-inspection would be too time-consuming. The GSD of 30 cm of the ADS40 imagery enabled to draw polygons around heavily damaged blocks of residential buildings using ESRI's ArcView. The polygons were linked with zip codes and street addresses, while field surveyors verified the damage assessments by spot-checking. To distribute the images among the variety of parties involved, the internet was used.

Although Hurricane Katrina was the costliest and one of the deadliest hurricanes in the history of the United States of America, on a global scale the Christmas 2004 tsunami disaster is unprecedented in terms of the amount of sudden loss of life, injured victims and demolished constructions. However, in terms of economic losses, this natural disaster, which did overwhelmingly affect nearly every coastal zone of the Indian Ocean, seems to be of a quite modest extent. Indeed, the ratio of actual damage and financial losses in the poor regions of the world is much more profound than in the richer ones. This seems to be an unavoidable social phenomenon, notwithstanding its unfairness.

14.7 Earthquake Monitoring

Countries situated in areas vulnerable to Earthquakes, such as Iran and India, are building GNSS networks for monitoring and studying crustal strain fields, velocities and displacements. In 1998 India launched a National GNSS Programme for Earthquake Hazard Assessment, especially to monitor the seismically active Himalayan region. To estimate regional crustal strains, and to identify and monitor seismically active regions, a network of about 700 semi-permanent GPS Stations is being set up covering the whole of India; the spacing between stations is about 40–60 km (Kulkarni, 2006). As far as possible, existing stations of the Great Trigonometrical (GT) Triangulation Network of India are included in the network.

Also Iran frequently suffers from heavy and disastrous earthquakes as the country is lying in one of the most tectonically active zones in the Alpine-Himalayan belt. In December 2003 and in March 2005 disastrous earthquakes took place in Bam (Fig. 14.4). The earthquakes are due to collision of the Eurasian plate in the north on the Arabian plate in the south, with shortening rate of 2–2.5 cm per year. To monitor surface displacement and measure velocity and strain fields, the National Cartographic Center of Iran has established a network of one hundred GPS stations: the Iranian Permanent GPS Network for Geodynamics (Djamour et al., 2006).

To properly study general tectonic movements, the 1,650,000-km^2 area of Iran actually needed to be subdivided into cells of 900 km^2 (30 km by 30 km) requiring 1,800 GNSS stations but just over a hundred GPS receivers were available when one designed the network. Prioritising of locating GNSS stations was on basis of seismic activity and population density. The network consists of two parts, a base network consisting of 41 stations covering the entire country (see Fig. 14.5), and three local networks in the most densely populated and active zones, built by using

Fig. 14.4   Buildings in Bam, Iran, demolished during the earthquake of December 2003

the remaining receivers. The stations collect and store raw GPS data and send them to Tehran processing centre on a daily basis for final processing and analysis.

Earthquakes are often followed by outbreaks of fire. In Japan such events are frequent and cities are very vulnerable to fire damage. The use of prior and real-time prediction systems could help to limit the numbers of people injured and houses destroyed. Geo-information technology enables to build such systems and The Japanese National Research Institute of Fire and Disaster has developed a real-time fire-spread simulation system as part of a more comprehensive fire-fighting programme (Shinohara and Sasaki, 2005). Shortly after any earthquake the system supports optimal fire fighting through predicting fire-spread by using data on each and every building and data measured on-site during the course of the fire, including wind speed and direction. Building data consist of an outlining polygon, number of storeys and floor area, as collected in advance from existing digital maps. Prior-simulation systems examine in advance (1) possible patterns of fire outbreaks and fire-fighting operations, (2) optimal use of the fire-fighting force for each possible pattern and (3) fire-fighting planning in case of an earthquake.

14.8 Disaster Management from Space

Space-programme research and development since the early 1970s has culminated today in the broad practical application of advanced Earth Observation (EO) technologies. Earth observation from space has proved of vital importance for disaster management; it can help to detect precursors of disaster and can contribute

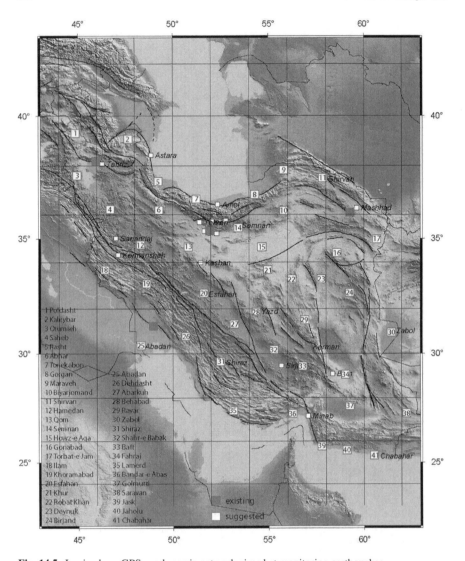

Fig. 14.5 Iranian base GPS geodynamic network aimed at monitoring earthquakes

significantly to relief efforts after natural disaster (Lemmens, 2005a). Such imagery provides comprehensive real-time coverage of large areas at frequent intervals.

14.8.1 Sichuan May 2008 Earthquake

It has often been proved that the rapid provision of high-quality radar and optical satellite imagery enables quick and detailed evaluation of disaster areas in the response stage (Lemmens, 2008b). One such proof was given in the emergency-response to the Sichuan Earthquake, which shook the steep eastern margin of the Tibetan plateau in Sichuan Province, China on 12 May 2008. The earthquake was so strong that office buildings in Beijing (1,500 km away) and Shanghai (1,700 km) swayed. The earthquake destroyed infrastructure over 270 km along a geological fault, created lakes and caused landslides and debris flows which buried populated areas and blocked streams; it left about seven million people home-less, killed over 70,000 and injured more than 360,000. It was the deadliest and strongest earthquake to hit China since the 1976 Tangshan quake, which killed at least a quarter of a million. Heavy rainfall and destroyed road networks hindered rescue work of rescue teams in the first few days.

In a joint attempt to help relieve the emergency situation in Sichuan, Infoterra GmbH and Spot Image SA have provided China with satellite imagery by tasking their respective satellites, the German radar satellite TerraSAR-X and the French optical satellite SPOT 5, to acquire data of the earthquake area. Seven hours after reception it was possible to deliver the first processed TerraSAR-X images. The post-earthquake TerraSAR-X imagery (StripMap mode, 3 m Ground Sample Distance (GSD), acquired 15 May 2008) was visually interpreted and compared with pre-earthquake reference data consisting of enhanced topographic maps. The enhancements were extracted as vector data from Spot5 imagery (panchromatic, 2.5 m GSD, acquired 2006) from which major roads, transport lines, bridges and urban areas were mapped. In this way the level of damage to towns, villages and infras-tructure could be assessed and potential threat from rivers blocked by landslides identified (Fig. 14.6). Over 30 of such river blockings could be pinpointed and when they would break or overflow many lives, both survivors and rescue workers, down-stream would be endangered. The detailed maps could be delivered the day after the recording of the TerraSAR-X post-earthquake imagery and they informed the Chinese authorities of the extent of the damage. The results were also made available through the internet to give other aid organisations all over the world access.

The experiences gained during the Sichuan May 2008 Earthquake was in line with earlier findings namely that space technology provides useful and timely information for disaster management. On 6 September 2008 China launched two satellites as the first components of a larger configuration of small satellites for monitoring the environment and disasters (Wang, 2010). Each satellite is equipped with two CCD cameras, capturing the visible and near infrared parts of the elec-tromagnetic spectrum in four bands (0.43–0.52 μm; 0.52–0.60 μm; 0.63–0.69 μm and 0.76–0.90 μm); GSD 30 m and swath width 360 km for a single camera and 700 km if two cameras are operating in tandem. One satellite is further equipped with a panchromatic camera, swath width 50 km and a sway of 30°, covering the spectral band 0.45–0.95 μm with GSD 100 m, while the other satellite has on board an additional infrared scanner covering two channels of the near infrared (0.75–1.10 μm; GSD 150 m and 1.55–17.5 μm; GSD 150 m) and two thermal infrared bands (3.50–3.90 μm; GSD 150 m and 10.5–12.5 μm; GSD 300 m). The wide-swath and short revisit-period has been proven beneficial to support governmental emergency responses. During the Yellow River ice flood in 2009 ice jam length

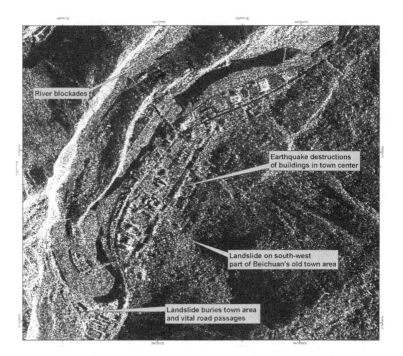

Fig. 14.6 Landslides visible on annotation map based on post-earthquake TerraSAR-X StripMap image, the landslide on the left blocks a river and the one on the right encounters the city of Beichuan; the landslide eventually slammed into Buryin, the old town area of Beichuan

and ice flood area could be determined. The images also enabled to monitor water reserves and to determine changes of cropland and ecosystem during the drought period scourging the south-west part of China in 2010. After the earthquake ravaging Kyushu in 2010 the first image was available within 5 h. The constellation was also beneficially employed during the wildfires in Victoria, Australia, in 2009. Through combining the multispectral images captured by the CCD cameras and panchromatic images, the 2010 spread of oil in the Gulf of Mexico, USA, could be assessed and predicted through time-series analysis.

14.8.2 Haiti Earthquake

On 12 January 2010 a disrupting earthquake stroke Haiti. Fortunately, there was no cloud cover so that optical satellite imagery became soon available. GeoEye succeeded to acquire Ikonos images (GSD 50 cm) within a few hours after the destructive shocks and licensing issues were temporarily set aside enabling Google to quickly disseminate the imagery via the internet. Within a few days the whole country was captured with airborne and spaceborne optical imagery as well as radar data and from these damage assessment maps could be rapidly created by manual interpretation carried out by many volunteers. To warrant all volunteers extract the same type of information comprehensive procedures had to be implemented while up-to-date and detailed reference geo-data were used as backdrop of the maps. If reference geo-data were insufficiently available lacking information was acquired by means of participatory mapping in relatively short time. Using mobile phones equipped with GNSS facilities and camera emergency workers and others communicated the situation on ground to coordination bodies directly or through media such as OpenStreetMap or Google Map Maker. The response to the Haiti disaster demonstrates that the formation of well-operating coordination bodies equipped with proper supporting tools is essential.

14.8.3 Conflicting Resolutions

The major benefits of satellite imagery are its ability to capture a wide area in one go, and to acquire up-to-date information. When blocked roads limit human access, and aeroplanes and helicopters are either unavailable or useless due to bad weather conditions, satellite sensors can still take images; active radar sensors, which operate in the microwave part of the electromagnetic spectrum, can do so by day and night and in all weather conditions. These images may be used to collate inventories for monitoring and assessing affected areas and to prepare emergency response in even the most remote parts. Satellite sensors can provide imagery with high temporal, spatial and spectral resolution, stereo-mapping facilities and weather-independent capture (SAR), but not by the same satellite at a time (Bhanumurthy and Behera, 2008). For example, the imagery from INSAT VHRR/CCD, METEOSAT and NOAA AVHRR have coarse spatial resolution but high temporal resolution. The Indian Remote Sensing (IRS) satellites, which are polar orbiting, have high spatial resolution but low temporal and spectral resolution. Events that change gradually over time, such as drought and land degradation, are easy to capture by EO satellites. Experts agree that for prompt assessment of disaster damage, satellite imagery information must be released no more than 3 h after the event. To enable quick response to highly dynamic disasters any area should therefore be captured at least once per day at a high level of detail.

An EO satellite can take images only at specific times and dates, depending on orbit, off-nadir viewing capabilities and swath width. This means a disaster area can be imaged only if the satellite is orbiting sufficiently close by. Time lag means that a satellite cannot always respond to disaster as quickly as needed. Furthermore, disaster areas are often subject to further devastation and ravage during the days following the calamity. For example, after the main tremor of the Sichuan quake, 52 major aftershocks were recorded within 72 h. Such dynamics imply a need for permanent, ongoing observation, yet at best satellites can provide only daily coverage. For example, the Japanese Advanced Land Observing Satellite, Daichi, launched in January 2006 and designed for mapping purposes, with as spin-off application disaster management, can take images of the same area only every 2 days. Four orbiting Daichi satellites would be required to capture the same area every 3 h.

In addition to the timelines, including revisit frequency and processing speed, GSD (spatial resolution) is of crucial importance. In sparsely populated rural areas GSDs of several metres may suffice for damage-assessment purposes, but this is quite inadequate when evaluating earthquakes affecting major cities. Here detailed images of at least 1 m GSD are essential for assessing damage to streets and buildings. Many tropical regions experience frequent cloud cover depending upon time of day, and up to recent times the satellites providing high-level detail have optical sensors, which are weather dependent and image at the same local time of day (usually around 10.30 a.m.). TerraSAR-X is a radar satellite that functions independent of weather conditions and has favourable GSD properties; in High Resolution Spotlight mode the GSD is 1 m. Nadir revisit time is, however, 11 days. At 2 days, the revisit time of the Synthetic Aperture Radar (SAR) on board Daichi is better, but its GSD is just 10 m. The features of the EO satellites are such that their use alone cannot cover all disaster-management needs. But the weaknesses of one system may be the other's strengths, so in tandem, three abreast or four-in-hand they are able to meet a great proportion of the needs. Another limitation is that present EO satellites have been designed to meet many applications; the instruments are general-purpose and not tailored to suit disaster management. The success of disaster management from space lies in harmonising EO characteristics of a great variety of current systems, such as spatial, spectral and temporal resolution, efficient data acquisition turnaround time, and development of standardised data products to properly serve decision makers.

14.8.4 Worldwide Collaboration

A constellation of satellites operating in tandem, combining high spatial, temporal and spectral resolution, all-weather capacity (radar) and stereo-mapping abilities is needed to serve the needs of decision-makers before, during and after disaster (Bhanumurthy and Behera, 2008). To address these needs requires collaboration on a global scale. With this aim, more than 70 governments and 40 organisations have joined forces to build a Global Earth Observation System of Systems (GEOSS). In 10 years' time, the thousands of observation instruments now operating in isolation will be linked together. They include in excess of 60 high-tech EO satellites; floating buoys for monitoring ocean currents, temperature and salinity, and air-quality and rainwater sensors. GEOSS is not focused solely on disaster mitigation, but also on water, ocean and marine resources, weather and air quality, biodiversity, sustainable land use, human health and wellbeing, energy resources and climate change. An example of co-operation at a more modest scale is the Disaster Monitoring Constellation (DMC) formed through an international partnership involving Algeria, China, Nigeria, Turkey, UK, Thailand and Vietnam. The DMC satellites can operate alone or in tandem.

Web sites can greatly facilitate dissemination of timely geo-data captured by orbiting satellites to emergency managers and thus stimulates quick responses. However, the storage of the diverse data sets on different servers often impedes quick availability. With respect to the coordination of data collection and dissemination, the European Commission Office for Humanitarian Aid (ECHO) acknowledged that the availability of a single portal to access information on any natural disaster would improve co-operation of all actors involved in disaster response (De Groeve, 2010). In 2004 this insight cumulated in the founding of the Global Disaster Alert and Coordination System (GDACS, http://www.gdacs.org) a UN framework aimed at strengthening cooperation among providers and users of disaster information worldwide. Today GDACS collects a diversity of data on volcanoes, earthquakes, floods, tsunamis and tropical cyclones from various sources distributed over the globe. This data are combined with number of inhabitants, vulnerability and other demographic and socio-economic data and analysed in a GIS environment. Once a sudden-onset disaster occurs alert notifications are automatically generated and disseminated to the concerning disasters managers while they also get access to tools to support response coordination.

14.9 Base-Map

Taking appropriate action in each of the five disaster management stages requires the right information at the right time and with the right persons. The above demonstrated how geo-information technology is able to collect, process and disseminate timely, accurate and detailed geo-information at any of the five stages and at all geographical scales. A great deal would already have been gained had more politicians acknowledged the suitability of geo-information technology as a means for efficient disaster management, taking into account that the appropriateness of geo-information depends on three factors:

1. The type of disaster (flood, fire, explosion).
2. For which of the five stages the information is needed.
3. Geographical scale of the disaster: local, zonal, national, regional, continental or global.

For example, airborne Lidar is very well suited for the generation of DEMs of river areas, urban conglomerates and coastal zones, both pre and post-disaster. A geodata type the significance of which can hardly be overlooked is the base-map. The function of a base-map is twofold. First of all the base-map acts as the geometric infrastructure to which all other geo-datasets are referenced in order to match a certain spot or feature in the dataset to its true location in the real world. A second function of the base-map relates to its thematic content. Topography, buildings and infrastructure provide information on the pre-disaster stage of the real world. By comparing the base-map with post-disaster data, an inventory of the damage can be made and decisions taken on appropriate action. The presence of a base-map, accompanied by its proper use, may thus rescue people and save goods of economic worth. In this context it is appropriate to note that geo-information should be made available free of charge for disaster management purposes.

14.10 Concluding Remarks

The experience of end-users faced with using geo-information in a disaster management context may vary from great to negligible. But in general they will be unaware of the special characteristics and limitations of such data. The approach to geo-data that takes into account its quality – by now second nature to the surveying and mapping community – is an unknown line of thought for the layman. This is, why not only delivery of the dataset should be at issue but also its proper use. A good understanding of the nature and magnitude of errors present in the data are essential for every user. Another need involves easily understandable visualisations. Since end-users are operating under super-stress, presentation of derived information should be simple and straightforward to understand; end-users in an emergency situation should be able to interpret a map in seconds. It is thus reasonable to assert that presentation is more important than content. To sum up, in addition to his role in making geo-data available, the analyst has another: equipping the end-user for its optimal use. This means disaster managers being familiarised with GIS and remote sensing capabilities, without going too deeply into technological issues.

In response to the need for more and better knowledge, methods and collaboration new research communities have been established which attempt at developing technology for solving urgent problems and develop long-term strategies. One such is Gi4DM (Geo-information for Disaster Management), a multidisciplinary symposium initiated by the Joint Board of Geospatial Information Societies (JB GIS) including ICA, FIG and ISPRS. The first Gi4DM was held in 2005, and has since become an annual event. The resulting proceedings (e.g. Zlatanova and Li, 2008; Nayak and Zlatanova, 2008; Konecny et al., 2010) aim at researchers, practitioners and students who work disciplines related to geo-information technology and disaster management.

References

Bhanumurthy V, Behera G (2008) Deliverables from space data sets for disaster management – present and future trends. ISPRS Int Arch Photogramm Remote Sens Spat Inf Sci XXXVII(Comm VIII/2):263–270

Corbley KP (2006) Rapid post-disaster mapping: airborne remote sensing and hurricane Katrina. GIM Int 20(8):19–21

De Groeve T (2010) Alerts and impact estimations. GIM Int 24(7):39

Djamour Y, Nankali HR, Rahimi Z (2006) Iranian permanent GPS network: using GPS to understand earthquakes. GIM Int 20(9):40–43

Klingseisen B, Leopold Ph (2006) Landslide hazard mapping in Austria: using GIS to develop a prediction method. GIM Int 20(12):41–43

Konecny M, Zlatanova S, Bandrova TL (eds) (2010) Geographic information and cartography for risk and crisis management: towards better solutions. Springer-Verlag Berlin Heidelberg

Kulkarni MN (2006) Indian GPS programme for earthquakes studies: permanent, semi-permanent and campaign-mode GPS Network. GIM Int 20(1):37–39

Landré M (2008) GIS in response time analysis: fire station, Zutphen, Netherlands. GIM Int 22(2):43–45

Lemmens M (2005b) Helpless Caesars of technology. GIM Int 19(2):11

Lemmens M (2007) Disaster response starts with a map. GIM Int 21(4):13

Lemmens M (2008a) Avoiding disaster surprises. GIM Int 22(1):11

Lemmens M (2008b) Disaster management. GIM Int 22(7):11

Lemmens MJPM (2005a) Abilities of airborne and space-borne sensors for managing natural disasters. In: Van Oosterom P, Zlatanova S, Fendel EM (eds) Geo-information for disaster management. Springer-Verlag Berlin Heidelberg, pp 355–364

Nayak S, Zlatanova S (eds) (2008) Remote sensing and GIS technologies for monitoring and prediction of disasters. Springer, Berlin

Shinohara H, Sasaki K (2005) Monitoring earthquake fires in Japan: real-time fire-spread prediction system. GIM Int 19(11):46–47

Sur U, Sokhi BS (2006) GIS in city fire hazard: petrol Station vicinity vulnerability assessment. GIM Int 20(8):50–51

Tibaijuka A (2008) Improving Slum Conditions through Innovative Financing, keynote address at the Opening Ceremony of the FIG Working Week 2008 in Stockholm, Sweden, 15 Jun 2008

Tran-Vinh P (2010) Early warning plea for Vietnam. GIM Int 24(5):29

Van Oosterom P, Zlatanova S, Fendel EM (eds) (2005) Geoinformation for disaster management. Springer-Verlag Berlin Heidelberg. ISBN 3-540-24988-5

Wang L (2010) Small satellites. GIM Int 24(9):55

Zlatanova S, Fabbri AG (2009) Geo-ICT for risk and disaster management. In: Zlatanova S, Fabbri AG (eds) Geospatial technology and the role of location in science. The GeoJournal Library, vol 96. Springer, Berlin, pp 239–266

Zlatanova S, Li J (eds) (2008) Geospatial information technology for emergency response. ISPRS Book Series, Taylor & Francis/Balkema, Leiden, The Netherlands

Chapter 15
Land Administration

The capacity needed for land administration includes on the one hand sound knowledge of technical subjects like data acquisition, database technology and data distribution, and on the other business subjects such as process design and workflow management, planning and control; all covered with a good sense of politics. Too many land-administration systems worldwide are slow, complicated and expensive due to the too narrow view adopted by land lawyers and land surveyors.

Prof. Paul van der Molen, the International Institute for Geoinformation Science and Earth Observation (ITC), University of Twente, The Netherlands.

Land administration is about registering land rights not only to secure these rights for the well-being of individual owners but also to support good governance and sustainable development and to eradicate poverty in developing countries (Williamson et al., 2010). How to register relationships between land and people? What is the value of land and what is its use? Who has what rights on which land? These are some of the questions related to land administration and to maintain land-administration systems or, in many countries, to set these up. Land administration is a millennium old activity aimed at securing land rights and stimulating good land management, an activity that also is increasingly receiving revived (academic) attention since the United Nations and other worldwide political organisations have recognised land administration as an important means in the combat against poverty (see Section 15.20). The relationships between land and people are based on rights codified in oral or written laws; without proper law system, land administration would lack any foundation. Land administration also entails identifying and documenting each parcel on which people have certain rights as an individual entity; the parcel should be registered and documented as an isolated part of the world. In addition to the rights and the land itself, the persons entitled should be identified and registered; land administration presumes on ground an adequate population registration system. This chapter provides an in-depth treatment of the tripod 'land $\longleftrightarrow$ rights $\longleftrightarrow$ people' with the emphasis on the land part of the equation.

M. Lemmens, *Geo-information*, Geotechnologies and the Environment 5,
DOI 10.1007/978-94-007-1667-4_15, © Springer Science+Business Media B.V. 2011

15.1 Land: The Most Valuable Resource

The solid part of the surface of the Earth, called land, covers around 30% of the Globe's surface and is essential for mankind's being on Earth. Land enables to produce food and raw material for making clothes and building shelter, and to provide other basic needs to safeguard human from starvation, diseases and poverty. As the basis for any agriculture production land is fundamental for securing enough food for an ever expanding world population. Land also offers mineral resources to ease living circumstances and to ascertain well-being and welfare. Land resources are the basis for human life: they provide soil, energy, water and the opportunity for all human activity (FIG, 2001).

Land and the constructions build on it are generally the major assets in any economy and in most countries, land accounts for between half to three-quarters of national wealth (Bell, 2006). For example, the value of the housing part of real property – also called real estate – in the Netherlands is estimated to be 1,250 billion Euros that is two times the Gross Domestic Product (GDP) of the country in which 16.5 million people are living. Land is thus not only vital for our living and well-being but for many it is also a source of wealth; for others, usually the ones who have only their unskilled labour to offer in the setting of developing countries – often euphemistically called the South – it an indispensable means to survive. In short, land is a fundamental resource as Larsson (1991) states while the UN ECE Land Administration Guidelines (UN-ECE, 1996) affirms that 'land is the ultimate resource, for without it life on Earth cannot be sustained. Land is both a physical commodity and an abstract concept in that the rights to own or use it are as much a part of the land as the objects rooted in its soil. Good stewardship of the land is essential for present and future generations'. Whether or not a region is economically successful, 'land security and land management are overriding imperatives for the new role of land administration in supporting sustainable development' (Williamson et al., 2010).

15.2 Human's Relationship to Land

The United Nations Economic Commission for Europe through its Working Party on Land Administration (WPLA) defines land administration as the processes of determining, recording and disseminating information about the ownership, value and use of land when implementing land management policies (UN-ECE, 1996). According to this definition, the aim of land administration is to support policies and activities within a society. Dale and MacLaughlin (1999) see a clear roll for land administration in resolving conflicts concerning the ownership and use of land and 'gathering of revenues from the land through sales, leasing, and taxation'. In the view of Van der Molen (2006), the concept of ownership should be understood as the mode in which people hold rights to land. Inherently, ownership defines a relationship between people and land and the types of relationship are based on consent and agreement within the society, which may be anchored in statutory law, common law, and customary traditions. The types of relationship acknowledged within a certain society depend on culture and historical development. For example, in many countries, especially those with a background as a former centrally planned economy, individual ownership was not allowed; all land belonged to all citizens of the country and was managed and controlled by the state on behalf of the citizens.

In other cultures land ownership is seen as a form of stewardship in which the present generation holds the land as trustees, connecting past generations with future generations. For example, the Ghanaian Customary Land Law contains the following statement: 'Land is believed to be God given, first to the ancestors and should be preserved and handed over to the numerous descendants. Land holders are only trustees and therefore cannot sell or dispose land in the way that will deprive the countless number of unborn children of its benefit'.

15.3 Value and Use

In addition to ownership, land administration is concerned with land value and land use within the overall context of land resource management. Land can have several types of value. The value determined for tax collection purposes may significantly differ from the market value, and this, in turn, may differ from the mortgage value, the loan a financial institute is willing to provide to the owner using the land as collateral. Value may also be seen from an insurance perspective, that is the costs of rebuilding a house if it were destroyed by a calamity such as flood, earthquake or storm. Value of land is determined by the wealth it creates and this, in turn, is anchored in its use or potential use. Location is the main determinant of land use and next comes the production capability of natural resources. Location is a primary determinant whether land is used for residential purposes or not. Irrespective of whether land may be productive in an agricultural sense or not a location close to an urban conglomerate may cause a (considerable) rise in value. The value of agricultural land will be primarily determined by what type of crops can be farmed, the yield per hectare, the price of the harvest on the local market or world market and the efforts necessary to cultivate the land. Presence of mineral resources underneath the surface of land may add a lot of value because it encourages transformation of agricultural, arable or wasteland into mine or oil fields.

The value of land does not necessarily correspond with the value of the right one has on the land. For example, many countries acknowledge the juridical construction that the owner of the land differs from the owner of the real property, that is the building erected on the land. This juridical construction is called long lease. Both the owner of the building and the owner of land, which will often be a national or local government, may sell their right to a third party but it will be clear that the owner of the building will receive less money when selling his right than his neighbour who has full ownership.

15.4 Recognition of Rights

Throughout history, the use of land has always been associated with rights and obligations, assigned either to individuals or groups of people by institutes acknowledged by all members of the relevant community as authoritative. Tax on land, which is in many countries a substantial source of income for (local) governments, is such an obligation and maybe even dates back to the time that mankind switched from a nomadic life style to permanent settlement. Written evidence shows that in 1400 BC the fertile land near the river Nile was divided into parcels amongst others for tax collection purposes. The relationship of human being with land is thus not only based on physically exploring the land for gaining benefits, but also on rights and obligations, which are granted, acknowledged and protected by a governing institution on behalf of the community, the sovereign or the divine.

The use of or gaining profit from land by individuals or groups of people, which are usually called *subjects* in the context of land administration, is thus based on rights that are recognised by others (Fig. 15.1).

Here subjects may concern an individual (man, woman), family, a group of individuals such as a clan, an organisation or a group of organisations or governmental organisations. Different subjects may have different types of rights on the same piece of land. For example, one subject may have a use right by owning the land, another exploits the land based on rent and a third one may have a security right based on mortgage. This situation, for example, occurs when the owner of a hacienda leases the land to farmers while he may have mortgaged simultaneously the land to build a factory to transfer maize into ethanol as car fuel in the wake of

Fig. 15.1 Relationship between man and parcel through rights (adopted from Henssen, 1995). Note that what is indicated as subject in the text is in this scheme named MAN and what is named Object in the text is called PARCEL

the 'Going Green' paradigm. In West Africa the Fulani people have unwritten rights based on a long tradition to pass farm land when driving their cattle from the North to the South when the dry season is approaching, and from the South to North when the rain season is coming.

A land administration system cannot functioning properly without well-functioning population registration and also not if an adequate legal system is lacking. Population registration and proper laws and legal protection facilities are prerequisites for the establishment of land-administration systems. Since also organisations such as firms may have rights on land, Chambers of Commerce plays also an important role in the land administration arena.

15.5 Fundamental Right

For more than 60 years now, the right to own land has been internationally acknowledged as a fundamental human right. Article 17 of the Universal Declaration of Human Rights – adopted and proclaimed by United Nations General Assembly resolution 217 A (III) of 10 December 1948 declares that (1) everyone has the right to own property alone as well as in association with others and that (2) no one shall be arbitrarily deprived of his property. Some governments attach so much value to the declaration of human rights that they chiselled them on square wide stone panels as the municipality of Linz, Austria, did (Fig. 15.2).

Governments of most countries concede this fundamental right and have anchored it in their constitution. For example, the 1999 Constitution of Nigeria states that every citizen of Nigeria shall have the right to acquire and own immovable property anywhere in Nigeria. The same constitution provides further that no such private property shall be compulsorily acquired by the state except on payment of prompt compensation and a right of access to court or tribunal for the determination of his interest in the property and the adequacy of compensation paid.

Fig. 15.2 Universal Declaration of Human Rights monumentalised at the Friedenplatz (Freedom Square) in Linz, Austria

15.6 Evidences of Right

Until man had taken his first step in advancing from a nomadic to a more settled existence, he had no need for proving land rights to other members of the community, nor did he have a need to record his claim to ownership. Ancient Egypt was one of the first regions where man took steps to become a food producer rather than a food gatherer. Evidence from the contents of tombs indicates that there was a form of public land registration and that the land courts would entertain no claim if the land were not registered. There is also evidence that surveying was used to set out the boundaries of individual parcels on a yearly basis to recover their beacons and boundaries after inundation of the annual flooding of the Nile. The corner beacons were set out or recovered by measuring from permanent markers above the flood line.

How do others in the community know of individual or group rights to land? To avoid conflicts it is essential that the types of right, the subjects who hold the rights and the objects themselves, i.e. the pieces of land, are clearly identified and registered orally or in writing. In small, tribal communities, or, more general, in close-knit communities, as Zevenbergen (2006) calls such societies, one may suffice with oral registration and recognition. In today's complex society, however, in which over 50% of the nearly seven billion inhabitants of this planet live more or less anonymously in (peri-)urban settings, disconnected from traditional, close-knit communities, registration which does not depend on the memory of the tribe seniors in a staunched social setting becomes essential to proof ownership and to be protected against claims from others. The buyer of land is no longer in close relationship with the seller, e.g. as son, nice or fellow villager, but is a stranger who has not the slightest clue of any century old unwritten ownership rights; the stranger wants the owner to prove his claim by official documents.

Throughout history, virtually all civilisations have devoted considerable efforts to defining rights to land and in establishing institutions to administer these rights (Bell, 2006). Also in Ancient Egypt surveying of land appeared necessary to register the boundaries, to record ownership in writing and to keep the documents in a public register. Interestingly, this system already shows the essential characteristics of today's land-administration systems in which the basic details of parcels, rights and subjects are collected in an official, public register and where this information can be easily consulted.

15.7 Security of Land Rights

15.7.1 Why Security?

The degree of recognition and guarantee of land rights determines the level of security. Security of land rights, also called security of tenure, is an important foundation for social and economic development (Bell, 2006). It is believed to increase the value of land, to encourage investments to improve, amongst other things, the productivity of agriculture, and to encourage economic participation of those who hold the rights (Barnes, 2003; Feder and Nishio, 1998). Security of land rights does not only benefit those who hold the rights but also society as a whole because it supports conservation and the sound use of natural resources. Furthermore, as Bell (2006) notes, land has been a cause of social, ethnic, cultural and religious conflict and many wars and revolutions have been fought over rights to land. Securing land rights may help to smooth the number, intensity and nature of conflicts relating to the use and transaction of land and may decrease the chance of their occurrences.

Securing land rights is above all germane for vulnerable groups such as indigenous groups, the poor and women. Dr. Anna Tibaijuka (2008), Under Secretary General and Executive Director UN-HABITAT estimates that only some 10% of land parcels in the world are registered and only some 5% of registered land is registered in a woman's name, i.e. worldwide land rights on just one in every two hundred parcels is in the hand of a woman. Users and holders of the 90% of land parcels, which are not registered, are likely confronted with some form of insecurity, now or in the future. Insecurity of land rights causes users and holders of land, whether rural or urban, feel that their rights on the land are at risk because they might be claimed by others, while the sources of evidence to proof their rights are weak. Also how long the right lasts may be indeterminate and would thus be a source of insecurity. A means to arrive at secure land rights is by formalising land rights and to record information about which subjects have what right(s) on which object in official registers. The institutions that administer who have what rights on which pieces of land (objects) are called land-administration systems.

15.7.2 Transfer of Land Rights

A key characteristic of rights on land is that they may be transferred from the one subject (seller) to another subject (buyer). This may sound trivial for earns grown up in Western countries, but it isn't. In Vietnam, a country in the past partly governed as a centrally planned and command-driven economy, the first Land Law, developed and adopted by the National Assembly in 1988, facilitated state allocation of agricultural land from agricultural co-operatives to households, for their permanent use. According to Professor Vo, former Vice-Minister at the Ministry of Natural Resources and Environment and chair of the Vietnam Association of Geodesy, Cartography and Remote Sensing this transformation from collective to individual rights was focussed on improving production of agricultural land, and households using land received Land Tenure Certificates, issued by local authorities on a temporary basis (Lemmens, 2008a). However, the first Land Law did not aim at securing rights; for that, 5 years later, in 1993 the second Land Law was adopted protecting the rights, interests and obligations of current land users by means of Land Tenure Certificates guaranteed by the state (Fig. 15.3). The second Land Law also gave five rights to households using land: exchange, transfer, lease, inheritance

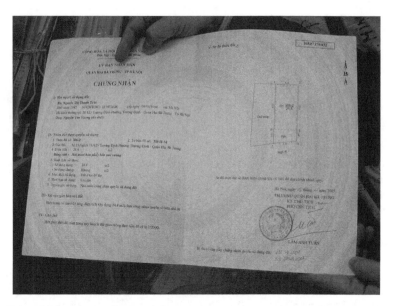

Fig. 15.3 Land Tenure Certificate (LTC) created by computer and capturing a land parcel in a locality in the Hanoi area, Vietnam. In addition to identification of owner, land right and parcel, the certificate contains a drawing at scale of the parcel

and mortgage. Furthermore, the state conceded the existence of price attached to land and defined this. A third Land Law, enacted in 2003, was necessary to facilitate the development of a land market.

Usually, the people in a community acknowledge that an individual possesses a good by custody of the good or by key keeping. The transfer (gift, sale) of the good from the 'old' owner to the 'new' owner is done through handing over the good itself. However, land, the constructions erected on it by humans, such as buildings, or natural features solidly grounded in the soil, such as trees, cannot be transported; they are immovable goods. The consequences are twofold. First, the seller cannot *prove* ownership, which gives him the right to sell, by custody or key keeping alone. He should demonstrate evidence by other mains. Second, the transfer of the right cannot be done by *handing over* the good itself, one has to look for other ways. Following Larsson (1991), Zevenbergen (2006) identifies two basic sources of transaction evidence, evidence in oral form and evidence in written form. Oral evidence is usually given by authoritative people in the community, such as local leaders, who witnesses, possibly underpinned by an oath, that the seller is the person entitled. Transaction evidence solely based on oral sources will usually occur in close-knit communities of which the members are illiterate and the number of transfers is insignificant. The transfer will often be accompanied or completed by a symbolic handing over of the land by giving the seller a product of the land (twig, grass) while standing on the land in the presence of other members of the community, so that those who may concern are getting aware of the transfer and will remember it.

15.7.3 Private Conveyance

Written evidence is provided by documents compiled by authoritative people in society, such as judges or notaries. The transfer of land rights takes place by compiling a document, often also called deed, describing which object is transferred from which previous owner to which new owner. In this, so-called, *private conveyance* system the deed is, together with all previous transfer documents, handed over to the new owner. The stack of documents demonstrates the chain of transfers starting from the very first owner and proves how the seller legally obtained his rights trough history and time. The very first owner could have gained his land right from an adjudication process or from a grant from government or a traditional leader. In Great Britain the timeline may possibly go back to William the Conqueror.

A main shortcoming of this system is that it is prone to fraud. For example, the owner may make a replica of the document and pretend that the duplicate is the original, enabling him to sell the land twice to two different persons, who are not aware of each other existence. The documents may also get lost due to fire or another disaster demolishing the essential link between subject and object and leaving the subject with deprivation of evidence to proof rights. The house, in which the documents are kept, may be target of burglary and the collared documents may be brought to the land market by the thief or his comrades. Also in case of theft the owner is confronted with a considerable diminution of evidences to proof rights.

15.7.4 Deeds Registration

Many countries have acknowledged that a clear-cut remedy to the above shortfalls is safeguarding the original documents by piling them up in the vault of an independent third party; depending on the country this may be a notary, lawyer, court, tax authority or others. Registration of a deed in a public repository provides security against loss, destruction and or fraud (Dale and McLaughlin, 1999). When the date and time of registration is recorded the system also provides a priority claim. One of the functions of registration of deeds is similar to the symbolic feat in the presence of the community members in oral agreements: it gives public notice that a transaction has occurred. To prevent a hotchpotch of ways how to prove rights on land, such a system of registration of deeds should be compulsory and the register should be the sole entry to the deeds. Unification is also necessary with respect to identification (description) of the land; this should not be left to the parties involved in compiling the deed. Unambiguous identification can be achieved by using an index map and assigning a unique number to the parcel and the use of both can be enforced by legislation. Easy retrieval of deeds, which is necessary in countries with well-developed land markets, should be warranted by introducing an entry register, which enables easy access to the pile of deeds.

15.7.5 Title Registration

In deed registration systems public agencies act as collectors and guardsmen of information related to rights on land and as doorkeeper, guide and possibly also as collector of admission fee for those who want access to that information. Although one of the basic principles of deed registration is providing security, the security is only related to preclude loss of the documents and prevention of falsification. In the words of Henssen and Williamson (1990), 'a deed registration system means that the deed itself, being a document which describes an isolated transaction, is registered. This deed is evidence that a particular transaction took place, but it is in principle not in itself proof of the legal rights of the parties to conduct that transaction and, consequently, it is not evidence of its legality. Thus before any dealing can be safely concluded, the prudent purchaser must trace his ownership back to a good root of title'.

In a deed registration system the state does not provide security that the one who is registered as owner is the actual owner. The role of the state is rather passive. In cases of dispute, the owner has to come into action and defence his claim himself possibly in front of a judge in court. That is because a deed is a document which provides evidence that transfer of land rights have taken place, but nobody warrants that it mirrors the situation on ground. If the data in a deed registration system are trustworthy massive disputes and uncertainty about ownership may be avoided. The term 'trustworthy' here implies that on the one hand data are reliable and accurate, and on the other hand it is acceptable to national and international stakeholders. A first requirement of trustworthiness is that all processes involved are transparent and all methodologies clearly defined and scientifically sound, meaning that the proprietors must trust the processes and institutions involved (Lemmens, 2006). But in many countries these conditions are not met (Zevenbergen, 2006).

A registration system showing imperfections leads to uncertainty of who has what rights on which piece of land and this in turn will affect the willingness of banks and housing finance companies to provide loans based on land as collateral to individuals or families. Such financial institutions feel the risk that the one who asks for the loan is not the real owner or, if he is the real owner, may have already vested a mortgage on the property. As a result investment in the land and enhancements ensued from these investments are blocked. The difficulty of legitimately proving land rights impedes not only access to mortgage finance but also the development of a land market and may result in excessive litigation, as is the case in New Delhi, the national capital of India. To eradicate the stumbling blocks New Delhi finalised a law by the end of December 2008 aimed at replacing the present deed registration system with a title registration system to establish 'without doubt' the ownership rights over land. 'A title registration system means that the deed is not registered but the consequence of that transaction, that is, the right itself' (Henssen and Williamson, 1990). The right itself (title) is created by the registration.

In a title registration system the government guarantees the rights. In course of time various types of title registration system have been developed. Well-known is the Torrens systems introduced in Australia in 1858 after a boom in land speculation.

In this system ownership needs not to be proved by retracing 'back to a good root of title' all the documents related to the rights on a certain land parcel. The register is the sole source of title information and it is thus not necessary to step beyond the Certificate of Title to review historical documentation (Dale and McLaughlin, 1999). Since all of the necessary information is on the Certificate of Title one does not need to lift the curtain to the past (curtain principle). The state is responsible for the veracity of the register and for providing compensation in the case of errors or omissions, thus providing financial security for the owners (insurance principle).

15.8 Land Administration

The UN-ECE (1996) definition of land administration given in Section 15.2 ('the processes of determining, recording and disseminating information about the ownership, value and use of land when implementing land management policies') may be criticised as being rather vague, broad and assembled from a cosmic perspective, but a more specific definition is hard to give because land administration may support many purposes in a society and the land policies, which are served by land administration, may differ from country to country and change over time. According to UN Habitat (unhabitat.org) land policies today should emphasise on protecting people from forced removals and evictions, reducing poverty and ensuring gender equality. With respect to land rights UN Habitat specifically stipulates:

> It is now well-recognised that secure land and property rights for all are essential to reducing poverty, because they underpin economic development and social inclusion. Secure land tenure and property rights enable people in rural and urban areas to invest in improved homes and livelihoods. They also help to promote good environmental management, improve food security, and assist directly in the realization of human rights, including the elimination of discrimination against women, the vulnerable, indigenous groups and other minorities.

The many and complex activities taking place in modern society requires that land and property rights should be easily identifiable and verifiable. For establishing these goals land-administration systems are indispensable. They are concerned with the recording and storing of information related to land as a natural resource to ensure its sustainable use and development (UN-ECE, 1996). A land administration system enables protecting rights against unlawful actions, and also that results of legal actions will be easy to predict.

The establishment and keeping in air of a land administration system should be accordingly shored by a broad pallet of buttresses including:

– Legal framework covering the tripod land rights, land policy and land use planning
– Institutional, Organisational and Administrative framework
– Technological framework focussing on the acquisition, storage, analysis and dissemination of land related (geo-) information

Land administration requires many levels of technology, which have the capability to manage millions of land parcels. Organisational structure for land administration varies widely between countries and regions throughout the world, reflecting local cultural and judicial settings (Enemark, 2006). In Estonia, for example, Siim Maasikamäe, Estonian University of Life Sciences, indicates that the main tasks of land administration are allocated to four ministries, county government and to local authorities (Lemmens, 2008b). The Ministry of the Environment is responsible for most land-administration tasks subsequently executed by the Estonian Land Board. The Ministry of Justice is responsible for registration of property rights in a title book. The Ministry of Internal Affairs administers land resources; two of its departments are responsible respectively for regional development and spatial planning, mainly orientated towards the prospective development of settlements and solving problems related to developing built-up areas. The Ministry of Agriculture provides conditions for sustainable and diverse development of rural areas in general. County governors represent the state in the land privatisation process and supervise municipalities in their land reform and planning activities. Three main municipal land-administration tasks are land reform execution, planning administration and assessment of land, and land taxation. Most are financed from the state purse. In the past the most important role of the Estonian Land Board was general guidance on land reform, the execution of which was mainly carried out by local authorities, county government and private surveyors, but other tasks gradually grew in substance, including central cadastral registration of land, exploitation of state land, and land assessment and analysis of land-market data.

15.9 Land Registration and Cadastres

Every land administration system should include some form of land registration that is the process of official recording of land rights through deeds or as title on properties. In place should be an official record (land register) of rights on land or of deeds concerning changes in the legal situation of defined units of land (Henssen, 1995). Land registration not only provides owner and (potential) purchaser with legal security but is also one of the three legs of the tripod supporting a properly functioning land market, consisting further of valuation and financial services (Dale and Baldwin, 2000; Fig. 15.4). Without a formal system in which land rights are registered, a modern market economy is unthinkable (De Soto, 2000). It is generally acknowledged that societies with a successful land market generate more wealth than those without; they are more stable and offer more economical opportunities to their members. As a result, the development of a well functioning land market is a goal of many governments and international agencies (Wallace and Williamson, 2006).

The second leg of a land administration system is the cadastre, which is 'an official record of information about land parcels, including details of their bounds, tenure, use and value' (McLaughlin and Nichols, 1989). A cadastre is similar to a

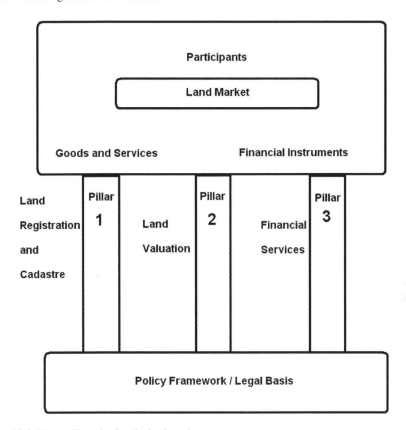

Fig. 15.4 Three pillars shoring the land market

land register in that it contains a set of records about rights on land, but rather than to duplicate each other land registration and cadastre usually complement each other. Nevertheless, often, the information overlap between Land Registry and Cadastre is 70% or more (Zevenbergen and Bogaerts, 2001).

The land register emphasises the relationship *subject-right*, whereas the cadastre emphasises the relation *right-object*. In a cadastre one may therefore expect to find more detailed information about the object (land parcel) than in a land register. In addition to what type of rights are vested on the land and who holds these rights, other types of information about the object may be stored, including attributes such as value (partly for tax collection purposes and also to serve the land market); use and details about the construction of buildings and population. Cadastres are not ends in themselves (Williamson, 1995). Improved security of land rights for individuals and groups of people is not the only reason why the basic details of parcels, rights and subjects are stored in land-administration systems. In addition to providing security, the information collected in land-administration systems can be beneficial for a wide spectrum of land related activities. When the information is properly organised, it provides the basis for handling broader social, economic, environmental activities, such as housing, transport, agriculture, fishery and forestry, sustainable economic development, and environmental management. It may also serve political stability and social justice. Rapid accessibility of data should be a key characteristic. In general terms, a cadastral system should be appropriately designed to serve the needs of the country.

15.10 Land Management

Land management encompasses all activities to put land and natural resources to good effect and in today's context this particularly means achieving sustainable development. To prove security of land rights one could suffice with keeping the original act, including a verbal or geometric parcel boundary description, in the office of public or private conveyancers at a safeguarded place and to provide a copy to the subject who holds the right. However, such an elementary system, which was used in some countries until very recently, would result in retrieving the information on rights only with considerable difficulty and would not be transparent. As a result it would impede good land management, which requires easy and rapid access to land-related data including land rights and land use. However, in some countries where land-administration systems are successfully implemented, the possibility is lacking to go beyond the retrieval of the features of individual parcels. For example, Siim Maasikamäe, complains that although Estonia has built up – since the country gained independence in 1991 – a modern cadastre in terms of registration of land rights of the whole country and information on nearly all individual parcels can be disseminated to institutions and the general public via the web, aggregated data on land and land use are missing, and the processes of land use change are a tabula rasa (Lemmens, 2008b). How much arable land has been transformed into residential land during the last decade? This is a question which cannot be answered. As a result the system does not support good land management practices.

Although land administration is primarily directed at protecting the interests of individual landowners, it is being seen increasingly as an instrument of national land policy, an aid in planning and in general a mechanism to support greater economic development. Consequently, cadastral information plays a key role in land management. From a technical point of view the associated activities often consists of combining land information, including the location parcel boundaries stored as coordinates of the corner points, with other types of information such as digital elevation models (DEM), land use maps, and database of road and railway infrastructure (Fig. 15.5). For example, when an architect is planning the construction of a building at a certain location, the municipality will have to give out a building permit and to check whether all conditions are fulfilled, precise parcel boundary information has to be combined with land use plans and maps of pipes, cables and many other features. For land management purposes it is therefore essential that

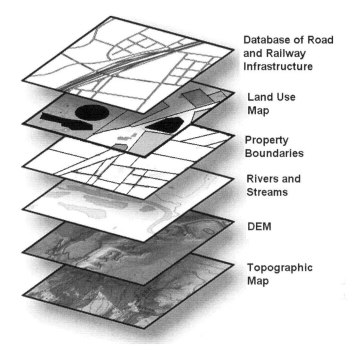

Database of Road
and Railway
Infrastructure

Land Use
Map

Property
Boundaries

Rivers and
Streams

DEM

Topographic
Map

Fig. 15.5 Integration of a wide variety of geo-information datasets is essential for good land management and requires all datasets to be geo-referenced in the same (national) geodetic system

boundaries – or more general, the spatial component – of land parcels are known in a local, national or regional geodetic reference system.

15.11 Adjudication

The initial establishment of a land administration system requires that the existing rights of land are officially recognised and recorded. This process of acknowledging existing rights by an authoritative institute is called adjudication. It is the first stage in the registration of land rights in areas where these rights are not yet officially known. Note that adjudication is not concerned with vesting land rights – they do already exist – but to *formalise* the rights. In general terms, adjudication refers to determining a binding resolution in a dispute or claim by a neutral third party, which has the authority to make a judgment; often this will be one or more representatives of the court system. Within the field of land administration adjudication represents the process through which existing rights with respect to a particular land parcel are finally and authoritatively ascertained (UN-ECE, 1996). It is the most common form of first registration and deals with the initial collection of the registers and therefore it is of great importance when no land administration system exists yet in an area (Zevenbergen, 2004). Before the registration can be affected, first the rightful claimants have to be determined and disputes and uncertainties about who owns what land have to be resolved. One of the main effects of a carefully carried out adjudication is creating security of rights for the people who use the land to make a living out of it while also the location of the boundaries is determined.

15.11.1 Sporadic and Systematic Adjudication

The determination of the owner of a parcel and the location of its boundaries may just commence when an owner submits a request to an authority to register the land, which may, for example, be necessary when the owner wants to sell the land. This is called sporadic adjudication, and the only provision that has to be taken by the authority is the establishment of a local registration office. Adjudication can be also carried out in a systematic manner. With systematic adjudication an area, such as a village or a local government area, is methodically approached. The selection in which order the areas should be covered in a region or country is determined by setting priority criteria, including whether the area is subject to rapid land development (e.g. transform from rural area to peri-urban area), level of economic development of the area, level of disputes and need for credit. In contrast to sporadic adjudication, systematic adjudication requires active involvement of both authorities and the community or communities living in the area. First of all the process relies on legislation that gives authority for the adjudication to be established. Also an adjudication commission consisting of officials and members of the community has to be put on ground. The holders of land rights within the community have to provide evidence for their claim. When all claims have been collected, the list – preferably accompanied by a map – is shown to the public, who may contest the claims. In case of contest, the one with the best evidence sources prevails. Finally, an official institution, usually the court, decides.

15.11.2 Mixed Strategy

The initiative of sporadic adjudication lies largely in the hands of holders of land rights and is actually voluntarily. Since the holders of land rights benefit from the registration, the approach permits that applicants pay a fee for the services provided by the registration office. By its very nature, systematic adjudication has to be compulsory. At the end of the process, the land rights will have been authoritatively acknowledged and therefore everybody who has a rightful claim should participate, else the rights of absentee owners may be infringed. However, citizens who take part in an activity forced by authorities – an activity they have not asked for and of which they may doubt viability – are unlikely to be willing to pay fees. To ensure cooperation from the community involved, systematic adjudication should therefore be paid from the national purse. Although sporadic adjudication may be cheaper in the short term, at least from the perspective of the national government, its ad hoc approach results in a patchwork of registered land parcels. The fragmentation obstructs good land management, which is becoming increasingly important in many parts of the world. The best approach for adjudication is a mixed strategy, with systematic adjudication in areas of main economic interest and sporadic adjudication elsewhere (Zevenbergen and Bogaerts, 2001).

15.11.3 Participatory GIS

Adjudication carried out in sub-Saharan countries is often criticised because of shortage of transparency and lack of participation of the main stakeholders, that is to say the land owners. To improve participation of community members in the adjudication process use can be made of participatory GIS (PGIS) (Lamptey, 2009). PGIS aims at acquiring and storing indigenous knowledge and is applicable in many domains for capturing the geographical knowledge of a community. In the context of land administration the method uses GNSS technology, Mobile GIS, satellite imagery and other easy to use geo-information technology to capture land rights by solidly involving local people (Fig. 15.6). PGIS appears to be an effective tool in eliciting indigenous knowledge about the changes over time of land ownership, use rights and land use.

Fig. 15.6 Participatory GIS is an effective tool for involving local people in the adjudication process

15.12 The Parcel

The basic building block in any cadastre is the parcel (FIG, 1995; Enemark, 2006). What constitutes a parcel depends on the purpose of the cadastre. In general a parcel may be defined as a continuous area of land within which unique and homogeneous interests are recognised. The interests may include ownership but also the same type of land use. Within the land administration framework a parcel is a unit of land over which homogeneous land rights are established (Cashin, 2003). Usually the boundaries of the parcels are surveyed by professional surveyors and mapped as closed lines (polygons) and indicated by a unique parcel identifier on a (digital) map (Fig. 15.7). The unique identifier is a key for retrieving the attributes belonging to the parcel and which are stored in records kept separate from the map. A cadastre may be considered as a geo-information system (GIS) but there is one fundamental feature that distinguishes the cadastre from other types of GIS systems, i.e. systems which provide information about geographical objects and their attributes. Data recorded in a cadastre have a social and legal meaning, and are based on accepted social concepts, and the rules derived from these concepts constitute the foundation of the system (Van der Molen, 2006). The consequence is that often the polygons stored in a cadastre have no physical counterpart in the real world, as they do in ordinary GIS systems. That is because they are based on legal and societal concepts

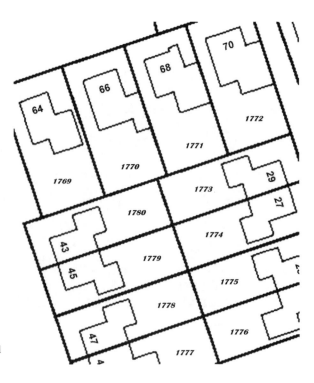

Fig. 15.7 Fragment of a digital Cadastre map, showing parcel boundaries, parcel numbers, outlines of buildings, house numbers and street names

and therefore the boundaries do not necessarily correspond to tangible boundaries in the physical world. Without the existence of a societal and legal framework the boundaries even would not exist and could not be drawn, or, at least it would be meaningless to draw them.

15.13 Boundary Description

When one is moving along a straight line over a land parcel in the same direction at some moment of time one will pass the boundary of the parcel. The boundary is the location, which may be visible or invisible, where one crosses the ownership of land and steps on the land owned by somebody else. Description of parcel boundaries, in whatever form, is an essential prerequisite for land administration. In many legal systems it is supposed that ownership of land extends from the centre of the Earth up to the infinite of the sky and includes all features natural and man-made attached to or beneath the surface of the Earth. As a consequence boundaries are infinitesimally thin vertical surfaces (UN-ECE, 1996). In practice, however, when talking about and measuring and reconstructing boundaries, they are treated as straight lines connecting corner points along the boundary of the parcel. Since the individual parcel is the key entity in any cadastre it is important that the manner in which rights to land are held includes a method of retracing boundaries of land parcels on the ground and locating them unambiguously. This is necessary to avoid disputes or when disputes between neighbours unfold their resolve.

When the aim would only be to retrace parcel boundaries in a later stage, one could suffice with either verbally describing the trace of the boundary, marking the boundary with man-made division features such as fences, hedges, walls, wooden posts, iron, steel or concrete markers, or letting the boundary coincide with physical features such as trees, streams or rock outcrops. When the boundaries are monumented well, for example, by walls, the parcels define themselves and an identifier is all what has to be recorded. Such an identifier is, for example, the postal address. If necessary the exact location of the boundaries can be recovered by inspection in the terrain.

The most basic recording of a boundary consists of verbal description. A common approach is to start the report at a certain, well-identifiable location, such as a rock or an oak, and describing the features one meets while walking along the boundary. The textbox shows an example from Iceland.

Boundary Description: Arnarbæli, Grímsnes

Corner monument: the ruin close to Heidrimakelda spring, south of Oddholtsmúla mound; from where there is a line of sight west to Hédinslækjabotnar hollow. From here the boundary follows Hédinslækur creek, and then Höskuldslækur creek to the Hvítá-river. To east of the above mentioned ruin close to Hedirimakelda spring the boundaries run south to verkelda spring, which runs from Galtatjörn pond .... (Source: National Archives of Iceland, 1884; see Ingvarsson et al., 2007)

When land is cheap and abundant, as was the case in the United States of America at its dawn, establishment and transfer of rights is often not recorded (Wolf and Ghilani, 2006). And if they were identified in writing, boundary descriptions were usually scanty or defective. The intersection of two parcel lines might have been written down as 'the place where John killed a bear' or 'the bend in a footpath from Jones' cabin to the river'. Still today land surveyors in the United States of America are confronted with many problems originating from such improper descriptions.

The description of a boundary in words only is associated with many disadvantages; the most important ones are as follows:

– Interpretation of the description requires extensive local knowledge about topography and nomenclature.
– In course of time the landscape may change under influence of natural forces or human activities impeding the identification of features because they are only partly present or have been shifted in location.
– The description is entirely focussed on retracing the boundary in case of dispute. As discussed above, today, land parcel information is used for many applications in the framework of good land management; often boundary information has to be combined with geo-referenced information and verbal description is unsuited for these purposes.

15.14 Fixed and General Boundaries

Parcel boundaries kept in a cadastre may be either 'fixed' or 'general'. The major difference between both types of boundary is that fixed boundaries provide confidence to owners where their property starts and ends since these boundaries are formally recognised, while for general boundaries, the location of the boundaries is left undetermined. The prefix 'fixed' can have two meanings: it can refer to either the accuracy with which the boundary is measured or refer to the fact that boundary locations may never change without legal documentation. In the first meaning of fixed, the boundary has been accurately surveyed and documented so that a surveyor can find any corner of the parcel from the recorded survey measurements, even when the boundaries themselves are not visible in the terrain, for example, due to removal of corners markers or when they never did exist. Fixed, in the meaning of unchangeable, has been adopted under the Torrens system: the location of the boundary cannot be changed without some document of transfer of land right. The boundary becomes fixed in space as a result of an agreement at the time of division of the land, irrespective whether the boundary has been (accurately) surveyed or not. In many systems, the agreement should be established between the adjoining owners and recorded as fixed. The recorded boundaries have precedence over what is on ground. In case of mismatch of the fence or hedge marking the boundary on ground with the recorded boundary, the latter prevails.

A general boundary, also called approximate boundary, is not as well defined in space as a fixed one. The boundaries as recorded are only indicative not definitive; the boundary monumentation on ground have precedence over what is recorded in the registers. So, when the fence or hedge marking the boundary on ground gets demolished and a new monumentation is erected of which the position does not correspond with the former position, the new situation prevails. In areas where withdrawal of natural resources, such as oil and groundwater, cause ground shifts that move corner monuments, the general boundary paradigm results in the prevalence of the corner monuments above any verbal or geometric description, irrespective how precise the location of the boundary has been determined. Ownership of land parcels can be registered without consultation of the neighbours and without agreement on the precise location of the boundary. This reduces the number of disputes in the short term but may give rise to problems in the longer term. The approach is often used where adjudication of title is undertaken sporadically. This undetermined nature is the basis of the English system and has often been the boundary system of choice for cadastres established in developing areas where time and financial constraints restrict the establishment of fixed boundaries (Cashin, 2003).

The use of general boundaries is well-suited in terrain, where it may be expected that future changes will be few. This will be, for example, the case in residential areas where houses will continue to exist for the coming decades and the gardens have been fenced. The method is also appropriate for rural areas, which have been cultivated for a long time so that land use patterns are well established and the likelihood of being guzzled by expanding urban conglomerates is minor. The establishment of general boundaries in rural areas can be done by photogrammetric means, using airborne or very-high-resolution satellite imagery, which is, in general, less costly and time-consuming than field based methods. Table 15.1 compares the advantages and disadvantages of fixed and general boundaries.

Table 15.1 Fixed versus general boundaries (compiled from UN-ECE (1996) and Cashin (2003))

Boundary type	Advantages	Disadvantages
Fixed	Boundaries are formally recognised and their location is precisely determined	Requires costly surveying
	Can be easily retraced in case of dispute	Danger of lost of boundary marks
General	Less demanding standards of survey and thus cheaper	Linear features may not (anymore) exist in the terrain
	As long as physical demarcated less chance of dispute	Reduced level of confidence

When one chooses fixed boundaries, one has to take in mind that precise boundary surveying is not only costly but also time-consuming. Sometimes the costs of identifying and measuring the land parcel are more expensive than the value of the land itself. In many parts of the world there is a lack of sufficiently skilled and training personnel. The deficiency of human capacity as well as financial resources may make happen that the putting on ground of a full-functioning land administration system may last for decades or centuries. Therefore, starting with general boundaries should be preferred when one starts with establishing or renewing a land administration system in a country. Upgrading of the general boundaries to fixed boundaries should be carried out in the initiative and at the expense of land owners whenever they feel the need (Zevenbergen and Bogaerts, 2001).

15.15 Geometric Boundary Surveys

Boundary descriptions can be done either by verbal means, geometrically or both. A geometric boundary survey, also called cadastral surveying, is a special type of boundary description. Cadastral surveying is that branch of land surveying which is concerned with the survey and demarcation of land for the purpose of defining parcels in a land administration system. Land surveying may be described as the process of measurement and delineation of the natural and artificial features (including parcel boundaries) of the Earth. Geometric description requires that the boundary is measured by an experienced or, in some countries, licensed professional surveyor. Geometric boundary descriptions can be grouped into two broad categories: stand-alone and embedded.

A stand-alone geometric boundary description is carried out by using the surveying method of closed traversing. Mathematically, the boundary of a parcel is treated as a polygon consisting of consecutive lines which connect the corner points of the parcel. The corners are mathematically linked to each other by measuring angles and distances. In its most basic form, a closed traverse can be measured by using compasses and measuring wheels. The compass readings provide the magnetic bearing and azimuth that is the horizontal direction between one corner and the other corner. The measuring wheel provides the distances between the corner points. One starts at a point of beginning (POB), which can be a man-made object, such as a fence post, or a natural feature which marks one corner of the parcel. Lengths and directions of the polygon lines are consecutively measured from this point of beginning. Although this method seems to be rather archaic in an era in which positioning by Global Navigation Satellite Systems (GNSS) has become ubiquitous, it is still in use, not only for cadastral surveying but also for other purposes, for example, to demarcate enumeration areas for census taking purposes as has been done in the Nigerian 2006 census (see Chapter 13). The stand-alone geometric description can be transformed into an embedded geometric description through measuring the distance and direction from POB to a reference point, this is appoint of which the coordinates are known in a geodetic reference system. Such a transformation requires conversion of directions and distances into coordinates, a calculation which is performed on computer using Least Squares Adjustment (see Section 11.5).

The results of a cadastral survey are stored on a map or in a digital database. Cadastral parcel information is not only helpful for parcel identification, parcel subdivision and boundary relocation, but also for land management, land use planning, valuation and facility management. Hence, boundary description by coordinates not only facilitates the relocation of lost and destroyed corner markers but also makes the positional relationship of the individual parcel with all other parcels explicit. In addition, embedment in a national geodetic reference system enables to combine land parcel information with a variety of other types of geo-information, using a GIS.

Three major categories of boundary surveys can be distinguished:

– Original surveys in areas which are not yet surveyed that still exist in many parts of the world, not only in developing countries but also in, for example, Alaska.
– Retracement surveys to recover previously established boundaries.
– Subdivision surveys aimed at establishing new smaller parcels of land within larger previously surveyed tracts and possibly join the subdivided parts with adjacent parcels.

The first category – original surveys – is closely related to adjudication. Adjudication and original surveys are both important elements to the development of land-administration systems. Surveying usually contributes highly to the overall costs of adjudication. Figure 15.8 shows an example of a subdivision survey using a total station, positioned on GCPs. From the GCPs the surveyor measures the directions and distances to the new boundary points and next calculates their coordinates using his measurements. In addition to the surveying and associated computation of the coordinates of the new boundary points, renumbering of the parcels involved has to be conducted.

15.16 Data Capturing Techniques

Various methods exist for the determination of coordinates of corner points of parcel boundaries. The oldest method makes use of direct measurements in the field. Today's typical measuring instrument is the 'total station' or electronic theodolite, which measures angles and distances to provide the basis for calculating the coordinates of boundary points. Global Navigation Satellite Systems (GNSS) – the best known system is the Global Positioning System (GPS) – is known by most people as a navigation system for cars, aircraft and ships, but it is also effectively used for cadastral surveying (Chapter 4). Used in combination with a GNSS infrastructure, GNSS enables to measure directly the coordinates of parcel boundaries in a national geodetic reference system with centimetre accuracy. Photogrammetry is another method used in cadastral surveying. The terrain is photographed either from the

Fig. 15.8 Part of a cadastral plan showing ownership parcels (*top*). The owner of parcel 1322 has sold parts of his property to his neighbours owning parcels 1319, 1320 and 1321 (*dashed lines*). To subdivide parcel 1322 and to join the parts with the other parcels, a land surveyor measured the new boundary points (A, B, C and D) with a Trimble 5600 DR Total Station (*centre*) positioned on GCPs (*bottom*). The method requires unobstructed view from GCPs to boundary points

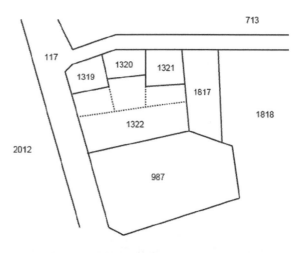

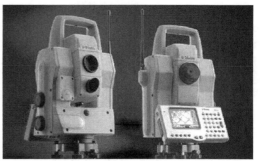

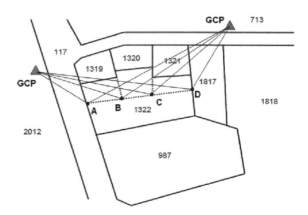

ground (terrestrial photogrammetry) or from the air (aerial photogrammetry) with fixed-wing aircraft or helicopters equipped with special metric cameras pointing downwards. The spatial resolution, usually called Ground Sample Distance (GSD), of several optical earth observation satellites ranges from 1 m to 50 cm. This high level of GSDs enables carrying out cadastral surveys from spaceborne imagery.

Although parcel boundaries may be delineated on airborne and satellite images, these images themselves cannot replace the map. This is because a well-made map is an accurate scale model in which features can be accurately identified. Airborne and satellite images have the appearance of a map but they miss integrity both with respect to geometry and content (see Section 6.6). Compared to an aerial or satellite image, a map has the following advantages:

- Geometric distortions are eliminated to a much greater degree.
- Through the use of conventional signs, contour lines and so on, the map can show all significant detail with greater simplicity and clarity.
- Non-visible information can be shown, also what is above or below the ground, and property boundaries which do not coincide with physical features in the terrain.
- Irrelevant detail can be omitted.

Therefore, accurate large-scale (digital) maps are the only sound basis for recording and analysing cadastral information. Since geo-information is increasingly stored in digital format also the map component of land-administration systems is getting stored in digital databases. Basically, one can arrive at a digital cadastre along three lines:

- Digitisation or scanning of existing documents.
- Surveying in the terrain using GNSS technology, total stations or other surveying devices.
- Extracting information from aerial and satellite images.

Whatever method(s) are used in a given setting, cadastral surveying is based on a number of sound land surveying principles. These are as follows:

- *Working from the whole to the part* by establishing a network of control points at a nationwide level, which is next split into smaller networks with points closer together; the resulting network is usually called national geodetic reference system (Fig. 15.9; see also Chapter 1).
- *Consistency*: once the higher order networks have been established, it is possible to work to less rigorous standards in the lower orders without affecting overall accuracy. The lower order points are connected to the higher order points and hence lower order points cannot be more accurate than the higher order reference points. So, it makes not much sense to spend a lot of money to achieve extremely high accuracy at lower orders; the high accuracy will be degraded by connection to the higher order points.
- *Economy*: since higher accuracy in general costs more money the surveyor should seek no higher accuracy than is necessary and sufficient for the task at hand.
- *Redundancy*: measuring more data than strictly necessary enables to carry out independent checks on the data (built-in quality control), for example, by measuring all three angles of a triangle even though the third angle measurement is

Fig. 15.9 Three monumented GNSS reference station of the first order Geodetic Reference Network located at the vertices of the Golden Triangle of Ghana (*left*). Any point within Ghana will be located within 100 km distance from a first order reference point. At the right side, easily accessible lower order point, protected with four stakes. See Poku-Gyamfi and Schueler (2008)

redundant. Also the redundancy obtained by measuring closed traverse provides means to check the measured angels and distances. The methodology to use and to cope with redundant data is based on least-squares adjustment.

– *Maintenance*: Changes take place all the time, the world is dynamic and therefore procedures and means must be put in place to cope with dynamics so that the dataset is kept as up to date as needed.

15.17 Land Policy

Since land is a fundamental and ultimate resource, governments of many countries mind how land in their jurisdiction is used and which rights have been given to whom. The array of targets set by the government to shape use of land rights is called land policy. Land policy comprises a whole complex of socio-economic and legal prescriptions that dictates how the land and benefits from the land are to be allocated (UN-ECE, 1996). The underlying tenets may originate in national insights and views on how land should be exploited and prevented from misuse but are also partly settled by international conventions, such as the Kyoto Protocol which came into force on 16 February 2005 and has been ratified by over 180 countries. According to Article 2, the ultimate objective is 'stabilization of greenhouse gas concentrations in the atmosphere at a level that would prevent dangerous anthropogenic interference with the climate system (. . .) to ensure that food production is not threatened and to enable economic development to proceed in a sustainable manner'. The effect of the Kyoto Protocol on national land policy may include encouraging farmers to replace crops aimed at food production by plantings suited for bio-fuel production. Other principles underpinning land policy include raising (gender) equality, social justice and economic development and preventing land use degrading the environment.

How people and governments deal with land is thus of major importance for the development of society. How to manage land is an issue of government policy in every country. Land policy determines values, objectives and the legal framework in relation to management of land as a legal, economic and physical object (Enemark, 2006). From the perspective of governments appropriate land policy is necessary to avoid conflicts within the boundaries of the nation, to warrant long-term economic development, and to promote confidence between the inhabitants, the enterprises and the government itself. Considered by many to be the mother of all cadastres, the essential idea shoring the French cadastre – instituted in 1807 by Napoleon Bonaparte – was legal security, not primarily as protection against violation of rights by other citizens but as a dam against unpredictable government ('L'etat cést moi'). In the words of Napoleon: 'The cadastre just by itself could have been regarded as the real beginning of the Empire, for it meant a secure guarantee of land ownership, providing for every citizen certainty of independence. Once the cadastre has been compiled [. . .] every citizen can for himself control his own affairs, and need not fear arbitrariness of the authorities', (Zevenbergen, 2003). Two centuries after the establishment of the French cadastre, many countries, especially in the developing and former communist countries do not have a well functioning land administration system and efforts continue to implement and improve it.

15.18 3D Cadastres

In countries that are heavily urbanised, such as Germany, Israel, Japan and the Netherlands one is struggling with scarcity of land. To cope with that problem their governments are stimulating multiple use of space that is simultaneously carrying out a great variety of activities within the same column of space. There are thousands of locations where different types of constructions are built on top of each other: underground railways, parking garages, shopping malls, bank offices, living quarters and many, many others (Figs. 15.10 and 15.11). Usually, the exploiter of any one of these constructions will differ from another who exploits the area of space above or below. How should one deal with vertical ownership rights? This is not a trivial question because in most countries ownership right is unlimited in the vertical dimension. So, theoretically any land ownership covers a pyramid, or cone, the apex of which is situated at the centre of the earth and the base floor of which expands into infinity. In practice, the owner takes many restrictions for granted, such as over-fly rights, mineral rights and other law-induced restrictions. In addition, land ownership is unlimited in time; it lasts forever. The description of the rights on objects with boundaries in the vertical space column is resolved by several legal instruments, including rights regarding servicing, condominium, building right and property in strata. The legislative detailing of these rights differs from country to country.

Fig. 15.10 Example of multiple uses of the same columns of space – buildings connected through passages high above the streets in the city of Modi'in, Israel

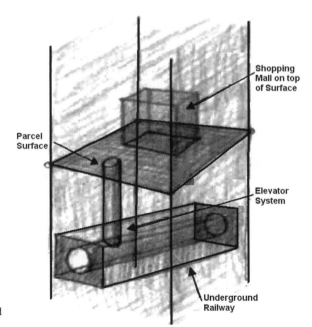

Fig. 15.11 Schematic impression of volumes built on top of each other with different ownership rights and users

The building of different types of construction on top of each other induces questions such as the following:

– Should the concept of land ownership be modified so that the extension of boundaries is defined not only horizontally but also vertically?
– How should the obligations and rights of vertical neighbours be defined?
– Should a more consistent and secure description of the rights on superimposed constructions result in a modification of land ownership?
– Should in addition of outlining of the parcel surfaces also the outlining of volumes be registered in land-administration systems so as to better serve the management of our complex environment?

Although some countries, such as Norway, Sweden, Queensland (Australia) and British Columbia (Canada), have properly addressed the legal issues involved in stratified ownership, these solutions are not complete in terms of 3D-cadastral registration since they are still not registered in their full three-dimensional extensions (Stoter and van Oosterom, 2006). The move from registering parcels as 2D objects to volumes is comprehensive as it requires redefinition of the cadastral concept. At the technical levels some of the issues are related to storage, querying and visualisation of volume objects and how to make sure that they do not overlap.

15.19 Core Cadastral Domain Model

The basic task of any cadastre in whatever country is to register the relationship (rights) between people and land, it is all about registering information on land property and giving others access to the information. How this task is conducted in various settings may differ substantially depending on culture and historical scenery. Indeed, both appearance and implementation of cadastral systems may vary greatly. One country may register general boundaries, the other fixed boundaries. The one cadastral system may have its roots in tax collection; in the other country securing land rights has been the predominant determinant. In some countries one has adopted the deed registration system, in other countries the system of title registration. In some countries the collection, custody and maintenance of data is centralised; in other countries, it dispersed over multiple agencies and organisations. In some countries cadastral surveys are carrier out by professionals employed by the cadastre itself; in other countries this is a task of private surveyors.

The basic service of cadastres is to allow others access to land information; others meaning here governmental agencies, non-governmental organisations, companies and citizens. These 'clients' can have domicile inside the territorial boundaries of the nation or are settled abroad. In the past the communication between cadastres and their clients proceeded along oral lines or paper channels. The omnipresent internet has not only triggered quicker and easier access to cadastral information but also attracted a broad pallet of new clients. In the framework of establishing

national, continent-wide and global geo-information infrastructures (GII) cadastral data play a pivotal role. A geo-information infrastructure, also called Spatial
Data Infrastructure (SDI), allows better addressing of pressing social, economic
and environmental issues. GII, properly embedded within the overall information infrastructure, is assumed to be crucial for improved electronic governance
(*e*-governance).

Meanwhile establishing and renewing land-administration systems all round the
world has been put high on the international agenda, not the least since the early
1990s as a result of the downfall of former centrally planned economies triggering
restoration of ignored cadastres. In addition, the important role land administration
can play for reducing poverty in developing countries has been widely recognised.
However, the great diversity of appearance and implementation of cadastral systems, each with their own taxonomy, impedes national and international exchange
of information. These hurdles also obstruct exchange of experiences and bringing
into effect aid from foreign donors. The terminology, technology and methods used
in land administration are not unified, harmonised or standardised all over the world.
This is increasingly felt as a deficiency and has bothered researchers already for
some time. Professional organisations, in particular FIG (International Federation of
Surveyors), picked up the emergency signals and are doing a great job in attempting
to abandon Babel and to arrive at a common cadastral language, practised all around
the world.

The language under development, which should bring together all professionals
active in the land administration domain, is called Core Cadastral Domain Model
(CCDM). The earliest developers of basic ideas and architects of the underlying
grammar are van Oosterom and Lemmen (2002). Basically, CCDM is a model
that covers land registration and cadastre in a broad sense and provides an extensible basis for establishing efficient and effective cadastral systems based on a
common language (Lemmen and van Oosterom, 2006). CCDM encourages and
supports the flow of information concerning land property between government
organisations mutually and between these organisations and citizens. More specifically, CCDM has been designed with two main goals in mind (van Oosterom et al.,
2006):

1. Enabling effective and efficient implementation of flexible and generic cadastral
   information systems based on a model driven architecture.
2. Providing the 'common ground' for data exchange between different systems in
   the cadastral domain.

Important considerations during the design process were that the model should cover
the most common aspects of cadastral registration all over the world and should
follow international ISO and OGC standards, whilst remaining as simple as possible so as to be useful in practice. CCDM is aligned with FIG Cadastre 2014, a
set of guidelines giving a start for implementing a cadastral model aimed at unambiguous exchange of information on how systems are implemented. The guidelines
are generic and implementation in a certain country requires further detailing and

adoption to local situations. In the meantime CCDM has been adopted by FIG as a standard approach. To communicate and promote CCDM the graphics oriented Unified Modeling Language (UML) has been embraced as common language (see Chapter 10). A specialisation of the CCDM, focussed on implementation in developing countries is in progress. Developing countries need a specialised approach because in these areas land administration is still in a stage of infancy. Furthermore, land administration is increasingly recognised as great help in the battle against poverty. The specialised model aimed at developing countries combating poverty is called Social Tenure Domain Model (Lemmen et al., 2007). In addition to FIG also UN-HABITAT plays an important role in bringing this model from the drawing table on the ground.

15.20 Land Administration and the Battle Against Poverty

In developed countries there is a high level of specialisation, a relatively low number of people involved in producing food and exploring mineral resources; the level of automation within the entire primary sector is high. Job numbers in this sector are lower than in the value-adding industry, where raw materials from agriculture and mining are transferred to, for example, cars, fast food, ready to wear clothes, TVs, computers and prefab houses. A country such as Japan has been able to create and sustain a developed society despite lacking mineral and energy resources within its own territorial boundaries, just by adding value to imported raw materials and energy resources. In contrast, a country like Nigeria, with an abundance of mineral and energy resources lying beneath its surface, does not add much value to exploited raw materials and so ends up with GDP per capita of below US $1,000. In developed countries infrastructures including transport of people, materials, energy, electronic and digital signals, and money are efficient because these activities are well organised and well managed. Family planning and birth control are well accepted and practised at virtually all levels of society. As a result, citizens enjoy a high standard of living, not only material well-being as measured by GDP per capita but also in terms of education, healthcare and life expectancy. Developing countries lack most of the above. The majority of people work in the primary sector: for example, in Nigeria 70% of the population is working in agriculture. Competition among labourers is high, leading to low wages. As a result, employers are not encouraged to invest in automating production processes, so that the overall level of automation in agriculture, mining and construction remains low and production rates per capita accordingly modest.

15.20.1 Millennium Development Goals

A world without poverty, hunger, pandemics and anguish, a world offering basic education for every child, equality, freedom and brother- and sisterhood; it is a world we might dream of. And also one far removed from reality. Perhaps the

United Nations (UN) had such a dream when it came together for the fifty-fifth time, from 6 to 8 September 2000, 'at the dawn of a new millennium'. All 189 Member States, 147 directly represented by their head of state or government, embraced the Millennium Declaration and Millennium Development Goals (MDGs). Following the UN example, other leading world organisations, such as the IMF (International Monetary Fund) and World Bank, welcomed the MDGs, consisting as they did of an array of eight goals to be achieved by 2015, with 1990 as reference datum. The goals are people-centred, time-bound and measurable.

The eight MDGs range from eradicating extreme poverty and hunger (Goal 1) and achieving universal primary education (Goal 2) to reducing child mortality (Goal 4) and improving maternal health (Goal 5). Beyond halfway the timeline the envisaged results have not even come into view. Let's take first Goal 1, to eradicate extreme poverty and hunger. This has as measurable target halving over the 25 years from 1990 to 2015 the proportion of people whose income is less than US $1 a day. According to MDG Report 2007, there were 1.25 billion people in developing countries living on less than US $1 a day in 1990. This number fell to 980 million in 2004. Expressed in percentage terms, this means the proportion of people living in extreme poverty across the globe dropped from 32% to 19% in nearly 15 years, and that is very promising indeed. If this rate of progress continues, the MDG target will be met.

But the progress shows geographical bias; the majority of the decrease results from rapid economic growth in China, India and south-east Asia. In contrast, poverty in sub-Saharan Africa fell only slightly, from 47% in 1990 to 46% in 1999, arriving at 42% in 2004. Imagine 42%! That means nearly half the population suffers from severe poverty and malnutrition. These are daunting percentages. But what do the absolute numbers say? There were 815 million hungry people in the developing world in 2002, nine million less than in 1990. However, in sub-Saharan Africa the number of people living in extreme poverty increased by 34 million over the same period, partly as a result of annual population growth rate of 2.3%, while numbers of extremely poor rose slightly from 296 million in 1999 to 298 million in 2004 (Lemmens, 2008c). Worse, sub-Saharan Africa enjoys the dubious privilege of being the only region in the world where the poor are getting poorer; the average income of people living on less than US $1 a day fell from US $0.62 in 1990 to US $0.60 in 2001. An UNDP/UNICEF report, published June 2002, states that should poverty reduction progress at this rate it will take until 2150 to halve extreme poverty. And that in a region where nearly half the population consists of children under the age of 14, this figure comes to a round 350 million, expected to reach over 400 million in 2015.

Speaking of children: how do things stand regarding enrolment in primary education for children living in sub-Saharan Africa? The target of Goal 2 is to ensure that by 2015 children everywhere, boys and girls alike, are able to complete a full course of primary schooling. The target is thus that 100% of children somewhere between the ages of 6 and 12 should be in school all working days, instead of helping their fathers in the field or carrying stones to a construction site for a coin or two. Globally, the net ratio of primary education enrolment in developing regions increased from 80% in 1990 to 88% in 2004. In sub-Saharan Africa the increase was even more spectacular, from 54% in 1990 to 70% in 2004 – but note that in 1990 only half of youngsters attended school. And 70% actually means that to date around two-thirds of children are enrolled, these statistics themselves probably providing too glowing a picture. Enrolment is an administrative matter and does not guarantee the physical presence of Promise in the classroom. Promise's parents may have other, higher priorities, and while her classmates do their spelling Promise is busy in the fields, reaping yam and cassava. (Some teachers and extramural others might surmise children are taking an extended summer break from school to enjoy a well-deserved holiday; however, it is no coincidence that the vacation coincides with harvest time). It will probably take sub-Saharan African countries until 2140 to achieve full primary school attendance for all their children.

In the UN MDG Report 2005, Kofi Anan, then secretary-general, stated unequivocally: 'Instead of setting targets, this time leaders must decide how to achieve them'. In the UN MDG report 2007, UN secretary-general, Ban Ki-Moon, reinforced Anan's words: 'There is a clear need for political leaders to take urgent and concerted action, or many millions of people will not realise the basic promises of the MDGs in their lives'. There is plainly a great deal of willingness to formulate goals and targets, but some reluctance to act on them. Let us keep our fingers crossed that the Millennium Development Goals will not go down in history as a gesture of humanitarian fervour inspired by no fewer than three noughts in the year, noughts which might turn out in retrospect to have been an omen sadly indicative of the final outcome.

15.20.2 Lack of Data

The modest progress of MDGs in sub-Saharan Africa, the only region in the world where not even a single country is on track is a matter of considerable concern to many influential people, including Robert Zoellick, president of the World Bank, UN secretary-general Ban Ki-moon himself, and representatives of the African Union and European Commission. It led to their meeting in New York and launching the MDG Africa Steering Group on 14 September 2007. The African Union sees the areas of agriculture, education, health, infrastructure and statistics as crucial for the African continent. The World Bank is to focus on the possibilities of increasing productivity in small-scale agriculture, but there is also huge need for investment in water and sanitation and, above all, in reliable and timely data. In addition to noting that promised international support had not been forthcoming, African Union Commissioner for Economic Affairs, Maxwell Mkwezalamba told the meeting that the issue of statistics could not be overstated. 'In Africa we have problems of data. Data are not reliable, are not timely. And therefore this is one area that we need to focus on. We do appreciate the fact that this meeting has also looked at developing statistical systems as one important area of focus', he added. Also the World Bank recognises that above all there is a huge need in Africa for reliable and timely data.

Furthermore, 'to arrive at sustainable development and social equity it is necessary to establish proper tenure systems', as Dr Anna Tibaijuka, UN under-secretary-general and executive director of UN-Habitat, told us (see Lemmens, 2007). At the crux of establishing such rigorous data-collection frameworks lies the use of detailed geo-information.

In the meantime Africa remains weakly mapped, and only few African countries have maps in place to support development. The United Nations has acknowledged the severe lack of geo-information in Africa over the years and the important role of maps in combating poverty.

15.20.3 Economic Apartheid

The battle against poverty in developing countries has for a long time been high on the agenda of Western countries. And that battle has always been associated with the issue of land. At first the focus was on improving the production capacity of a country by converting natural environments into agricultural land. Next, land productivity was increased by introducing mechanisation, fertilisers and insecticides, and by making crop species more disease resistant. The success rate varied; a number of projects failed, partly or entirely because of too rigid an implementation of solutions which worked well in the western context but appeared under different socio-economic circumstances to be counterproductive. But overall the Green Revolution can be classed a success, from the food-production perspective.

However, man lives by more than bread alone, and an essential part of this 'more' is the striving of people for power, the drive to rule over others. Improvements induced by external initiatives appeared in the long run to destabilise the socio-economic context, especially in countries where illiteracy is high: it enabled the powerful to gather more power, the rich to collect more money, whilst the poor remained poor or got even poorer. Particularly those living in rural areas suffered from a changing balance of power. In an attempt to relieve misfortune many initiatives were developed by non-profit organisations to encourage small farmers to earn money by themselves being active on the world market. However, these solutions were driven primarily by ideology and the products accordingly found a main market among those willing to pay high prices for low quality; not a permanently sustainable outcome.

To keep pace with the developments of technology changing ways of agricultural food production and produce food in an economically feasible manner, the farmer must regularly invest in cultivators, fertilisers, drainage works and so on. For most farmers it will be impossible to make such investments from savings; farmers, as so many entrepreneurs, need access to credit but before it will provide a loan a bank requires a guarantee and land is key collateral for this. However, one has to show official documents to prove that one is the legal owner. So that if one has no formal property rights one is excluded from formal credit and investments; one is faced with 'Economic Apartheid'.

15.20.4 Land as Collateral

A relatively new weapon in the battle against poverty is facilitating the poor to own land and to secure property rights via effective laws and solid land-administration systems. In effect, attention is shifting from harvesting land produce to the land itself. Securing land tenure enables to reduce poverty by providing farmers and other entrepreneurs with the opportunity to translate their property into collateral, giving them access to credit at reasonable rates of interest. This is asserted today by so many so often that this statement has become a generally accepted and undeniable truth. Experts have understood the close correlation between security of tenure and poverty eradication for more than three decades. But it was the seminal work of Hernando De Soto (2000) which made these ideas easily accessible and understandable to policy and decision-makers. So articulate is De Soto in his presentation of the straight line of reasoning linking security of tenure with poverty eradication that he has affected many a mindset. The mechanism is propelled forward like a three-stage rocket: first property is transferred into collateral, enabling access to cheap credit, credit is then translated into investment, and finally investment is harvested as increased income.

The line of thinking may be clear-cut, but implementation is no piece of cake. At the first stage of the money generating-rocket is property. How do people, particularly banks, ascertain that a poor farmer in a sub-Saharan country legitimately owns a particular piece of land? Announcements such as 'This land is not for sale nor for lease' are a common sight, chalked on walls or fences as proof that there are folk who do not stick to selling what belongs to them. Prerequisites for securing a person's right to a piece of land are proper laws, a well-functioning land administration system and good governance. Indeed, the last decade has witnessed augmented activity in establishing land administration system in developing countries all over the world, often supported by the World Bank, IMF, United Nations and the European Union. However, why should a poor farmer register his land when everyone in the village knows he is the rightful user, and registration will cost money, lots of it; money that is not available? This is the first hurdle.

The second stage of the rocket involves land accepted by a bank as collateral being translated into money. How should property be valued? There are two basic methods. The first, and common practice in developed countries, is by determining its market value. An appraiser adjusts the value by comparison with similar objects recently transferred. However, in many developing countries the land market is not as well developed as in western countries, and therefore it is widely advocated that high priority should be given to establishing a land market in these countries. A second method is by assessing the productive capacity of land and the market value of the harvest, but this demands other expertise, and the world market for agricultural products is unstable. A loan comes at a price. Interest rate is basically determined by two main parameters: official rate, linked to level of inflation, and risk. A bank requires not only security of tenure but also a guarantee that the farmer has sufficient present and future income to incrementally repay loan and interest. Although any default on repayment metamorphoses the bank from money provider into owner, it will not appreciate the role reversal; these are bankers, not farmers. A bank wants to be pretty sure that a farmer can settle his debts.

In the third stage the money is invested so that earnings minus interest and progress payments are higher than the farmer's previous income. The money enables him to increase his harvest by improving production methods and buying fertilisers and more disease-resistant seeds. He can invest in techniques to enable him to pound and dry yam or cassava himself, allowing his wife to sell the products at market months after harvest, when the prices are likely to be higher. The loan might help him buy a milling machine, and the value so added might provide higher income per kilogram of harvest.

Poor farmers and their wives are only human. They realise that there is a long way to go from winning security of tenure to becoming wealthier. It may take years and years, a decade even or more – while a shorter route exists, at least apparently. Borrowing from the bank makes it possible to spend more; the money can be used (some erroneously say 'invested') to buy, for example, the status symbol of a 100-cc Chinese-made motorbike. Satisfaction is immediate. Why take the long road? A loan does not make a farmer richer. It might, if properly invested, be the engine for earning more money in the future. Used for consumption, however, it only makes him poorer. His income fails to increase, while he is loaded down with debt. Consumption increases GDP per capita, one of the main indicators of the Millennium Development Goals; this is the definition of GDP. As a result, the weird situation may arise of poverty being statistically shown as successfully addressed, while the poor become poorer.

15.20.5 'Aid Should Continue'

Since acknowledgement of the MDGs by the United Nations, many countries, especially developing ones, have been engaged in attempts to reform their land administration system, sometimes on their own but recurrently with donor support. However, such reforms have seldom resulted in satisfactory systems. It has been estimated that as many as 90% of rural and 50% of urban property rights in developing countries are still not formally protected, either by registration or in another form. What is the reason for these failures? Is it due to failed implementation of advanced technology? Is it because of hesitation on the part of individuals to formalise their rights? Is the present process of formal registration to blame, involving as it does – due to altering goals of land administration – too many governmental agencies, resulting in fragmentation and lengthy and expensive processes? Is it because many property rights in developing countries are not individual but customary rights? Is it because indigenous customary tenure institutions do not meet good governance objectives especially in peri-urban areas as they tend to be weakened by statutory institutions operating in parallel (Arko-Adjei et al., 2010)? All of the above may contribute to failure. However, the main problem is that conditions vary from country to country. Thus the use of a template derived from systems which work well elsewhere – an approach which has been adopted all too often in the past – is inappropriate.

How helpful it would be to establish a successful land administration system simply by making an inventory of the factors which characterise a well-functioning one. Then the most suitable system would emerge from straightforward analysis. Indeed, there is nothing more practical than a good theory. However, a land-administration theory is non-existent. Many are working on the problem of arriving at insight into the factors which determine the well-functioning of land administration. One such offshoot is the work of Augustine Mulolwa (2002). He has studied land administration in Zambia and made a number of suggestions for improvement. This work is highly recommended to sharpen the insight of those involved in establishing land-administration systems in developing countries.

Every intervention creates non-projected side effects and disturbances. For example, many tourists visiting developing countries believe that by giving large tips they are supporting the local economy. They do not realise that all that money may increase inflation, making even poorer those poor people not involved in the tourist industry. So in thinking we do a good turn, in fact we might do a bad one. Although the ultimate and laudable goal is to arrive at a better world in which poverty has been eradicated, it is questionable whether the projected result of reducing poverty by implementing land-administration systems will be achieved in the long run. Will the ownership of land not end up in the hands of the few powerful and knowledgeable enough to understand the nitty-gritty of the systems and laws, and smart and affluent enough to (mis)use their power? And so once again the powerful may become more powerful and the rich richer, whilst the poor remain poor.

Professor Easterly (2006), who spent most of his career as an economist at the World Bank, is a profound criticaster of foreign aid, delivered by the rich part of the world to the poor part. He argues that aid does not work because aid is provided in the context of ambitious plans based on generosity: something has to be done, even when it does not work. 'A planner thinks he already knows the answers', Easterly writes. The approach of planners is top-down, while the poor need bottom-up solutions, which can be better provided by searchers rather than planners. The poor themselves should come up with solutions because they know the best what they need and next aid should be provided by piecemeal interventions. Up to know the sum of all plans is negative, he states, and that is because aid money actually does not reach the poor, and when goods bought from the money reach them they are not properly used. They are given for free and are used for a wide variety of purposes except for the intended ones. Take, for example, insecticide-treated mosquito nets, which cost a few dollars; they can prevent most malaria infections from which hundreds of millions of people suffer in Africa. They are used as fishing nets or wedding veils.

Dr. Anna Tibaijuka, Under Secretary General and Executive Director UN-HABITAT, communicated her feeling of the claim that financial aid provided to Africa over the past half century has had no effect at all, and when it has this has been a negative one (Lemmens, 2007):

> I think there is no factual basis for drawing such a drastic conclusion. In order to be able to answer this question, which is actually an academic one, you have to be able to show how Africa would have been today without having received any aid. After WWII the Europeans

got assistance via the Marshall Plan and it is obvious that when your development is held back for historical reasons you need support. This is a matter of decency and a part of international commitment. The problem is that the Africans have not received sufficient co-ordinated aid to make a difference. As an African I would say that we are looking for improvement in the situation. Better knowledge exchange and harmonisation between the parties involved is crucial in this respect. In conclusion, I would say that the observation is misplaced. There is no scientific basis for it and, in my opinion, aid which has gone to Africa has been useful and should therefore continue.

15.21 Future

What will be the tasks of the cadastral surveyor 30 years hence? What are the underlying societal and technological dynamics? To keep it simple, let us answer these questions by limiting ourselves to three major ongoing demographic developments: urbanisation, advantage of scale in agriculture and globalisation.

The concept of the parcel emerged when farming was the main way of life for the majority of people on this planet. Many urban conglomerates, such as Istanbul, Turkey, Tehran, Iran, and Lagos, Nigeria, have grown massively over the last three decades; to date more people live in cities than in rural areas. In a megalopolis one does not see 'parcels' but 'objects', including high-rise buildings, dwellings, petrol stations, railways and roads. Each (part of an) object is a nucleus to which a wide variety of data can be assigned, not only to secure tenure but also to enable managers and planners to collect tax, guarantee safety and sufficient drinking water, prevent people from starving and respond to emergencies. 3D property boundaries should be measured and registered at cm-level. The attributes assigned to the object should be detailed and include ownership, use and value, number of storeys and so on.

While cities expand at an unprecedented rate, depopulation of rural areas continues. Goods and data move from one part of the world to the other, money even in a split second. Air transport enables sugar-snap peas grown in Kenya to be offered for sale all over the world. To stay competitive, farmers have to produce food on a large scale using heavy machinery. It is widely believed that eradication of poverty can be achieved by formally registering land belonging to small farmers, enabling them to invest through a mortgage. This assumption may however be challenged, since small farmers have a low production capacity which will continue to fall as globalisation progresses. Within one or two generations, adjacent farmlands now owned by hundreds of small farmers will probably be swept together into one big property parcel. The children will move to the cities and the small farmer become extinct. So it does not make a lot of sense to invest great effort in improving security of tenure in areas which will always remain rural. In stark contrast is the situation at urban fringes, where the city meets the countryside. Here farmers face the threat of ejection from their land with little or no compensation, and it is of the utmost importance that tenure security is established here without delay.

Another issue recently put forward is as follows: What role should cadastres play in the climate-change debate? (Van der Molen, 2009, 2010). In answering this question, Paul van der Molen states (see Lemmens, 2010c):

I see two aspects. Firstly, adapting to and mitigating climate change requires reforestation, grazing-land management, cropland management, and re-vegetation. The resultant increase in government intervention in private property rights would be impossible without knowing who owns what and where. Secondly, the Kyoto Protocol provides for a commercial market for carbon credits based on the use of greenhouse sinks: carbon sequestration in soil and vegetation. Such emission rights are unbundled property rights, which constitute a tradable title. Tenure and market security by registration are also needed here.

Cadastre 2014 is an influential publication produced by a FIG Commission 7 working group between 1994 and 1998. The task was to develop a vision for the modern cadastre of 20 years hence. Authored by working-group chairperson Jürg Kaufmann and secretary Daniel Steudler, this review of the strengths and weaknesses of cadastral systems of 20 years ago and vision for the future were presented at the FIG Congress in Brighton in July 1998. As Prof. Ian Williamson, then chair of FIG Comm. 7, noted in his foreword, the vision recognised many ongoing changes, including the role of government and surveyors in society, relationship of humankind to land, the growing role of the private sector in cadastre operation, and dramatic influence of technology on cadastral reform.

It is now relevant to ask to what degree the objectives of Cadastre 2014 have been accomplished. Those involved in land administration also need to signal societal and technological dynamics that may affect the practice of land administration worldwide over the coming 20 years. Bennett et al. (2010) isolated six design elements relating to the role and nature of future cadastres:

– move from approximate boundary representation towards survey-accurate boundary representation
– focus shift from purely parcel-based systems towards systems of layered property objects
– expansion from 2D approaches to include the third (height) and fourth (time) dimensions
– updating and accessing of cadastral information in real time
– making national and state-based cadastres interoperable at regional and global levels
– inclusion in property interests, now designed around strict bearings and distances or Cartesian coordinates, of modelled organic natural environment by enabling fuzzy and dynamic boundary definitions

To encourage discussion on the suitability of the proposed six design elements the present author invited leading experts and practitioners to make known their own views and vision (Lemmens, 2010a, b). The following international experts voiced their views and opinions: Dr. Clarissa Augustinus, UN-HABITAT; Keith Clifford Bell, World Bank; Dorine Burmanje and Dr. Martin Salzmann, Kadaster, the Netherlands; Dr. Mohamed El-Sioufi, UN-HABITAT; Jürg Kaufmann, co-author of Cadastre 2014; Jarmo Ratia, National Land Survey of Finland; Daniel Roberge, Quebec cadastre, Canada; Dimitris Rokos, Ktimatologio S.A., Greece; Daniel Steudler, co-author of Cadastre 2014. Paul van der Molen, Twente University (ITC), the Netherlands. From the ten replies received the following can be concluded. The six chosen design elements proposed by Bennett et al. (2010) emerged from considering highly urbanised areas in developed countries where societal needs can be summed up in three key words: accuracy, detail (3D, 4D, rights-restrictions-responsibilities (RRR)) and real-time. Further, globalisation forces adjustment of cadastral content based on transnational interoperability criteria, while a shift is proposed in modelling boundaries of natural phenomena such as rivers, shores and forest, from crisp to fuzzy. However, completely different societal needs arise in developing countries, and design elements for these areas cannot be drawn up with anything like such steady hands.

How can cadastres contribute to eradicating poverty (a main Millennium Development Goal) and corruption? How can they enforce sustainable development of land? There is no such thing as 'one-size-fits-all'. There are urban areas and rural areas. There are developed and developing countries. In developed countries the needs of society seem clearer and finding solutions a matter of organising scarce resources, properly applying technology and anticipating technology to come. Much more challenging tasks face the cadastre in developing countries, where there is an abundance of complicating issues. Here it is not just a matter of cadastre aiming to support a relatively frictionless society, but also enabling creation of a better one; eliminating malnutrition, gender inequity, illiteracy, corruption and the immense gap between the haves and have-nots. So that is at least four sizes, each requiring specific approaches and solutions; summarised in Table 15.2.

Clarissa Augustinus confirms the above, stating in her reply: 'The challenge facing the land industry is to design tools for the whole range of global society, not just the developed world'. From a global perspective, the six design elements are far from comprehensive. She thus seamlessly joins her argument to that of Keith Bell, who challenges Bennett et al., asking 'Are real-time, spatially accurate cadastres more important than water, sanitation and nutrition?' Daniel Roberge too recognises that the 'survey accuracy' design element is more a thing for developed countries, as 'developing countries, where the need for land-rights infrastructure is primary and resources are scarce, require light and low-cost solutions' creating exact rather than accurate data. Paul van der Molen also has an eye for the gap between developed and developing countries, and again sharpens the focus: 'Do people invest more in their

Table 15.2 Urban and rural areas, developed and developing countries, each need vision, approach and solutions of their own

	Urban	Rural
Developed	Objects (3D, 4D)/survey accuracy and RRR	Parcels/survey accuracy and RRR
	Exploiting advanced technology	Exploiting advanced technology
Developing	Determination of role in society	Determination of role in society
	Supporting good governance	Supporting good governance
	Getting cadastres off the ground	Getting cadastres off the ground
	Objects (3D)/survey accuracy	Parcels/general boundaries
	Society first, technology next	Society first, technology next

land, and do new landowners have better access to credit?' His point is that issues of corruption and lack of return on investment must be resolved before the six design elements have any chance of getting off the ground in developing countries.

References

Arko-Adjei A, de Jong J, Zevenbergen JA, Tuladhar AM (2010) Customary tenure institutions and good governance. XXIV FIG international congress 2010: facing the challenges: building the capacity, Sydney, Australia, pp 1–21, 11–16 Apr 2010

Barnes G (2003) Lessons learned: an evaluation of land administration initiatives in Latin America over the past two decades. J Land Use Policy 20(4):367–374

Bell KC (2006) World Bank support for land administration and Management: responding to the Challenges of the Millennium Development Goals. XXIII FIG congress, Munich, Germany, 8–13 Oct 2006

Bennet R, Rajabifard A, Kalantari M, Wallace J, Williamson I (2010) Cadastral futures; building a new vision for the nature and role of cadastres, FIG Congress 2010: Facing the Challenges: Building the Capacity, Sydney, Australia, 11–16 April, pp 1–15

Cashin SM (2003) The application of high-resolution imagery and geographical information systems in cadastral mapping: a case study of the republic of Moldova. PhD thesis, Department of Geography, University of Cambridge, UK

Dale P, Baldwin R (2000) Lessons learnt from the emerging land markets in Central and Eastern Europe. Proceedings FIG working week, Prague, pp 4–12, 21–26 May 2000

Dale P, McLaughlin J (1999) Land administration. Oxford University Press, Oxford, 169 p. ISBN 019 823390-6

De Soto H (2000) The mystery of capital: why capitalism triumphs in the West and fails everywhere else. Bantam Press, London

Easterly W (2006) The white man's burden: why the West's efforts to aid the rest have done so much ill and so little good. The Penguin Press, New York, NY, 436 p

Enemark S (2006) Understanding the land management paradigm: need for establishing sustainable national concepts. GIM Int 20(1):12–15

Feder G, Nishio A (1998) The benefits of land registration and titling: economic and social perspectives. Land Use Policy 15(1):25–43

FIG (1995) The FIG Statement on the Cadastre. International Federation of Surveyors, Australia

FIG (2001) FIG Agenda 21, Agenda for implementing the concept of Sustainable Development in the activities of the International Federation of Surveyors and its member associations. FIG Publication No. 23

Henssen J (1995) Basic principles of the main cadastral systems in the world, Modern Cadastres and Cadastral Innovations, Proceedings of the one day seminar held during the Annual Meeting of Commission 7, Cadastre and Rural Land Management, of the International federation of Surveyors (FIG), Delft, the Netherlands, May 1995, pp 5–12

Henssen JLG, Williamson IP (1990) Land registration, Cadastre and its interaction: a world perspective. FIG XIX international congress, Helsinki, Finland, pp 14–43

Ingvarsson TM, Barry T, Hauksdóttir M (2007) Reform of Icelandic cadastre: well-laid foundation ensures strong future. GIM Int 21(3):51–53

Lamptey F (2009) Participatory GIS tools for mapping indigenous knowledge in customary land tenure dynamics: case of Peri-urban Northern Ghana. MSc thesis, International Institute of Geo-information and Earth Observation (ITC), Enschede, The Netherlands

Larsson G (1991) Land registration and cadastral systems: tools for land information and management. Longman Scientific and Technical, Harlow, 175 p. ISBN 058 208952

Lemmen CHJ, Van Oosterom PJM (2006) FIG core cadastral domain model version 1.0. GIM Int 20(11):43–47

Lemmen CHJ, Augustinus C, van Oosterom PJM, van der Molen P (2007) The Social Tenure Domain Model – Design of a First Draft Model, FIG Working Week, Hong Kong SAR, China, 13–17 May 2007

Lemmens M (2006) Carving up a country. GIM Int 20(1):11

Lemmens M (2007) Secure land tenure key. GIM Int 21(11):7–9

Lemmens M (2008a) Land administration in Vietnam. GIM Int 22(9):7–11

Lemmens M (2008b) Land administration in Estonia. GIM Int 22(11):7–9

Lemmens M (2008c) Land rights in Sub-Saharan Africa. Prof Surveyor 28(2):54–58

Lemmens M (2010a) Towards Cadastre 2034. GIM int 24(9):41–49

Lemmens M (2010b) Towards Cadastre 2034: Part II. GIM int 24(10):37–45

Lemmens M (2010c) A career devoted to cadastres: interview with Paul van der Molen. GIM int 24(12):6–9

McLaughlin JD, Nichols SE (1989) Resource management: the land administration and cadastral systems component. Surv Mapp 49(2):77–86

Mulolwa A (2002) Integrated land delivery: towards improving land administration in Zambia. PhD thesis, Delft University Press, The Netherlands. ISBN: 90-407-2347-8

van Oosterom PJM, Lemmen CHJ (2002) Towards a standard for the cadastral domain: proposal to establish a core cadastral data model, COST workshop 'twards a cadastral core domain model'. Delft, the Netherlands

van Oosterom PJM, Lemmen CHJ, Ingvarson T, Van der Molen P, Ploeger H, Quak W, Stoter J, Zevenbergen J (2006) The core cadastral domain model. Comput Environ Urban Syst 30: 627–660

Poku-Gyamfi Y, Schueler T (2008) Renewal of Ghana's geodetic reference network. 13th FIG symposium on deformation measurement and analysis, UNEC, Lisbon, 12–15 May 2008

Stoter JE, van Oosterom PJM (2006) 3D Cadastre in an international context: legal, organizational and technological aspects. Taylor & Francis, CRC Press, Boca Raton, FL, 323 p. ISBN: 0-8493-3932-4

Tibaijuka A (2008) Improving slum conditions through innovative financing, Keynote address at the Opening Ceremony of the FIG Working Week 2008 in Stockholm, Sweden, 15 June 2008

UN-ECE (1996) Land Administration Guidelines, European Commission for Europe, United Nations Publication, No E.96. II.E.7. ISBN 92-1-116644-6

Van der Molen (2006) Tenure and tools, two aspects of innovative land administration: background paper 'evening lecture' RICS 13 December 2006. London

Van der Molen P (2009) Cadastres and climate change. Proceedings of the FIG working week: surveyors key role in accelerated development, Eilat, Israel, 3–8 May 2009, 12 p. ISBN 978-87-90907-73-0

Van der Molen P (2010) Cadastres and climate change: multi-purpose land administration systems. GeoInformatics 13(2):34–39

Wallace J, Williamson I (2006) Building land markets. J Land Use Policy 23(2):123–135

Williamson I (1995) Appropriate cadastral systems. Proceedings of modern cadastres and cadastral innovations, FIG Commission 7, Delft, The Netherlands, May 1995

Williamson I, Wallace J, Enemark S, Rajabifard A (2010) Land administration for sustainable development. Esri Press, Redlands, CA

Wolf PR, Ghilani CD (2006) Elementary surveying, an introduction to geomatics, 11th edn. Pearson Prentice Hall, Upper Saddle River, NJ, 916 p. ISBN 0-13-148189-5

Zevenbergen JA (2003) Systems of land registration – aspects and effects. PhD thesis, Netherlands Geodetic Commission, Delft, The Netherlands. ISBN 90 132 277 4

Zevenbergen J (2004) A systems approach to land registration and cadastre. Nord J Surv Real Estate Res 1(1):11–24

Zevenbergen J (2006) Slowly towards trustworthy land records of pre-exiting land rights. XXIII FIG congress, Munich, Germany, 8–13 Oct 2006

Zevenbergen J, Bogaerts Th (2001) Cadastres in the 21st Century (2), different approaches available for successful systems. GIM Int 15(2):38–41

Name Index

M. Lemmens, *Geo-information*, Geotechnologies and the Environment 5,
DOI 10.1007/978-94-007-1667-4, © Springer Science+Business Media B.V. 2011

Subject Index